D1661651

THE INTERNATIONAL SERIES OF
MONOGRAPHS ON CHEMISTRY

THE INTERNATIONAL SERIES OF MONOGRAPHS ON CHEMISTRY

1. J.D. Lambert: *Vibrational and rotational relaxation in gases*
2. N.G. Parsonage and L.A.K. Staveley: *Disorder in crystals*
3. G.C. Maitland, M. Rigby, E.B. Smith, and W.A. Wakeham: *Intermolecular forces: their origin and determination*
4. W.G. Richards, H.P. Trivedi, and D.L. Cooper: *Spin-orbit coupling in molecules*
5. C.F. Cullis and M.M. Hirschler: *The combustion of organic polymers*
6. R.T. Bailey, A.M. North, and R.A. Pethrick: *Molecular motion in high polymers*
7. Atta-ur-Rahman and A. Basha: *Biosynthesis of indole alkaloids*
8. J.S. Rowlinson and B. Widom: *Molecular theory of capillarity*
9. C.G. Gray and K.E. Gubbins: *Theory of molecular fluids. Volume 1: Fundamentals*
10. C.G. Gray and K.E. Gubbins: *Theory of molecular fluids. Volume 2: Applications* (in preparation)
11. S. Wilson: *Electron correlation in molecules*
12. E. Haslam: *Metabolites and metabolism*
13. G.R. Fleming: *Chemical applications of ultrafast spectroscopy*
14. R.R. Ernst, G. Bodenhausen, and A. Wokaun: *Principles of nuclear magnetic resonance in one and two dimensions*
15. M. Goldman: *Quantum description of high-resolution NMR in liquids*
16. R.G. Parr and W. Yang: *Density-functional theory of chemistry*
17. J.C. Vickerman, A. Brown, and N.M. Reed (editors): *Secondary ion mass spectrometry: principles and applications*

Secondary Ion Mass Spectrometry

Principles and Applications

Edited by

John C. Vickerman

Director of Surface Analysis Research Centre
Department of Chemistry
UMIST

Alan Brown

Sims Product Manager, VG Microtrace Winsford

and

Nicola M. Reed

Surface Analysis Research Centre
Department of Chemistry
UMIST

CLARENDON PRESS·OXFORD

1989

Oxford University Press, Walton Street, Oxford OX2 6DP
Oxford New York Toronto
Delhi Bombay Calcutta Madras Karachi
Petaling Jaya Singapore Hong Kong Tokyo
Nairobi Dar es Salaam Cape Town
Melbourne Auckland
and associated companies in
Berlin Ibadan

Oxford is a trade mark of Oxford University Press

Published in the United States
by Oxford University Press, New York

British Library Cataloguing in Publication Data
Secondary ion mass spectrometry
1. Secondary ion mass spectrometry
I. Vickerman, John C. II. Brown, Alan
III. Reed, Nicola M. IV. Series
543'.0873
ISBN 0-19-855625-X

Library of Congress Cataloging in Publication Data
Secondary ion mass spectrometry: principles and applications/John C. Vickerman, Alan Brown, and Nicola M. Reed.
p. cm. — (The International Series of monographs on chemistry; 17)
Includes bibliographies.
1. Secondary ion mass spectrometry. I. Vickerman, J.C. II. Brown, Alan. III. Reed, Nicola M. IV. Series.
QD96.S43S4 1989 543'.0873-dc19 88-38479
ISBN 0-19-855625-X

Set by Colset Private Limited, Singapore
Printed by St Edmundsbury Press,
Bury St Edmunds, Suffolk

PREFACE

Since the early 1970s, secondary ion mass spectrometry (SIMS) has developed from the ion probe, primarily used for the bulk analysis of solids, into a number of sophisticated techniques for surface and near-surface analysis. *Static SIMS* using gentle bombardment conditions has provided the analyst with a technique able to characterize extensively the chemistry of the top surface of almost all types of materials whilst causing negligible surface damage. *Scanning or imaging SIMS* using highly focused ion beams has been able to provide detailed *chemical* images with the spatial resolution associated with scanning electron microscopy (less than 100 nm). Concurrent with these developments, *dynamic SIMS*, the successor of ion probe, has become the most sensitive elemental analysis technique applicable to solid state analysis, able to provide concentration–depth profiles with depth resolutions of a few nanometres and elemental sensitivities down to 10^{14} atoms cm^{-3}.

Quantifying data from surface analysis techniques is never really straightforward. The physics of the process probably makes SIMS data more difficult tham most in this respect. However, in the 1980s new developments in which the neutral particles emitted from the surface are post-ionized by electrons or laser photons have provided methods, known generally as sputtered neutral mass spectrometry, which will potentially make quantification easier.

SIMS is thus not a single technique but a very versatile suite of techniques able to provide a wide range of information regarding the surface region. Increasingly materials scientists and engineers, academic and industrial, are recognizing its very considerable powers. In particular SIMS is beginning to compete with, and certainly complement, the established techniques of electron spectroscopy in industrial materials research and development and quality control.

This book seeks to *introduce* SIMS and the account provided does not profess to be exhaustive. The aim is to enable those wanting to exploit the technique themselves to approach the practical use of SIMS or its data with understanding. The bibliographies will enable the interested reader to delve deeper. All the contributors are or have been associated with the Surface Analysis Research Centre at UMIST and are active SIMS practitioners. They are experienced in basic research into the phenomenon of SIMS and in its application to surface analytical problems, both fundamental and applied.

In the first three chapters, John Vickerman introduces the phenomenon of

SIMS and the theories which have been developed to understand it. John Eccles provides a general description of the instrumentation required for the main SIMS variants in Chapter 4.

Chapters 5 to 8 outline the application of SIMS in its various forms. These chapters are strongly application oriented. Again they do not profess to cover the whole of an analytical area but are personal accounts which illustrate what is possible. Practical examples are used extensively to demonstrate analytical procedures and the results which can be expected. In Chapter 5, the use of dynamic SIMS to provide elemental depth profiles is illustrated by David McPhail, with a detailed account of the problems to be addressed in the analysis of dopant concentrations in electronic materials. Basic surface science has been slow to recognize the benefits of static SIMS. John Vickerman seeks to redress the balance in Chapter 6. The application of static SIMS to the full surface chemical characterization of materials is *the* great growth area of SIMS. Nicola Reed outlines some of the many exciting areas of analysis in Chapter 7. The ability to define detailed chemistry with spatial resolutions of less than 100 nm is a major advance in surface analysis. In Chapter 8, Paul Humphrey describes the application and power of scanning or imaging SIMS. Finally, in Chapter 9, John Vickerman briefly reviews the advances which have been made in developing a quantitative SIMS by sputtered neutral mass spectrometry, SNMS. This chapter also summarizes developments in the SIMS-related techniques of laser ablation mass spectrometry and plasma desorption mass spectrometry.

There is a need for standard procedures in SIMS analysis, particularly when tackling difficult, electrically insulating materials. An attempt has been made to provide operational protocols for static SIMS and scanning SIMS in Chapters 7 and 8. Finally as a further aid to the SIMS analyst a set of Appendices have been provided containing a variety of frequently used information.

We hope this book will prove to be an effective introduction to what is developing into an exciting technique of major importance.

Manchester and Winsford
October 1988

John Vickerman,
University of Manchester Institute of Science and Technology, UK.

Alan Brown,
VG Microtrace, Winsford, Cheshire, UK.

Nicola Reed,
University of Manchester Institute of Science and Technology, UK.

CONTENTS

List of contributors xiii

1 Introducing Secondary Ion Mass Spectrometry **1**

John C. Vickerman

1.1 What is SIMS? 2
1.2 Historical development of SIMS 3
References 8

2 The SIMS phenomenon—the experimental parameters **9**

John C. Vickerman

2.1 Sputter rate 9
2.1.1 Monolayer lifetime 10
2.1.2 Dependence on primary beam parameters 10
2.1.3 Dependence on target parameters 12
2.2 Ionization probability 17
2.2.1 Dependence on ionization potential 17
2.2.2 Dependence on electronic state of target material (matrix) 18
2.2.3 Dependence on primary particle reactivity 20
2.2.4 Secondary ion energy distributions 22
2.3 Sensitivity 23
2.4 Surface charging 24
2.5 Generation of cluster or molecular ions 26
2.5.1 Inorganic cluster ions 27
2.5.2 Organic molecular-cluster ions 28
2.5.3 Matrix effects in molecular-cluster ion emission 30
2.5.4 Cluster emission and structural damage 31
References 32

3 SIMS—the theoretical models **34**

John C. Vickerman

3.1 Sputtering models 35
3.1.1 Linear collision cascade theory 36
3.1.2 The spike regime or prompt thermal sputtering 47

3.1.3 Molecular dynamics (MD) models 48
3.1.4 The binary collision approximation 56
3.1.5 Other cluster emission models 58
3.1.6 Electronic sputtering 61
3.2 Ionization 62
3.2.1 The perturbation model 62
3.2.2 The surface excitation model 63
3.2.3 The bond-breaking model 63
3.2.4 The molecular model 64
3.2.5 Desorption ionization 68
3.3 Quantitative analysis 69
3.3.1 Local thermal equilibrium theory 69
3.4 Conclusions 70
References 71

4 SIMS instrumentation 73
A. J. Eccles

4.1 Introduction 73
4.2 Vacuum systems 73
4.3 Ion guns 74
4.3.1 General principles 75
4.3.2 Ion gun components 79
4.3.3 Types of ion source 81
4.3.4 Primary beams for insulating samples 88
4.4 Mass spectrometers 91
4.4.1 General principles 92
4.4.2 Mass spectrometer components 93
4.4.3 Types of mass spectrometer 94
4.5 Conclusions 101
References 103

5 SIMS depth profiling of semiconductors 105
D. S. McPhail

5.1 Introduction 105
5.1.1 Quantification 107
5.1.2 Depth resolution 109
5.1.3 Parameters influencing quantification and depth resolution 110
5.2 The experimental arrangement 110
5.2.1 The primary ion beam 110
5.2.2 Secondary ion collection 112

5.2.3 Data logging and instrumental control 114

5.3 Quantification of the data 115

5.3.1 Quantification of the depth scale 115
5.3.2 The concentration scale: the relationship between the secondary ion count-rate and the dopant–impurity concentration 117

5.4 Description of a typical depth profile 119

5.4.1 Choice of experimental conditions 119
5.4.2 Quantification of the data 123

5.5 Further examples of SIMS depth profiles 126

5.5.1 Stationary beam analysis—a useful diagnostic technique 126
5.5.2 Mass interferences in SIMS depth profiling 127
5.5.3 The depth resolution in SIMS depth profiling 130
5.5.4 Distinguishing between uneven etching during analysis and diffusion during growth 140
5.5.5 Charging effects—how to detect and overcome them 140
5.5.6 Multi-layer analysis 146

References 148

6 The application of static SIMS in surface science **149**

John C. Vickerman

6.1 Metal surface characterization 150

6.2 Studies of thin metal films 152

6.2.1 Overlayer growth mode 155
6.2.2 Overlayer coverage 156

6.3 Oxidation of metals 156

6.3.1 Initial oxidation of chromium and copper 157
6.3.2 Initial oxidation of a copper–zinc alloy 159

6.4 Studies of chemisorption 160

6.4.1 Molecular or dissociative adsorption? 162
6.4.2 Surface coverage measurements 166
6.4.3 The energetics of adsorption 169

6.5 Studies of adsorbate structure 169

6.5.1 CO adsorbate structure 171
6.5.2 Angle-resolved SSIMS 173
6.5.3 Adsorbate–adsorbate interactions 173

6.6 Surface reactivity 175

6.6.1 Deuterated ethene adsorption and reaction on ruthenium 175
6.6.2 Oxidation of CO over Pd(111) 178
6.6.3 Other reactions 180

6.7 Studies of non-metal surfaces 181

6.8 Adsorption studies on organic surfaces 181
6.9 Conclusions 182
References 184

7 Static SIMS for applied surface analysis 186
Nicola M. Reed

7.1 Introduction 186
7.2 Surface potential 187
7.2.1 Control of surface potential—electron beams 188
7.2.2 Control of surface potential—neutral primary beams 189
7.2.3 Electron neutralization vs. ESIE 190
7.2.4 Consequence of incomplete control of surface potential 191
7.3 Comparison of ToF and quadrupole instruments 193
7.4 Primary beam effects 194
7.4.1 Fluence 194
7.4.2 Primary particle mass and energy transfer 196
7.4.3 Atom vs. ion beam damage 199
7.5 Spectral interpretation 201
7.5.1 MS-MS 201
7.5.2 Fingerprint spectra—chemical structure—organic materials 204
7.5.3 Fingerprint spectra—inorganic materials 208
7.6 Applied polymer analysis 213
7.6.1 Polymer surface—end group analysis 214
7.6.2 Surface composition—copolymer analysis 214
7.7 Adhesion 216
7.7.1 Optical coatings 216
7.7.2 Composite materials 217
7.7.3 Polymer surface treatment 219
7.8 Pharmaceutical applications 221
7.8.1 Characterization of polymer delivery systems 221
7.8.2 Drug release 221
7.9 Semiconductor surface characterization 225
7.9.1 Surface cleanliness 226
7.9.2 Microelectronics analysis 229
7.10 Oxide analysis 231
7.10.1 Surface state and reactivity of oxides 232
7.10.2 Surface segregation 234
7.11 Catalyst analysis 236
7.11.1 Catalyst preparation 236

7.11.2 Cr–Silica catalysts 238
7.11.3 Surface reactivity 239
7.12 Conclusions 240
Appendix 240
References 241

8 SIMS imaging **244**
P. Humphrey

8.1 Introduction 244
8.2 Experimental 245
8.2.1 Ion microprobe 245
8.2.2 Ion microscope 246
8.2.3 Data systems 246
8.3 Sensitivity vs. damage 248
8.3.1 Sensitivity 248
8.3.2 Damage 249
8.4 Contrast mechanisms 252
8.4.1 Introduction 252
8.4.2 Topographic contrast 253
8.4.3 Material contrast 255
8.4.4 Crystallographic contrast 256
8.4.5 Voltage contrast 256
8.4.6 Primary beam contrast 258
8.5 Specimen charging and image acquisition 259
8.5.1 Introduction 259
8.5.2 Acquisition protocol 259
8.6 Three-dimensional imaging 262
8.7 Applications of SIMS imaging 263
8.7.1 Metallurgical samples 263
8.7.2 Semiconductor devices 264
8.7.3 Non-conducting samples 267
8.8 Future developments 269
References 270

9 SIMS-related techniques **272**
John C. Vickerman

9.1 Sputtered neutral mass spectrometry 272
9.1.1 Electron bombardment post-ionization 273
9.1.2 Electron plasma post-ionization 279
9.1.3 Laser-induced post-ionization 282
9.1.4 Summary 290

9.2 ^{252}Cf plasma desorption mass spectrometry 290
9.3 Laser desorption mass spectrometry 295
References 298

Appendices **299**

1. Physical constants and useful relations 299
2. Atomic weights of the elements based on the ^{12}C standard 301
   2.1 Calculation of isotope distribution patterns 304
3. Erosion rates and monolayer lifetimes as a function of sputtering parameters 305
4. Ion velocities 307
5. Standard static SIMS spectra and their acquisition 310
6. Commonly observed fragment ions in SSIMS spectra 323
7. Sputter yields under positive ion bombardment 327
8. Relative secondary ion yields due to Cs^+ and O^- 329
9. Effects of primary ion beam energy and incidence on sputter yields and secondary ion yields from semiconductor materials 333

Index 337

Plates fall between pp 258 and 259 of the text

CONTRIBUTORS

ALAN BROWN VG MicroTrace Ltd, Nat Lane, Winsford CW7 3QH, UK

JOHN ECCLES VSW Scientific Instruments Ltd, Warwick Road South, Manchester M16 0JT, UK

PAUL HUMPHREY Surface Analysis Research Centre, Department of Chemistry, UMIST, PO Box 88, Manchester M60 1QD, UK

DAVID S. McPHAIL Department of Physics, University of Warwick, Coventry CV4 7AL, UK

NICOLA M. REED Surface Analysis Research Centre, Department of Chemistry, UMIST, PO Box 88, Manchester M60 1QD, UK

JOHN C. VICKERMAN Surface Analysis Research Centre, Department of Chemistry, UMIST, PO Box 88, Manchester M60 1QD, UK

1

INTRODUCING SECONDARY ION MASS SPECTROMETRY

The electron spectroscopies have dominated surface and near-surface analysis during the period from the mid-1960s. X-ray photoelectron spectroscopy (XPS) has been the technique of choice for quantitative elemental analysis above the 0.1 per cent of a monolayer level. Auger electron spectroscopy (AES) is somewhat more sensitive than XPS and the use of a focused electron beam as the exciting radiation has allowed high spatial resolution to be obtained, down to 20 nm in favourable cases. The particular sensitivity of both spectroscopies to the surface layers arises from the fact that the electrons emitted from the orbitals of the atoms comprising the solid can only travel very short distances before being scattered. Thus they can escape from the solid only if they originate within two to twenty layers of the surface, and the information observed arises from the top layer plus a number of sub-surface layers.

The ability to-monitor XPS chemical shifts in electron binding energy due to changes in the chemical state of surface atoms has been particularly valuable. However, in many cases these shifts are very small and the resulting chemical state information can be rather imprecise.

Increasingly there has been an analytical demand for very high elemental sensitivities down to parts-per-million (p.p.m.) and parts-per-billion (p.p.b.) levels. The requirement for top surface analysis has also grown. Thus although electron spectroscopy is now well developed and mature there are increasing analytical demands which are difficult to satisfy.

Amongst the techniques of bulk chemical characterization mass spectrometry has been spectacularly successful over the last three or four decades. Its ability to provide both elemental composition data and detailed chemical structure information in the analysis of volatile materials is well known and has revolutionized organic and bio-organic chemistry. The mass spectrometric analysis of involatile materials has, however, always been more difficult. The methods developed for transferring the atoms and molecules of interest from a solid surface into the electron bombardment source for ionization—spark source and field desorption—are difficult, require the solid to be in a particular form, and require special operator skills. Whilst these methods can yield useful data about the bulk of the solid, their inherent destructiveness meant that mass spectrometry was not able to feature as a technique in true surface analysis.

It is in the area of involatile materials analysis and the surface analysis of solids that secondary ion mass spectrometry (SIMS) has developed into a major technique over the last ten to twenty years. As we shall see, it is able to provide the sensitivity and surface chemical specificity lacking in the electron spectroscopies.

1.1. What is SIMS?

SIMS is the mass spectrometry of ionized particles which are emitted when a surface, usually a solid, though sometimes a liquid, is bombarded by energetic primary particles. The primary particles may be electrons, ions, neutrals, or photons. The emitted (so-called 'secondary') particles will be electrons; neutral species, atoms, and molecules; atomic and cluster ions. It is the secondary ions which are detected and analysed by a mass spectrometer. It is this process which provides a mass spectrum of a surface and enables a detailed chemical analysis of a surface or solid to be performed. In a development of SIMS those particles initially emitted as neutrals may be post-ionized and contribute to the analysis.

Figure 1.1 schematically outlines the mechanism of SIMS as it is at present understood. The detailed theory will be discussed in Chapter 3. In the case of ion or atom bombardment the energy of the primary beam is transferred to the atoms in the solid by a collision, a billiard-ball-type process. A cascade of collisions occurs between the atoms in the solid; some collisions return to the surface and result in the emission of atoms and atom clusters (a process known as sputtering), some of which are ionized in the course of leaving the surface and can be analysed with a mass spectrometer. At first sight the process seems conceptually very simple. Readers familiar with mass

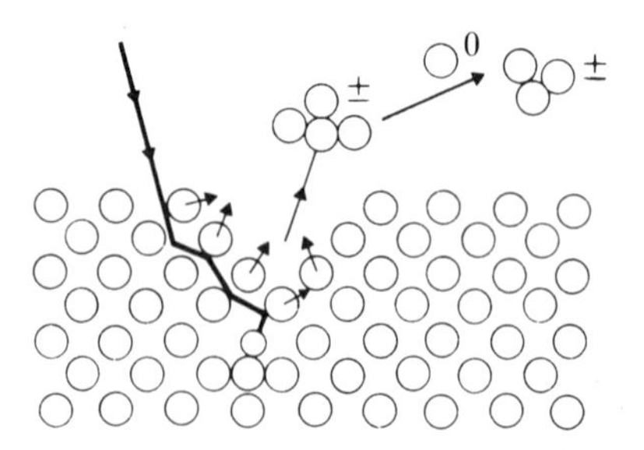

FIG 1.1. Diagram of the SIMS phenomenon indicating the collision of the primary particles with a solid surface and the emission of secondary particles. Reproduced with permission of The Royal Society of Chemistry, from J. C. Vickerman (1987). *Chem. in Brit.*, 969–74.

spectrometry will appreciate that the technique apparently has considerable potential. Nature, however, frequently insists on payment for her benefits! SIMS is no exception.

We will see in Chapters 2 and 3 that the fact that ionization coincides with sputtering introduces considerable problems in understanding the mechanism of the process and means that careful calibration is necessary when quantifiable data are required. Furthermore, particle bombardment results in the removal of atoms and clusters from the solid surface. This will introduce sample damage. How can useful surface analysis data be generated? This could be a problem, but Chapters 6 and 7 in particular will show how detailed chemical state data can be obtained which are complementary to, and in many cases extend those accessible from the mature techniques of electron spectroscopy.

The basic arrangement for the SIMS experiment is shown in Fig. 1.2. The primary particle source is capable of producing a beam of ions or atoms in the 0.5–50 keV energy range. The emitted secondary ions are collected by an ion transport optics and directed into a mass spectrometer which may be of the magnetic sector, quadrupole, or time-of-flight type. The instrumentation used is described in some detail in Chapter 4.

1.2. Historical development of SIMS

It was Sir Joseph John Thomson who first observed and identified in 1910 the emission of positive secondary ions when primary ions bombarded a metal surface in a discharge tube.[1] However, it was not until 1931 that Woodcock published the first negative secondary ion spectra obtained by bombarding NaF and CaF_2 with 500 eV Li^+ ions. As Hönig has pointed out, it was probably Vieböck in Vienna who, in 1949, under the supervision of Herzog,

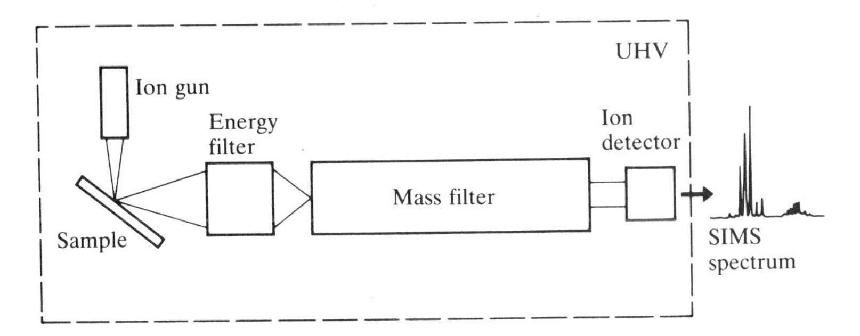

FIG 1.2. Schematic diagram of the main components of the SIMS experiment. Reproduced with permission of The Royal Society of Chemistry, from J. C. Vickerman (1987). *Chem. in Brit.* 969–74.

first built a modern SIMS instrument and recorded secondary ion mass spectra from metals and oxides.[2,3] This instrument used electric fields to accelerate the primary ions to the target and an electric field parabola spectrograph to separate the secondary ions.

In the early 1950s Hönig and co-workers at the RCA laboratories developed a SIMS instrument which was used quite extensively. However, it was in the 1960s that really rapid progress began. This was enabled by the advances made by Liebl and Herzog in ion optical design.[4] The first spectrometers were used in a mode of operation known as **dynamic SIMS** (see Chapter 5). In order to obtain a very high yield of secondary ions a high flux of primary ions is directed at the material surface. The surface is eroded very rapidly and it is possible to monitor changes of elemental composition with depth and thus a **depth profile** may be generated (see Fig. 1.3).[5]

Two types of SIMS instrument were developed: in 1960 Castaing and Slodzian produced their prototype ion microscope (see Chapter 4).[6] This uses a broad ion beam to bombard the sample, but the secondary ion collection optics operates like a microscope to give about 1 μm spatial resolution. In 1967 Liebel introduced his ion microprobe.[7] This uses a finely focused ion beam which is raster-scanned over the surface to erode a rectangular crater for depth profiling or to produce a chemical map. Lateral resolution of 1 μm was obtained using a duoplasmatron ion source.

In the 1980s much-improved spatial resolution has been obtainable using the liquid metal ion sources developed by Prewett and co-workers.[8] These sources operated by field ionization of metal ions such as gallium from a

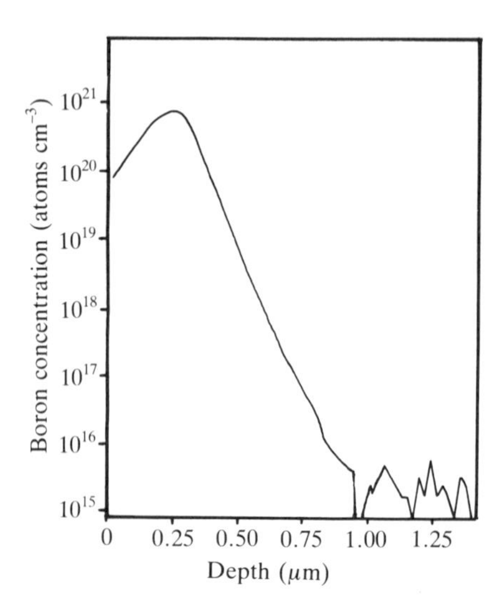

FIG 1.3. SIMS depth profile of a 70 keV boron implant in silicon.

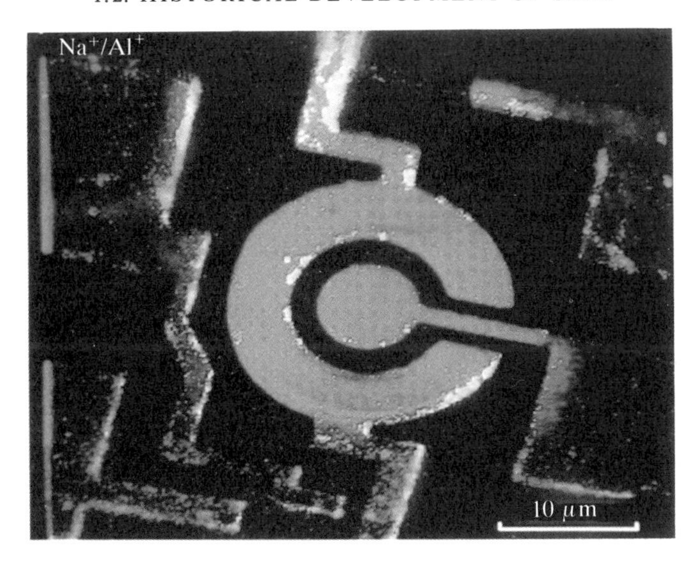

FIG 1.4. SIMS image of sodium contamination (grey) on the aluminum (white) tracks of semiconductor device.

sharp metal tip. Beam diameters of 50 nm or below are possible with high beam brightness. By raster-scanning the beam across an area of surface and collecting the secondary ions at each point a chemical image can be generated (see Fig. 1.4). This has promoted the exciting development of **imaging SIMS**, a 'scanning chemical microscopy'.

Clearly, analytical conditions using highly destructive primary beam densities are not suitable for surface analysis. At the beginning of the 1970s true *surface* analysis by SIMS was made possible by a modification of the technique demonstrated by Benninghoven.[9] **Static SIMS (SSIMS)** uses a primary beam whose current density is maintained at a *very low* level so that secondary ions are emitted from areas not previously damaged and the surface monolayer lifetime is many hours, well in excess of the time required for analysis. Good sensitivity is realized because of efficient collection of ions and the development of highly sensitive single particle ion counting detectors. The importance of SSIMS for surface analysis lies in the possibility it offers of studying not only the elemental composition but also the *chemical structure* of surfaces. This is because the surface mass spectrum includes cluster ions as well as elemental ions (see Fig. 1.5). These cluster ions reflect the surface chemistry in a detailed way. Initially, most static SIMS was performed using quadrupole mass analysers; however, it was realized in the mid-1980s that the more efficient ion collection of time-of-flight spectrometers would yield a considerable improvement in performance and these are now being increasingly used.

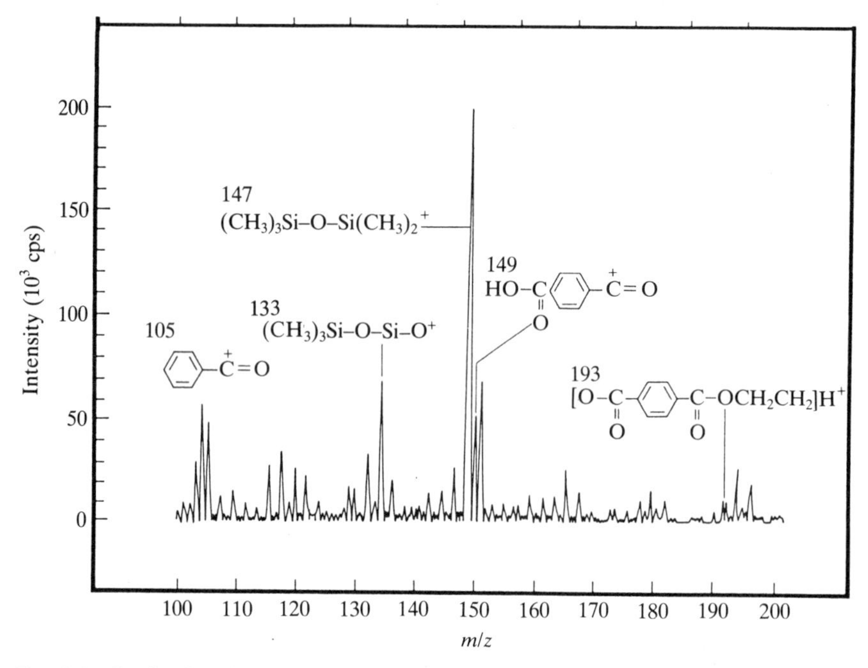

FIG 1.5. Static SIMS spectrum of silicon contamination of a polyethylene terephthalate (PET) fibre. (FABMS using 2 keV Ar^+, at 10^9 particles cm^{-1} s^{-1}.) Reproduced with permission of the Royal Society of Chemistry, from J. C. Vickerman (1987) *Chem. in Brit.* 969–74.

In common with many techniques of surface analysis which use a charged particle as the probe species, ion bombardment of a poorly conducting surface can result in surface charging which in the SSIMS experiment can result in spectral loss or instability. Flooding the surface with low-energy electrons can be a solution although stable neutralization of surface charge is not always easy. In addition, electron bombardment can have a deleterious effect on delicate surfaces. Although neutral beams had been used by earlier workers to generate secondary ions,[10] in the period 1979–81 Vickerman and co-workers showed that the charging problem could be greatly reduced by using a neutral primary beam instead of an ion beam and, in consequence, insulator materials could be routinely analysed.[11,12] This modification of SSIMS is sometimes known as **fast atom bombardment (FAB) SIMS**. Although the initial development of the technique was for surface analysis, where it is now extensively used (see Chapter 7), in the early 1980s it was spectacularly applied in organic mass spectrometry by Barber and co-workers for the analysis of involatile organic and bio-organic compounds.[13]

In most applications of SIMS to date it is the secondary ions emitted in the bombarding process which are detected. In other words, the ionization process occurs during particle emission from the surface. Thus the yield of secondary ions of a particular type from the surface of a solid is not only dependent on the ionization potential of the species but also strongly dependent on the electronic state of the solid. This means that quantitative analysis requires careful calibration procedures for each solid matrix investigated. In the SIMS process only a small proportion of the particles emitted are in fact ions; most are neutrals. In the early 1980s a good deal of study went into ways of decoupling the emission and ionization processes by post-ionizing the sputtered neutrals. **Sputtered neutral mass spectrometry (SNMS)**,[14] using electron plasmas or laser-induced multi-photon ionization, allows quantitative elemental analysis without the matrix calibrations. Discussion of the main features of the technique and its instrumentation will be covered in Chapter 9. The main application of SNMS at present is as a variant of dynamic SIMS. However, using laser-induced post-ionization there are prospects for SNMS in SSIMS.

In parallel with these developments in SIMS a number of alternative ion desorption methods appeared in the mid-1970s. These will also be briefly reviewed and referred to in Chapter 9. First, Macfarlane and co-workers investigated the use of fission fragments from the decay of ^{252}Cf as a sputter source.[15] These fragments have energies in the MeV region. They found that bombarding the rear of a thin foil covered with an organic sample generated molecular and fragment cluster ions which enabled the characterization of the chemical structure. In fact, the spectra generated have subsequently been shown to be qualitatively very similar to those obtained using SIMS. The method is known as **^{252}Cf plasma desorption** and has been very successful in generating molecular ions from very large bio-molecules ($m/z > 40\,000$). A development of this approach is to use very heavy, high-energy ions, e.g. 90 MeV $^{127}I^{14+}$, using a tandem accelerator. The resulting data are very similar to plasma desorption mass spectrometry (PDMS).[16]

Laser desorption is the other main method which began to develop rapidly at about the same time. The use of UV, nanosecond pulsed lasers to vaporize and ionize micron volumes of solid led to a rapid method of elemental analysis. Initially, the laser was fired from the rear of the sample, so thin samples were necessary and only bulk analysis was possible. More recently, front surface desorption with lower laser powers has permitted surface layer analysis. At high laser power elemental analysis predominates, whereas at lower powers cluster ions are also observed, yielding cluster ion spectra similar to SSIMS and PDMS.[17]

SIMS and its many variants has brought the analytical power of mass spectrometry to surface and near-surface analysis. The following chapters seek to provide a general understanding of the SIMS process and an insight into the analytical potential of the technique.

References

1. Thomson J.J. (1910). *Phil. Mag.*, **20**, 252.
2. Hönig, R.E. (1986). In *Secondary Ion Mass Spectrometry, SIMS V*, Springer Series in Chemical Physics, Vol. 44 (ed. A. Benninghoven, R.J. Colton, D.S. Simons, and H.W. Werner), p. 2. Springer Verlag, Berlin.
3. Herzog, R.F.K. and Vieböck, F.P. (1949). *Phys. Rev.*, **76**, 855L.
4. Liebl, H.J. and Herzog, R.F.K. (1963) *J. Appl. Phys.*, **34**, 2893.
5. Zinner, E. (1983) *J. Electrochem. Soc.*, **130**, 199C.
6. Castaing, R. and Slodzian, G. (1962) In *J. Microscopic.* (1962), **1**, 395.
7. Liebl, H.J. (1967). *J. Appl. Phys.*, **38**, 5277.
8. Prewett, P.D. and Jefferies, D.K. (1980). *J. Phys. D: Appl. Phys.*, **13**, 1747.
9. Benninghoven, A. (1970). *Z. Physik*, **230**, 403.
10. Devienne, F.M. (1973). *Vide*, **167**, 193.
11. Surman, D.J. and Vickerman, J.C. (1981). *Appl. Surf. Sci.*, **9**, 108.
12. Surman, D.J., van den Berg, J.A., and Vickerman, J.C. (1982). *Surf. Interface Anal.*, **4**, 160.
13. Barber, M., Bordoli, R.J., Sedgwick, R.D., and Tyler, A.N. (1981). *Nature*, **293**, 270.
14. Oechsner, H. (1984). In *Thin Film and Death Profile Analysis*, Topics in Current Physics, Vol 37, p. 63. Springer Verlag, Berlin.
15. Macfarlane, R.D. and Torgerson, D.F. (1976). *Science*, **191**, 920.
16. Sunqvist, B., Hedin, A., Hakensson, P., Kamensky, I., Salepour, M., and Sawe, G. (1985). *Int. J. Mass Spectrom. Ion Proc.*, **65**, 69.
17. Hillenkamp, F. (1983). In *Ion Formation from Organic Solids (IFOS II)*, Springer Series in Chemical Physics, Vol 25 (ed. A. Bennighoven), p. 190. Springer Verlag, Berlin.

2

THE SIMS PHENOMENON — THE EXPERIMENTAL PARAMETERS

Two processes are involved in the SIMS phenomenon: the emission of the particle or sputtering, and particle ionization. There is a wealth of experimental data concerned with sputtering but we are still some way from a basic understanding of the mechanism by which the impact of a high-energy particle with a solid surface gives rise to the emission of low-energy secondary ions from the surface. As our theoretical considerations will show, there is still considerable discussion as to whether the sputtering and ionization are coincident or consecutive.

It is generally accepted that on impact of the primary particle with the surface atoms, energy is transferred as the primary particle penetrates the surface and is brought to a standstill, and collision sequences between atoms in the near-surface region are initiated. Some of these collision cascades result in the dissipation of energy into the bulk whilst others return to the surface, causing the emission of secondary ions, atoms, or molecules. The range of penetration of the primary particle will be dependent on the particle mass, its energy, its angle of incidence, and the properties of the impacted solid. A number of computer programs have been developed based on models of the sputtering process to calculate the primary particle range (see Chapter 3).

Two features of this process favour the application of SIMS to surface analysis: it will be clear that, in general, the emitted particles are released from a point remote from the initial impact zone where structural disruption may be greatest; and the cascade of collisions which result in emission will dissipate much of the original energy imparted to the surface and thus secondary particles will be lifted off rather gently.

In this chapter we will briefly describe the basic experimental parameters underlying the SIMS process and in Chapter 3 we will consider some of the mechanistic models which attempt to describe the sputter emission of ions.

2.1. Sputter rate

In static SIMS we are concerned to have high surface sensitivity with minimal surface damage, whereas in dynamic SIMS layers are removed rapidly to generate depth profiles with high sensitivity for trace elements. The surface removal rate is important in both cases and is dependent solely on the sputter rate, whereas sensitivity is also a function of ionization probability.

2.1.1. Monolayer lifetime

A crucial parameter is the lifetime of the surface monolayer

$$t_m = N_s/(6.2 \times 10^{18} I_p Y) \qquad (2.1.1)$$

where the density of atoms in the surface is N_s cm^{-2} and may be taken to be 10^{15} cm^{-2}; Y is the number of secondary particles removed per primary particle impact or sputter rate, which usually lies between 0.1 and 10;[1] and $6.2 \times 10^{18} I_p$ is the number of primary particles incident at the surface cm^{-2} s^{-1} A^{-1} for a primary ion current I_p. Thus when I_p is 1 nA ($\sim 10^{10}$ particles cm^{-2} s^{-1}) and $Y=1$, then $t_m \sim 10^5$ s. These would seem to be good conditions for SSIMS. However, *each* primary particle impact is thought to affect an area of about 10 nm^2. Thus it would only require 10^{13} impacts to influence the whole of a 1 cm^2 sample area: $[10^{13} \times (10 \times 10^{-18}\ \mathrm{m}^2)] = 10^{-4}\ \mathrm{m}^2 = 1\ \mathrm{cm}^2$. If a 1 nA primary beam current is used, the surface structure might begin to be noticeably affected after about 1500 s. Clearly, the conditions for SSIMS could be rather stringent and will depend greatly on the material being analysed and it's sensitivity to bombardment-induced damage.

For dynamic SIMS we are more interested in the rate of removal of material or the time required to complete a depth profile

$$t_{dp} = d/(2 \times 10^{-10})\,(N/Y6.2 \times 10^{18} I_p) \qquad (2.1.2)$$

where d is the depth to be profiled and the approximate depth per layer is 2×10^{-10} m. Thus if we wish to profile a 1 μm depth in about 10 000 s we will need a primary beam current of about 5×10^{-5} A cm^{-2}. If the depth profiling is performed by scanning a focused primary beam over an area about 200 μm square, the total area is 4×10^{-4} cm^2, and thus the current required in the beam to carry out this depth profile is 20 nA.

Clearly, the magnitude of Y is important and it has been shown to be dependent on the primary beam parameters as well as on the target properties.

2.1.2. Dependence on primary beam parameters

In view of the collisional nature of the process it is not surprising to find that, in general, Y increases with beam energy and with the primary particle mass although the increase is far from linear in either parameter. At high energies (>10 keV) a maximum between 1 and 10 is reached; see, for example, Fig. 2.1 for aluminium.[2] Similar curves are obtained for all the elements and are summarized by Andersen and Bay in Ref. 1. The sputter rate is also a

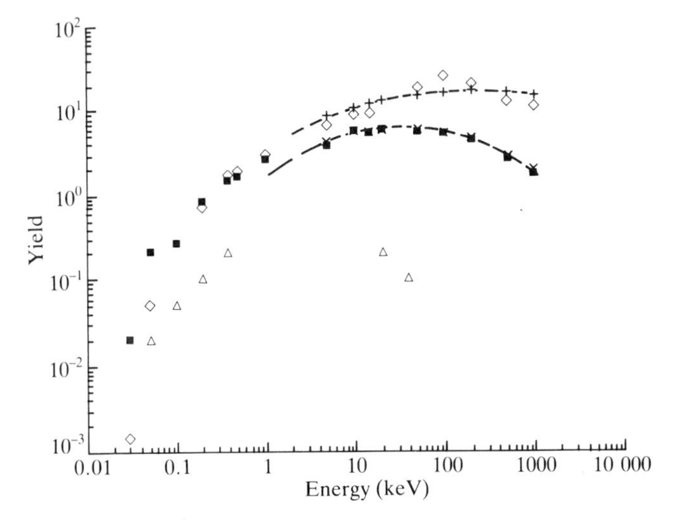

FIG 2.1. Experimental sputter yield data for aluminium for a number of primary ions as a function of primary ion energy. Δ, He; ■, Ar (experimental); ◇, Xe (experimental); ×, Ar (theoretical); +, Xe (theoretical). The figure summarizes data from a wide range of sources.

function of impact angle and reaches a maximum at about 70° to the surface normal as a consequence of depositing most of the impact energy in the surface layers (see Fig. 2.2).[3]

It is frequently assumed that the sputter rate is independent of the charge

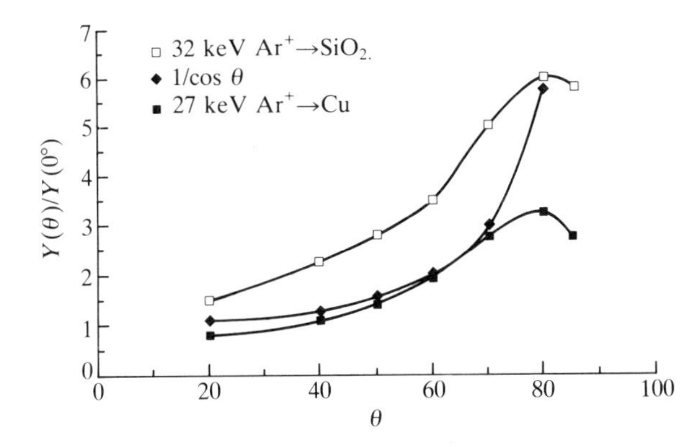

FIG 2.2. The variation of sputter yields as a function of the angle of incidence of the primary beam to the surface normal. The $1/\cos\theta$ curve represents the theoretical calculations by Sigmund (see Chapter 3). Reproduced with permission from Ref. 3.

state of the primary particles. Whilst this is almost certainly so in the case of metal targets, where the charge on the incoming ion will be neutralized before impact, it is unlikely to be the case with poor conductors or insulators. There is good evidence now for an electronic contribution to sputtering. Table 2.1 shows that there is a significantly larger sputter yield when poorly conducting materials are bombarded with an ion beam than with a neutral beam.[4]

TABLE 2.1

Measured sputter yields for 10 keV ion and atom bombardment at normal incidence

	Sputtering yield		
Material	Ions	Atoms	Ratio
Au	4.5	4.5	1.0
Ta_2O_5	3.5	2.2	1.6
Si	1.7	0.7	2.4
GaAs–GaAlAs	3.6	2.4	1.5
Glass	2.9	1.1	2.6

2.1.3. Dependence on target parameters

The physical state of the target has a significant influence on the sputter rate.

(a) *Crystallinity*. The sputter yield is sensitive to the crystallinity of the surface, and there is a large variation in yield from one surface plane to another (see Fig. 2.3). The yield is greatest for the close-packed hexagonal (111) face of copper. The more open (100) and (110) faces are less easily sputtered whilst the polycrystalline sample shows an intermediate yield but a displaced energy maximum.[5] Channelling of the incoming beam and preferred emission along close-packed directions are the reasons for this behaviour.

(b) *Topography*. Although these effects are not seen on polycrystalline materials, surface topography can have a strong influence on the average sputtering yield. Roughness which is comparable with or larger than the dimensions of a cascade generally leads to higher yields than from a flat surface.[6] Pore structure may lead to retrapping of sputtered material and hence to a lowering of the sputtered yield. At high doses new surface topographies are generated (see Fig. 2.4), which will in turn affect the sputter rate.[7,8]

(c) *Atomic number*. The variations in sputter yield at a specified primary beam energy as a function of the atomic number of the target can be derived from data of the type assembled by Andersen and Bay (see Fig. 2.5). The

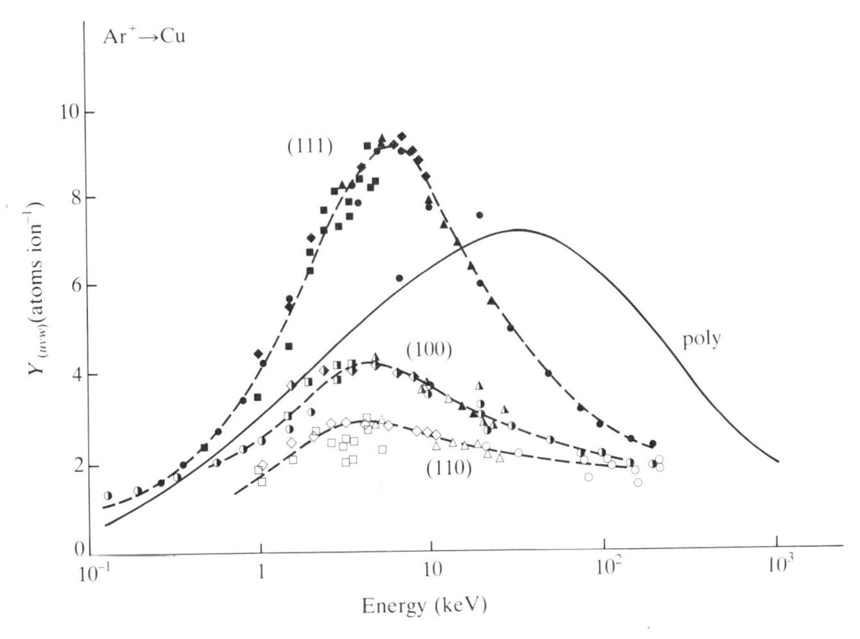

FIG 2.3. The energy dependence of the sputtering yields for Ar^+ bombardment of polycrystalline copper and the (110), (100), and (111) planes of copper. Reproduced with permission from Ref. 5.

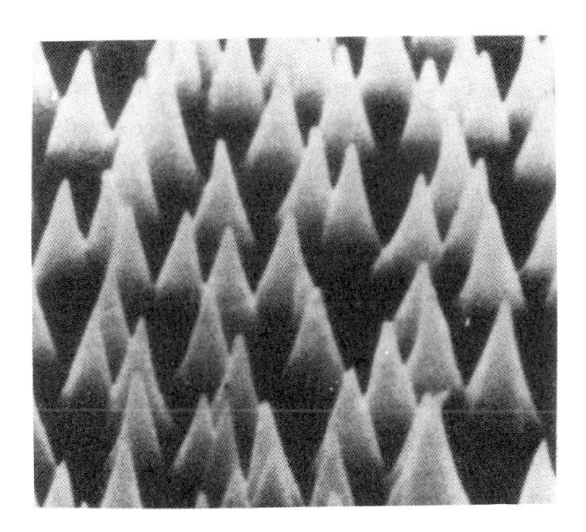

FIG 2.4. Dense pyramid array developed on a copper single crystal with orientation (11,3,1) after a bombardment dose of 10^{19} of 40 keV Ar^+ ions cm^{-2}. Magnification is $\times$ 10,000. Reproduced with permission from Ref. 8.

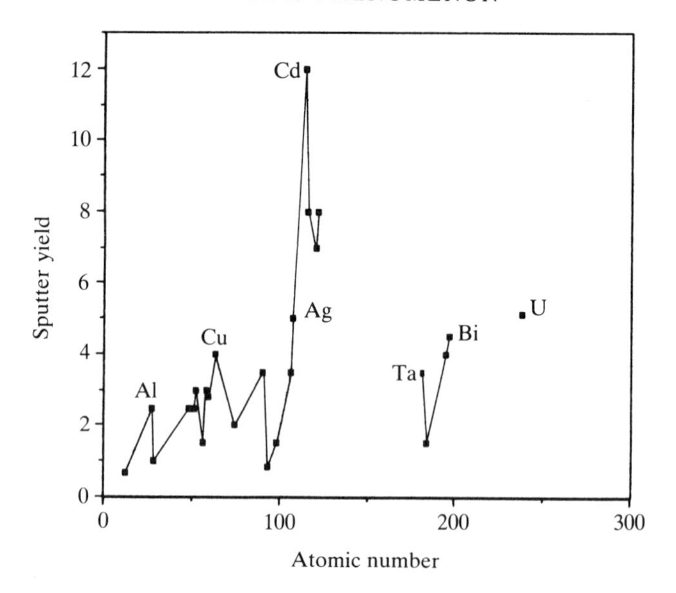

FIG 2.5. The variation of sputter yield as a function of atomic number for bombardment by 2 keV Ar^+. The data are derived from the sputter yield plots in Ref. 1.

yield varies by about a factor of three to five through the Periodic Table and is related to the surface binding energies of the elements.

(d) *Kinetic energies of sputtered particles.* The sputtered atoms are emitted with a range of kinetic energies. The distribution peaks at around 1–5 eV but has a high-energy tail extending up to 100 eV or more (see Fig. 2.6).[9]

(e) *Preferential sputtering.* Sputtering from multi-element targets results in the observation of preferential loss of one of the components. In general, the experimental data show that for most alloys and compounds the observed surface enrichment is in agreement with the sputtering yields of the pure elements, i.e. the component with the lower elemental yield becomes enriched under ion bombardment. The earliest observations seemed to suggest that surface enrichment was correlated with mass such that the lighter species were lost preferentially; for example, oxygen was lost from many oxides. Simple collisional considerations would make this appear reasonable. However, Kelly lists alloys which preferentially lose the lighter component, e.g. Al–Au or Cu–Pd; alloys which preferentially lose the heavier species, e.g. Au–Ni or Pb–Sn; and finally alloys such as Au–Cu and Cr–Pd, whose composition is unchanged by sputtering.[10,11] It is clear that significant changes in structure and surface composition do occur and some semi-empirical rules are possible for some compounds. For example, it has been found that the oxides lying above PbO in Fig. 2.7 preferentially lose oxygen

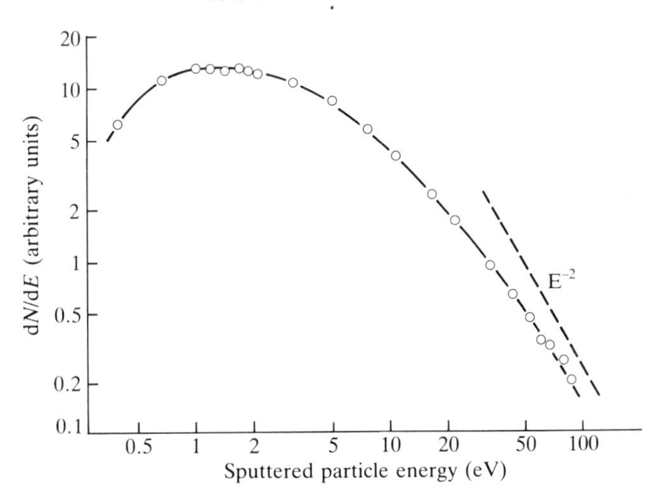

FIG 2.6. The energy distribution of sputtered Au atoms due to the impact of 15 keV Ar^+ on polycrystalline gold. Reproduced with permission from Ref. 9.

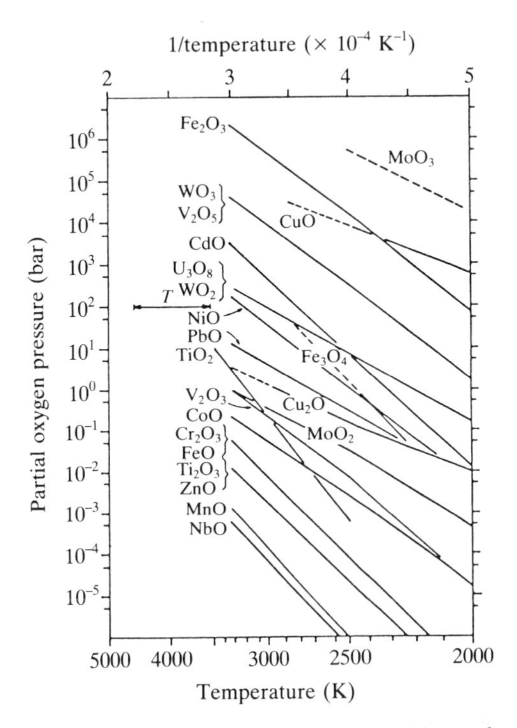

FIG 2.7. The partial pressure of oxygen above the oxide at decomposition as a function of temperature for a range of oxides. The oxides lying above PbO show bombardment-induced loss of oxygen whereas most of the others do not. Reproduced with permission from Ref.10.

whereas those below generally do not. Clearly, there would seem to be a dependence on binding energy involved. This will be considered further in our review of the theory of sputtering. However, a very useful listing of the observed preferential sputtering effects is given in Ref. 10.

(f) *Sputter-induced mixing*. Experimental data show that compositional changes do occur over much larger depths than are attributable to preferential sputtering. Changes occur in what is called the altered layer due to various transport processes which are initiated by the bombardment event. These include thermal diffusion, radiation-enhanced diffusion, recoil implantation or cascade mixing, thermal surface segregation and radiation-enhanced segregation. It is often difficult to determine which process is responsible without extensive theoretical modelling. The effect can be seen in Fig. 2.8, where the surface composition of a Pd–Ag alloy is monitored by AES during Xe^+ bombardment.[12] The alloy is first bombarded with 0.5 keV Xe then the energy of the beam was step changed to 5 keV. It can be seen that there is a transition region where there is a rapidly changing concentration which is initially rich in Ag. Thus although there was a steady concentration in the surface layer under 0.5 keV bombardment, just ahead of the beam in the sub-surface layers the concentration was rather different due to the various mixing processes. After equilibrium is reached again, now under 5 keV bombardment, a step change to 0.5 keV is made and a different altered layer is revealed which is apparently deeper and in this case enriched in Pd. Analysis of these data shows that the depth of the altered layers are comparable with the calculated range of the primary ions at 0.5 and 5 keV.

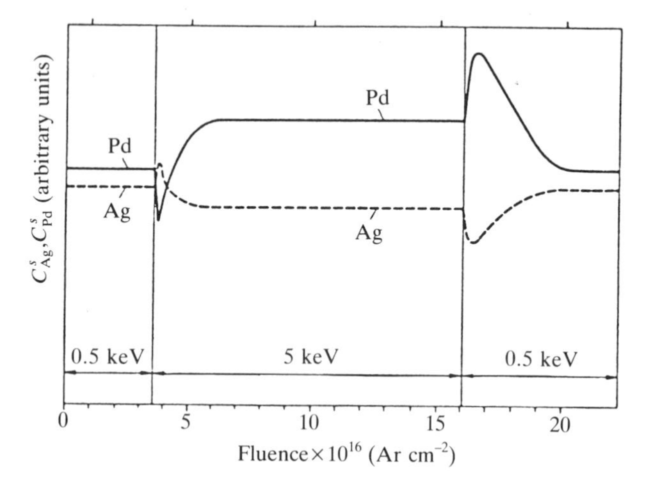

FIG 2.8. Changes in the Ag and Pd surface concentrations, C^s_{Ag} and C^s_{Pd}, (measured by AES) with fluence and Xe^+ energy for an 80 at. % Ag–Pd alloy. Reproduced with permission from Ref. 12.

Such bombardment-induced effects are important in depth profiling studies because they may distort concentration changes and broaden interfaces.

(g) *Emission of cluster particles.* All the observations referred to are concerned with the yield of atomic particles from metallic targets. In SSIMS we are also interested in the sputtering of molecular or cluster particles from multicomponent inorganic and organic materials. Other than for so-called chemical sputtering,[13] where the primary beam is a reactive gas, there has been little systematic study of the experimental dependence of the emission of multi-atomic neutral species on primary particle or indeed target parameters, although, as we shall see in the next section and ensuing chapters, there is a considerable body of data on the emission of cluster *ions*.

Oechsner has suggested from SNMS studies that clusters from inorganic materials may be generated by either recombination of atomic species above the surface or by direct emission of clusters by fragmentation of the lattice. In the former case, the elements in a cluster may not arise from nearest neighbours in the solid whereas in the latter they would (see Section 3.1.3).

2.2. Ionization probability

Whilst the yield of sputtered particles is important, as far as SIMS analysis is concerned it is the yield of secondary ions which is crucial

$$i_s^M = I_p Y \alpha^+ \theta_M \eta \qquad (2.2.1)$$

where i_s^M is the secondary ion current of an element or species M, I_p is the primary particle flux, α^+ is the ionization probability of M, θ_M is the fractional concentration of M in the surface layer and η is the transmission of the analysis system.

Clearly, sensitivity to M is governed by α^+ and η. The transmission of the analysis system is fixed and will be considered in the next section.

2.2.1. *Dependence on ionization potential*

It is well known that the ionization potential of the elements follows the regular group variations of the Periodic Table. The positive ion yields of different elements sputtered from a *common matrix* also exhibit a regularity as a function of atomic number (see Fig. 2.9), which implies an inverse dependence of ion yield on ionization potential. In fact, there is an inverse exponential dependence on the ionization potential of the sputtered atom (there is some evidence for a similar dependence of negative ion yield on electron affinity).[14]

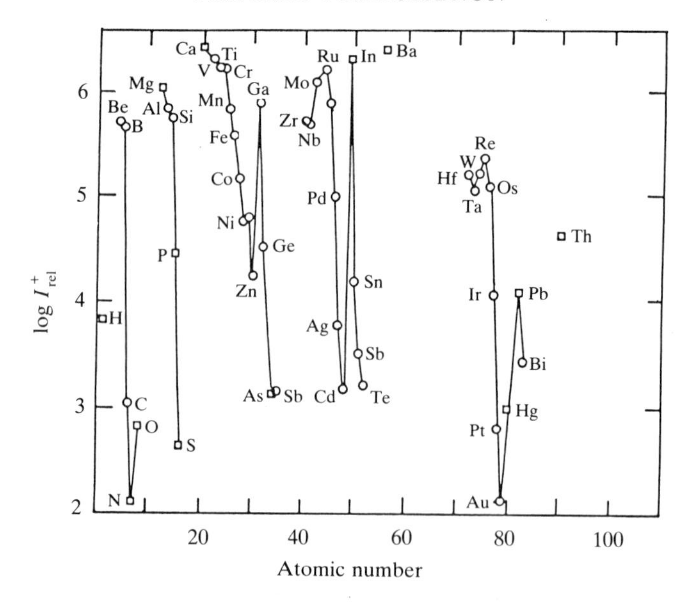

FIG 2.9. The variation of positive ion yield as a function of atomic number for 1 nA, 13.5 keV O^- bombardment. ○, from elements; □ from compounds. Reproduced with permission from Ref. 19.

Figure 2.9 shows that the variation in ion yield across the Periodic Table is very considerable. Thus, although the sputter yields in Fig. 2.5 only varied by about a factor of five, the ion yields vary by more than 10^4. The sensitivity of SIMS to the detection of Ca will be 10^4 greater than to Au. SIMS is a technique which requires careful calibration if quantitative data are to be obtained. However, the influence of the electronic state of the matrix makes this even more complicated.

2.2.2. Dependence on electronic state of target material (matrix)

Positive ion yields are greatly enhanced in the presence of oxygen or other electronegative species at the surface. The adsorption of a complete coverage of carbon monoxide on a transition metal is sufficient to raise the yield of M^+ by one to two orders of magnitude (see Fig. 2.10).[15] This adsorption is usually accompanied by an *increase* in work function. Thus the increase in positive ion yield is associated with an increased probability of ion escape from the surface as the work function increases.

If a metal surface is oxidized it is also found that the positive ion yield increases.[16] Figure 2.11 shows the increase in the Si^+ yield as a silicon surface

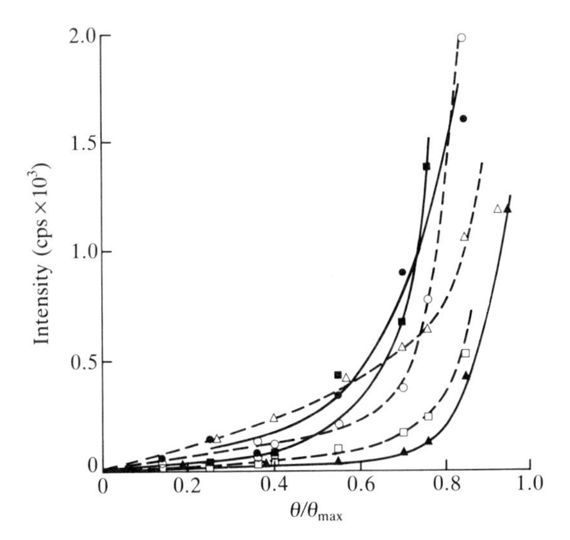

FIG 2.10. The variation of Pd_n^+ and Pd_nCO^+ ions as a function of the relative coverage of CO on Pd (111) at 300 K. ○, Pd^+; ●, Pd $CO^+ \times 10$; □, $Pd^+{}_2$; ■, $Pd_2CO^+ \times 10$; △, $Pd_3{}^+ \times 10$; ▲, $Pd_3CO^+ \times 10$. Reproduced with permission from Ref. 15.

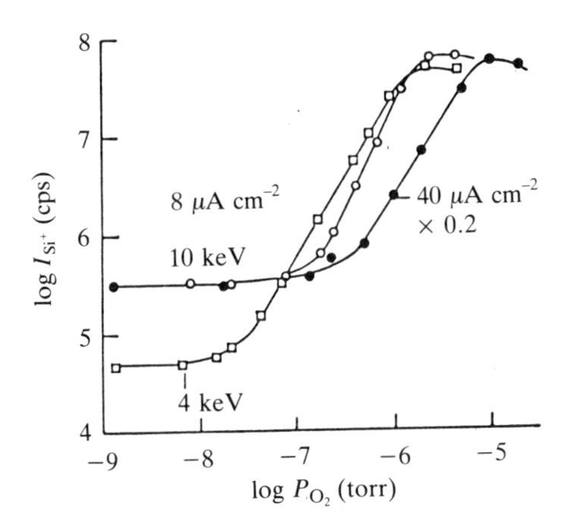

FIG 2.11. The increase in the Si^+ yield with increasing partial pressure of oxygen in the analysis chamber during Ar^+ bombardment of silicon under various energy and fluence conditions. □, 4 keV and 8 $\mu A\ cm^{-2}$; ○, 10 keV and 8 $\mu A\ cm^{-2}$; ●, 10 keV and 40 $\mu A\ cm^{-2}$. Reproduced with permission from Ref. 17.

TABLE 2.2

Secondary ion yields from clean and oxidized metal surfaces (from Ref. 18)

Metal	Clean metals M^+ yield	Oxide M^+ yield
Mg	0.01	0.9
Al	0.007	0.7
Si	0.0084	0.58
Ti	0.0013	0.4
V	0.001	0.3
Cr	0.0012	1.2
Mn	0.0006	0.3
Fe	0.0015	0.35
Ni	0.0006	0.045
Cu	0.0003	0.007
Ge	0.0044	0.02
Sr	0.0002	0.16
Nb	0.0006	0.05
Mo	0.00065	0.4
Ba	0.0002	0.03
Ta	0.00007	0.02
W	0.00009	0.035

is oxidized by flowing oxygen over the surface during analysis.[17] Table 2.2 indicates the extent of the increase in ionization probability which results from oxidation. Obviously, since the increases are often of the order of 10^2–10^4 this will lead to greatly increased sensitivities.[18] However, it can be seen that the extent of the increase varies between the elements, which complicates further the calibration procedures required for quantification. There is evidence that the effect is related to the increase in work function which occurs for most metals on oxidation.

2.2.3. Dependence on primary particle reactivity

The ionization probability will not increase with the mass of the primary particle alone. Thus although the secondary ion current will increase when Xe^+ is used instead of Ar^+, this is a consequence of the increase in Y shown in Fig. 2.1. If a reactive species is used in the primary beam, following from Section 2.2.2, a change in the ionization probability can be obtained. Thus oxygen primary beams, O^- or O_2^+, are frequently used where high sensitivity is required. If a high beam flux is used, the area around the bombardment zone becomes oxidized and hence the ion yields are of the same order as for an oxidized surface.

If Cs^+ is used in the primary beam it is found that the negative ion yield for some elements is increased. Generally, it is those elements whose positive ion yield was not great under oxygen bombardment which display a significant negative ion yield under Cs^+. Figure 2.12 shows the varying negative ion yield with atomic number.[19] Again, the variation in sensitivity between the elements is considerable, with Au amongst those having the highest yield and Ca having the lowest.

Obviously, in the analytical situation, bombardment with such reactive ions would not be carried out when *surface analysis* is required. Oxygen and caesium ion bombardment are only used in dynamic SIMS when the highest sensitivities are required. In these circumstances Fig. 2.13 is useful in enabling the analyst to decide which primary ions should be used for maximum sensitivity for the element of interest. This figure plots the ratio of negative ion yield (M^-) under Cs^+ bombardment to positive ion yield (M^+) under O^-. Hence where $M^-/M^+ > 1$ Cs^+ should be used, and where it is < 1 O^- should be used.

In SNMS, where ionization of sputtered neutrals is effected *after* emission, the elemental ion yields are mainly dependent on the ionization

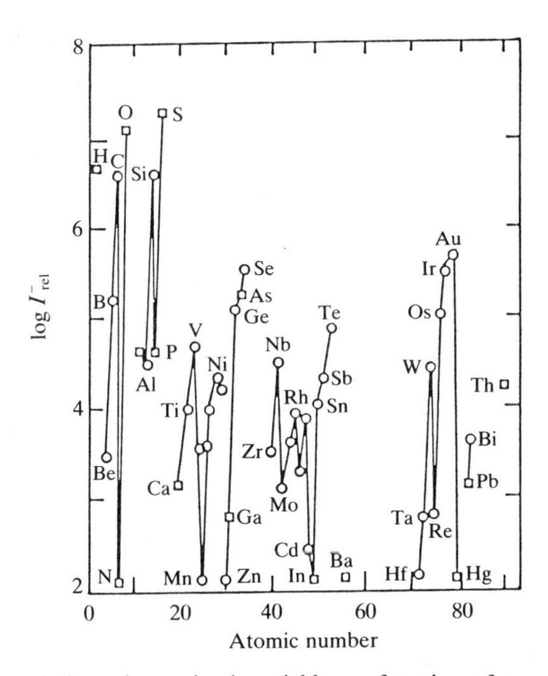

FIG 2.12. The variation of negative ion yield as a function of atomic number for bombardment by 1 nA, 16.5 keV Cs^+. ○, from the elements; □, from compounds. Reproduced with permission from Ref. 19.

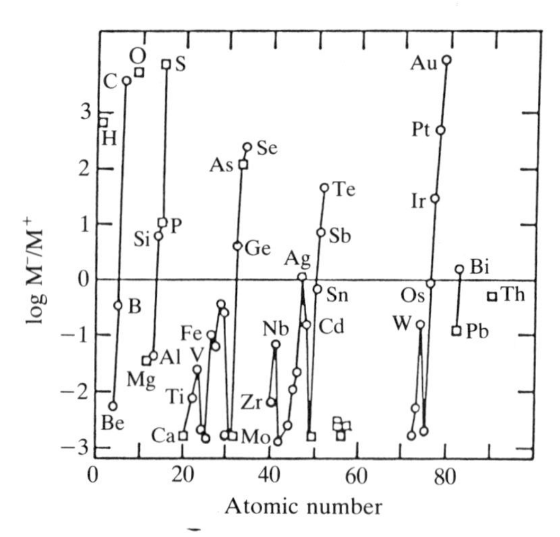

FIG 2.13. The ratio of the negative ion yield (M^-) under Cs^+ bombardment to positive ion yield (M^+) under O^- bombardment as a function of atomic number. Reproduced from data in Ref. 19.

potential of the element whose values vary by less than an order of magnitude through the Periodic Table. The relationship between concentration and secondary ion current is relatively simple and is not affected by matrix effects. However, the ionization process is not as efficient as in SIMS and elemental sensitivities are usually lower (see Chapter 9).

2.2.4. *Secondary ion energy distributions*

It has been found that the vast majority of the ions emitted arise from the first two surface layers. The final sub-surface collisions which give rise to the emission of secondary species will have a range of energies which are reflected in the observed kinetic energy distribution of the secondary ions. This distribution, which generally peaks at rather low energies (1–10 eV), is independent of the energy of the primary beam. The energy distributions display fairly low-level tails up to several hundred eV and are significantly broader for elemental ions than for cluster ions. In fact, the more complex the cluster the narrower the distribution tends to be and the lower the energy of the peak maximum.[2,20]

The differing energy distributions of elemental ions and cluster ions can be exploited in analysis. It is possible to discriminate against the lower-energy cluster ions using an energy analyser and collect mainly elemental ions, thus simplifying the spectrum for elemental analysis (see Fig. 2.14).

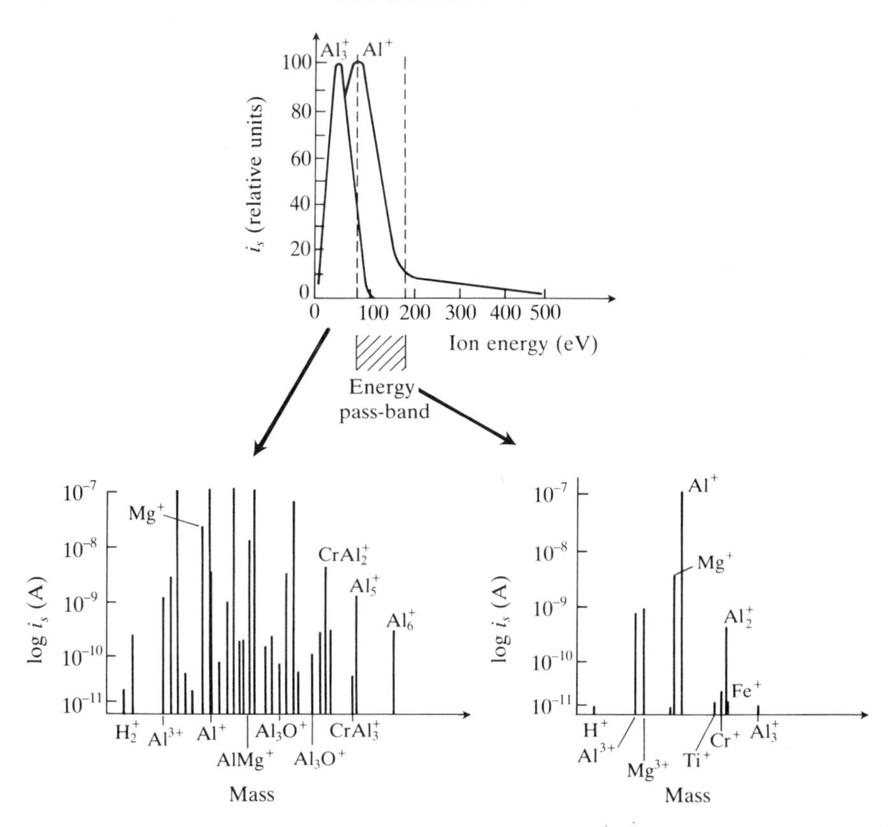

FIG 2.14. The energy distribution of secondary ions and spectra obtained by varying the energy acceptance of the mass analyser. Reproduced with permission from Herzog, R., Poschenrieder, W., and Satkiewicz, F. (1972). *Proc. Int. Conf. Ion Surface Interaction* (ed. R. Behrisch and W. Heiland), p. 173. Gordon and Breach, London.

2.3. Sensitivity

We have considered the sputter yield Y, and the ionization probability, α^+, both of which will be important in determining the minimum amount of an element which can be detected by the technique. The sensitivity of SIMS is also dependent on the transmission of the analytical system, η. This is an instrumental parameter which varies only with the configuration and type of mass analyser used. These matters will be discussed in detail in Chapter 4. However, in order for us to estimate the sensitivity range of static and dynamic SIMS it is sufficient for us to know that η usually lies between 10^{-3} and 10^{-1}. Using eqn (2.2.1), Table 2.3 summarizes the minimum detectable

TABLE 2.3

Analytical sensitivity in static and dynamic SIMS. Fractional surface composition detectable for a range of instrumental transmission (η) and secondary ion yield ($Y\alpha^+$) values

$Y\alpha^+$		Static SIMS I_p = 1 nA cm^{-2}		Dynamic SIMS I_p = 1 μA cm^{-2}	
	η	10^{-3}	10^{-1}	10^{-3}	10^{-1}
10^{-4}		10^{-2}	10^{-4}	10^{-5}	10^{-7}
1		10^{-6}	10^{-8}	10^{-9}	10^{-11}

concentrations under static and dynamic conditions if we assume that the minimum detectable current of secondary ions is 10^{-18} A or ~10 cps. The table takes two examples of secondary ion yield: the first a very low value, where $Y\alpha^+ = 10^{-4}$, which would be typical of argon ion bombardment of iron (see Table 2.2); the second, a high value, where $Y\alpha^+$ is 1, which might be typical for oxygen ion bombardment of chromium. We also consider two instrumental arrangements such that η is either 10^{-3}, characteristic of a quadrupole analysis system, or 10^{-1}, which may be obtainable with a time-of-flight system (see Chapter 4).

The sensitivity of *static* SIMS ranges from 10^{-2} to 10^{-8} of a monolayer or from 10^{21} to 10^{15} atoms cm^{-3} dependent on the element studied and the instrument used, whilst for *dynamic* SIMS using 10^3 higher current density the sensitivity range is that factor higher. These are calculated estimates of what should be possible; a detection sensitivity of 10^{12} atoms of Cr cm^{-3} has yet to be demonstrated!

2.4. Surface charging

In common with all methods of surface analysis involving either incoming or outgoing charged particles, surface charging of insulators can be a problem. The SIMS process usually involves bombardment with primary positive ions, which gives rise to the emission of secondary ions and neutrals and large numbers of secondary electrons. The predominant species will usually be electrons and neutrals, thus with incoming positive ions insulating surfaces usually charge positive. This is illustrated in Fig. 2.15, which shows the rise in surface potential with time of a simulated insulator (a metal surface insulated from instrument earth through a 250 V, 0.1 μF capacitor) under four different primary beams having a beam flux of 6.2×10^9 particles s^{-1}

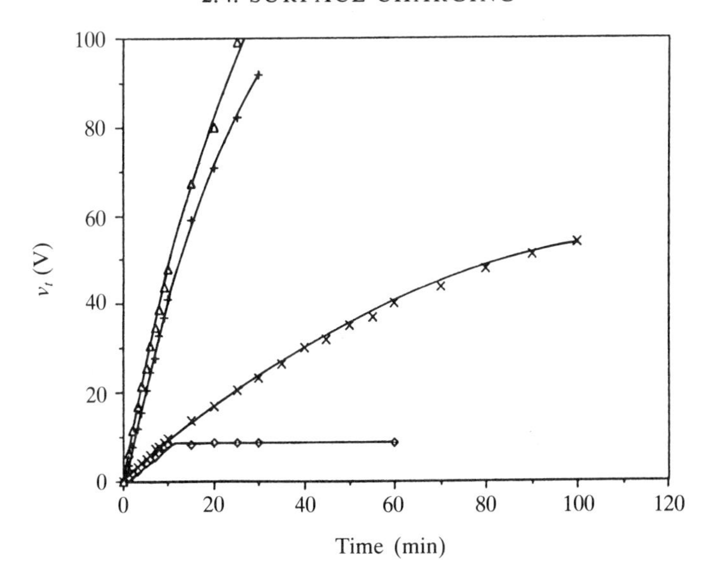

FIG 2.15. The variation of surface potential (v_t, volts) of an insulating sample as a function of bombardment time under different primary particle beams. The beam flux is the same in each case, equivalent to 1 nA or 6.2×10^{18} particles s^{-1}. △, 10 keV Ga^+; +, 2 keV Ar^+; ◇, 2 keV Ar^0; ×, a mixture of 2 keV Ar^+ and Ar^0.

(equivalent to 1 nA). Bombardment with 10 keV Ga^+ and 2 keV Ar^+ leads to a rapid rise in surface potential to in excess of +100 V in 20 min due to incoming positive ions and outgoing secondary electrons. The measurement was terminated at 100 V, but it was found that the rate of rise was doubled if the beam flux was doubled, indicating that the charging was almost entirely due to incoming positive ions. In the range measured the rate of potential rise is also dependent on the energy of the beam.[21] The limit will be determined by the capacitance of the material.

As far as SIMS is concerned, the main effect of this charging is that positive ions are given further energy, accelerating them beyond the acceptance energy of the analyser. This is particularly a problem for quadrupole-based systems (see Chapter 4), which have a narrow energy acceptance window, typically 5 eV. In the case of negative secondary ions a positive surface potential will inhibit or totally suppress their emission.

Surface charging can be reduced by the use of neutral primary particles—fast atom bombardment (FAB).[22] Although some charging does occur under FAB it is only due to secondary electron emission. Under keV Ar atom bombardment it has been shown that charging is very much reduced and reaches equilibrium between 10–30 V, for example, Fig. 2.15. This behaviour is to be expected since secondary electron emission will be mainly responsible for the development of a positive surface potential and as this rises the emission of the low-energy electrons will be suppressed. In practice this means that the analyser ion optics can be easily adjusted to collect positive secondary ions; however, the emission of low-energy negative ions may still be inhibited.

Surface potential may also be influenced by concurrent electron bombardment of the sample surface. Usually the secondary electron emission coefficient under high-energy electron bombardment is >1, so that charging to a positive potential would still be expected. Furthermore, high-energy electrons can have a deleterious effect on surface chemistry. The use of low thermal energy electrons would seem to be appropriate; however, they are easily deflected by stray electric fields. In practice the use of 500–700 eV electrons has been successful, although it is thought that neutralization may be effected by the charged surface trapping low-energy secondaries emitted from the surface or from surrounding hardware.

It is in the detection of negative cluster or molecular ions that the control of surface potential is most crucial. These ions are chemically very informative but they have low kinetic energies and are thus easily suppressed by a positive surface potential.[23] The detailed effects of charging on the experimental acquisition of SIMS data are considered for depth profiling in Section 5.5.5; for static SIMS in Section 7.2; and for SIMS imaging in Section 8.5. The instrumental requirements are described in Section 4.3.4.

2.5. Generation of cluster or molecular ions

Cluster or molecular ions are potentially of most interest in *static* SIMS for surface chemical structure characterization. However, they are sometimes used in dynamic SIMS if a cluster characterizes a phase of interest or an element is more easily identified when in the cluster state (see Chapter 5).

TABLE 2.4

The statistical and experimentally observed yields of Ni^+, Ni_2^+ and Ni_3^+ ions from the (110), (100) and (111) planes of Ni

	(110)	(100)	(111)
No. of ways of producing Ni_2^+*	2	4	6
Obs. Ni_2^+/Ni^+**	1	1.8	3.2
No. of ways of producing Ni_3^+*	4	24	48
Obs. Ni_3^+/Ni^+*	1	6	38

* Considering top monolayer only.
**Normalized to the (110) face.

2.5.1. Inorganic cluster ions

Cluster ions are observed from almost all surfaces during SIMS analysis. On simple single-component metals M_2^+, M_3^+, etc., are detected and their yield relative to M^+ have been shown to be sensitive to surface crystallinity[24] (see Table 2.4). Thus the relative yields M_2^+/M^+ from the three low index faces of Ni (111), (100) and (110) are in the ratio 3:2:1, which corresponds to the relative geometric probability of finding two nearest-neighbour metal atoms on each of the three surfaces. Adsorption on metals gives rise to the emission of cluster ions. For example, oxidation generates MO^+, MO_2^-, $M_xO_y^\pm$. The relative yield of MO^+ to M^+ has been shown to be dependent on the bond dissociation energy of the related oxides (see Fig. 2.16).[25] Furthermore, the relative yields of other cluster ions, $M_xO_y^\pm$, from oxides seem to be related to the oxidation state of the metal atom. It will be shown in Chapter 7 that the so-called valence theory postulated by Benninghoven and Plog quite accurately mirrors the effect of metal atom valence on the yields of these ions.[26]

Some workers suggest that the process of ion generation from oxides may proceed either by direct lattice fragmentation or recombination of atomic species above the surface. There are strong reasons to suggest that negative ions are almost always produced by lattice fragmentation (see Section 3.2.4).

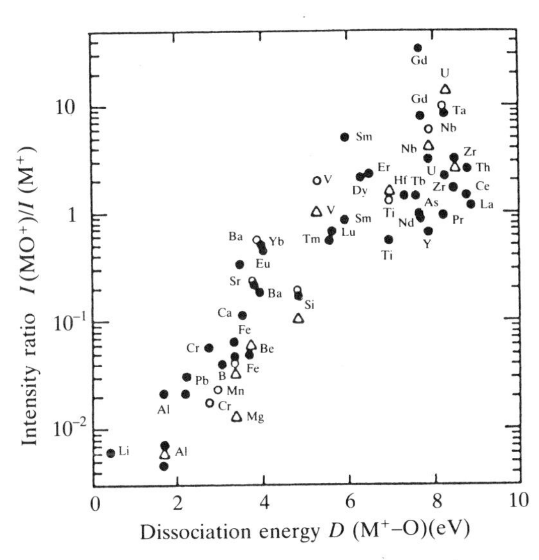

FIG 2.16. The relative intensity of oxide ions $I(MO^+)/I(M^+)$ as a function of bond dissociation energy $D(M^+ - O)$ for oxygen sputtering of metals and inert gas sputtering of oxides or metals in an oxygen atmosphere. △, $O_2^+ \rightarrow$ metals; ●, metal dopants in oxides; ○, metals + O_2.

The adsorption of CO yields MCO^+, M_2CO^+, and M_3CO^+, and these clusters can be related to the coverage and structure of absorbed CO (see Chapter 6).

In studies of inorganic frozen molecular solids such as Ar, N_2, CO_2, NO, etc., it has been found that, even for such simple solids, large clusters can be generated.[27] The spectra from these solids are composed of some simple elemental ions and of clusters built up of a central ion (not necessarily identical with the molecular ion of the solid, e.g. for N_2 it may be N^+ or $N_3{}^+$ or N_2^+) plus a number of 'solvating' molecules (again not necessarily identical with the molecule of the solid, e.g. for NO it is N_2O_3). However, the SIMS cluster spectra of mixtures of, say, O_2–N_2 are quite different from those of compounds of the same elemental composition, NO. These materials are insulators, as are many of the materials for which cluster ion generation is important in analysis. Charge can be generated in the primary impact and cause further molecular and intermolecular disruption. In the present case, the cluster ions are thought to arise from reactions between simple elemental or molecular ions generated by the primary ion impact and the molecules and fragments in the impact zone. It seems that similar processes may occur during the emission of multi-molecular clusters from organic solids. However, it is likely that the larger the molecule the more energy can be dissipated and 'reactions' in the impact area may become less likely.

2.5.2. Organic molecular–cluster ions

Extensive studies of organic materials have shown that cluster ions are emitted whose identity and yield are related to the chemical structure of the materials from which they are derived. In common with organic mass spectrometry, molecular or quasi-molecular ions are observed along with ions which may arise by fragmentation, rearrangement, decomposition, or reaction of the constituent molecules of the material. Most of the observed ions are thought to arise as a consequence of surface desorption, reaction in the so-called selvedge region or fragmentation in the gas phase.

It is noteworthy that very frequently the cluster ions generated in the spectra for a particular material are essentially the same whether SIMS of a solid surface, liquid FAB, or ^{252}Cf PDMS are used. For example, the spectrum of thiamine generated by each of these techniques is dominated by the benzylic cleavage products at m/z 122 and 144. This suggests that the internal energies deposited in these techniques are comparable even though the initial bombardment processes are rather different. Cooks has tabulated the possible ion formation processes in molecular SIMS (see Table 2.5).[28] These are equally applicable to SIMS–FABMS from solid or liquid matrices. There are at least three types of molecular ion.

TABLE 2.5

Ion formation processes in molecular SIMS

Reaction	Probable region of occurrence	Comments
$M^0(s) \rightarrow M^0(g) \xrightarrow{\pm e^-} M^{\pm}(g)$	selvedge	electron ionization
$C^+(s) \rightarrow C^+(g)$	surface	direct desorption
$M^0(s) \rightarrow M^0(g) \xrightarrow{+C^+} (M + C)^+(g)$	selvedge	cationization
$M^0(s) \rightarrow M^0(g) \xrightarrow{+A^-} (M + A)^-(g)$	selvedge	anionization
$M^0(s) \rightarrow M^0(g) \xrightarrow{+H^+} (M + H)^+(g)$	selvedge	cationization by protonation
$(C_mA_n)^{\pm}(s) \rightarrow (C_mA_n)^{\pm}(g)$	surface	direct cluster desorption
$C^+(s) + nS^0(1) \rightarrow CS_n^+(g)$	surface (liquid)	solvent attachment
$C^+(g) + nS^0(g) \rightarrow CS_n^+(g)$	selvedge	cationization of solvent
$CS_n^+(g) \rightarrow C^+(g) + nS^0(g)$	free vacuum	desolvation
$M_1^0(s) + M_2^0(s, 1) \xrightarrow{\pm e^-} M_1^{\pm}(g) + M_2^{\pm}(s, 1)$	selvedge	electron transfer
$M^0(s) \rightarrow F(g) \xrightarrow{+C_n^{\pm}(g)} (C_n + F)^{\pm}(g)$	surface → selvedge	"beam damage"
$[C^0(s) + Ad(g)] \rightarrow (C + Ad)^{\pm}(g)$	surface	intact emission
$Ad(g) \xrightarrow{+C^+} (C + Ad)^+(g)$	selvedge	cationization
$nC(s) \rightarrow C_n^{\pm}(g)$	surface	intact cluster emission
$M_1^0(s) + M_1^0(s) \xrightarrow{+H^+} (2M_1 + H)^+(g) \rightarrow M_2^+(g)$	surface	surface reaction
$\rightarrow F$	free vacuum	unimolecular fragmentation

M = molecule, C^+ = cation, A^- = anion, C = atom or molecule, S = solvent, Ad = adsorbed species, F = fragment, m,n = positive integers

These may be formed by electron ionization, direct desorption of precharged ions, or ion/molecule processes such as cationization or anionization. In the disrupted surface region, or selvedge, ions may react with other ions to form other clusters. Thus in liquid matrices solvent molecules may be cationized. In the main, it is thought that fragment ions are formed by unimolecular decompositions of desorbed cluster ions or neutrals after leaving the surface. This suggestion is supported, although not proved, by the observation that the tandem mass spectrometry, MS/MS, spectrum of the intact molecular cation from a pure material generated in SIMS is almost identical to the standard MS mass spectrum. This emphasizes the commonality of the fragmentation mechanisms in the gas-phase MS/MS experiment and the sputter experiments.[29]

The yield of 'molecular' or cluster ions rises with the amount of energy brought in by the primary particle. Thus Standing *et al.* found that molecular ion yield from simple amino acids bombarded by ions ranging from Li^+ to Cs^+ having energies from 1 to 16 keV increased with the total energy deposited.[30] Similar observations have been reported for polymer materials, where it has been shown that as inert gas ion energy or mass is increased the yield of high-mass clusters increases (see Section 7.3).

2.5.3. *Matrix effects in molecular–cluster ion emission*

The matrix which comprises or supports an organic material has a very significant influence on the identity and yield of cluster ions. Yet because the matrix not only absorbs the primary excitation, but is also the source of the desorbing species, *and* mediates the ionization process, it is extremely difficult to unravel the processes involved. There are a number of possible matrix effects on cluster ion generation.

(1) Enhanced ionization can occur through cationization or anionization. Impurity alkali metals such as sodium or potassium can have a very marked effect in this respect. A metal substrate can play a role in ionization by providing cations (C^+) which complex with sputtered neutral molecules to yield the observed $(C+M)^+$ ions. This procedure has been used effectively by Benninghoven *et al.* to ease the analysis of complex organic molecules by depositing monolayers of them on a silver substrate. The silver then cationizes the molecule (see Section 6.4). In some cases, it seems that the process of cationization occurs during emission, whereas in others the molecules adsorb onto the metal and are chemisorbed to the metal atom. This latter process can be extended to produce *preformed* ions in the matrix using Brønsted acids or bases. These ions are then simply desorbed in the emission process.

(2) The dilution of a molecule or compound of interest in another solid or liquid matrix may reduce molecular ion fragmentation. Dilution of organic

compounds in liquid glycerol or similar low vapour pressure solvents for FABMS is primarily to maintain a rapidly renewable surface for compound analysis. However, it is found that dilution of molecules in solid matrices such as ammonium chloride can be beneficial in SIMS analysis of organic compounds. It reduces intermolecule interactions and the internal energy of sputtered molecules and thereby reduces the extent of fragmentation.[29]

(3) Dilution of a compound in a mixture or solid solution can also lead to enhanced absolute ion yields. Thus the intact cation signal in the SIMS spectrum of a neat pyrilium salt was three times smaller than for the salt diluted 1000 times in NH_4Cl.

(4) Increased dilution also reduces the possibility of intermolecular reactions. This frequently leads to the increased relative intensity of the molecular ion.

2.5.4. *Cluster emission and structural damage*

Naturally, the emission of structurally related cluster ions is dependent on the maintenance of that structure. Preferential sputtering or mixing effects can produce an altered layer resulting in a change in the cluster ion emission. In Section 2.1.3(e) the possibility of preferential loss of oxygen from oxides was mentioned; this will result in the reduction of the surface layer and hence loss of $M_xO_y^{\pm}$ species. The ion dose required to reduce these cluster ions will depend on the stability of the oxide.

Many organic materials are easily damaged by particle bombardment. As we saw in Section 2.1.1, the primary dose conditions for SSIMS for such materials may be below 10^{13} particles cm^{-2}. In the analysis of polymer materials it has been found that primary particle doses in excess of 10^{12}–10^{13} cm^{-2} can begin to generate damage which reveals itself in the loss of structurally related ions and their replacement by ions characteristic of a carbonaceous layer (see Section 7.4). Figure 2.17 shows the negative ion spectrum from poly(methylmethacrylate) after a dose of 10^{12} ions cm^{-2} and after 5.5×10^{13} ions cm^{-2}.[31] The structurally significant ions at $m/z = 85$, 141, and 185 have been almost completely lost in the latter spectrum. Clearly, this has serious implications for SSIMS analysis of these materials and will be addressed in more detail in Chapter 7.

As was indicated in Section 2.1.2, the ion-induced sputter yield in insulators is higher than for atom bombardment. This leads to a more rapid generation of damage in organic materials and underlines the fact that the use of atom beams in the analysis of the surface chemical structure of insulating materials is beneficial not only because it reduces the charging problem but also because it reduces the damage generated in the solid. It is clear that the definition of *static* primary particle bombardment conditions needs to be tailored to the type of material being studied.

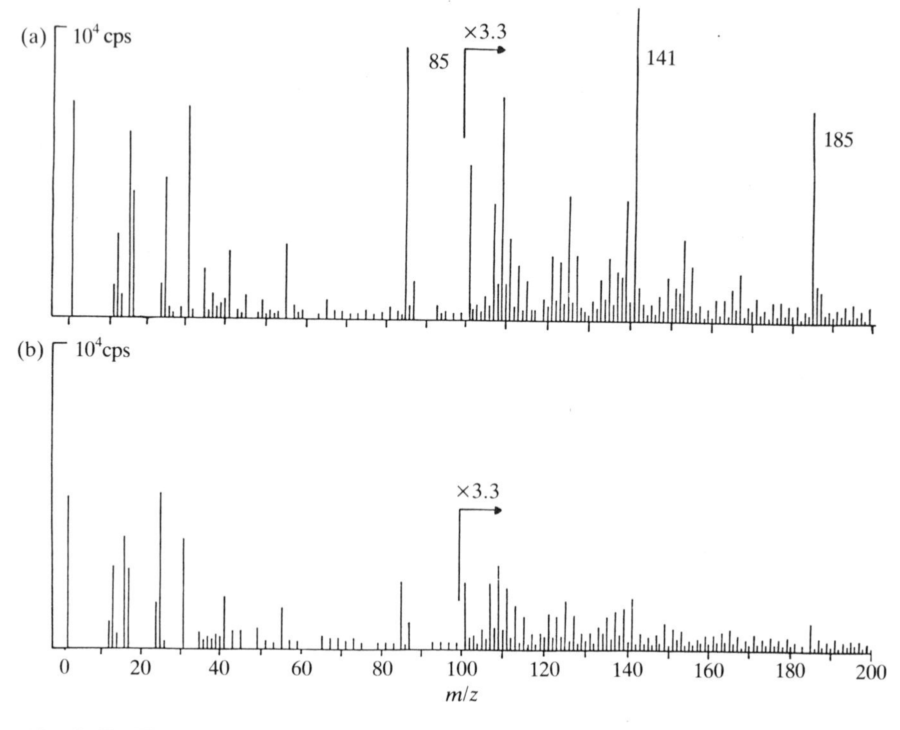

FIG 2.17. The negative ion spectrum from poly(methylmethacrylate) after a dose of 10^{12}, 4keV Xe^+ ions (a), and after a dose of 5.5×10^{13}, 4 keV Xe^+ ions (b). Reproduced with permission from Ref. 31.

References

1. Anderson, H. H. and Bay, H. L. (1981) In *Sputtering by Particle Bombardment I*, Springer Series Topics in Applied Physics, Vol 47 (ed. R. Behrisch), p. 145. Springer Verlag, Berlin.
2. Wittmaack, K. (1975). *Surf. Sci*, **53**, 626.
3. Oechsner, H. (1973) *Z. Physik*, **261**, 37.
4. Eccles, A. J., van den Berg, J. A., Brown A., and Vickerman, J. C. (1986). *Appl. Phys. Lett.*, **49**, 188.
5. Roosendaal, H. E. (1981). In *Sputtering by Particle Bombardment I*, Springer Series Topics in Applied Physics, Vol 47 (ed. R. Behrisch), p. 219. Springer Verlag, Berlin.
6. Littmark, U. and Hofer, W. (1978). *J. Mater. Sci.*, **13**, 2577.
7. Auciello, O. (1984). In *Ion Bombardment Modification of Surfaces: Fundamentals and Applications* (ed. O. Auciello and R. Kelly), p. 1. Elsevier, Amsterdam.

8. Whitton, J.L., Tanovic, L., and Williams, J.S. (1978). *Appl. Surf. Sci.*, **1**, 408.
9. Hucks, P., Stocklin, G., Vietzke, E., and Vogelbruch, K. (1978). *J. Nucl. Mat.*, **76/77**, 136.
10. Betz, G. and Wehner, G.K. (1983). In *Sputtering by Particle Bombardment II*, Springer Series Topics in Applied Physics, Vol 52 (ed. R. Behrisch), p. 11. Springer Verlag, Berlin.
11. Kelly, R. (1980). *Surf. Sci.*, **100**, 85.
12. Betz, G., Opitz, M., and Braun, P. (1981). *Nucl. Instrum. Methods* **182/183**, 63.
13. Roth, J. (1983). In *Sputtering by Particle Bombardment II*, Springer Series Topics in Applied Physics, Vol 52 (ed. R. Behrisch), p. 91. Springer Verlag, Berlin.
14. Andersen, C.A. and Hinthorne, J.R. (1972). *Science*, **175**, 853.
15. Brown, A. and Vickerman, J.C. (1983). *Surf. Sci.*, **124**, 267.
16. Slodzian, G. and Hennequin, J-F. (1966). *Compt. Rend. (Paris)*, **B263**, 1246.
17. Maul, J. and Wittmaack, K. (1975). *Surf. Sci.*, **47**, 358.
18. Benninghoven, A. (1975). *Surf. Sci.*, **53**, 596.
19. Storms, H.A., Brown, K.F., and Stein, J.D. (1977). *Anal. Chem.*, **49**, 2023.
20. Blaise, G. and Slodzian, G. (1973). *Rev. Phys. Appl.*, **8**, 105.
21. Humphrey, P. (1988). PhD thesis, UMIST.
22. Surman, D., van den Berg, J.A., and Vickerman, J.C. (1982) *Surf. Interface Anal.*, **4**, 160.
23. Brown, A. and Vickerman, J.C. (1986). *Surf. Interface Anal.*, **8**, 75.
24. Barber, M., Bordoli, R.S., Vickerman, J.C., and Wolstenholme, J. (1977). In *Proc. 7th Int. Vacuum Congr. and 3rd Int. Conf. on Solid Surf.* (Vienna, 1977), p. 983.
25. Wittmaack, K. (1979). *Surf. Sci.*, **89**, 668.
26. Plog, C., Wiedmann, L., and Benninghoven, A. (1977). *Surf. Sci.*, **67**, 565.
27. Michl, J. (1983). *Int. J. Mass Spectrom. Ion Phys.*, **53**, 255.
28. Pachuta, S.J. and Cooks, R.G. (1985). *Desorption Mass Spectrometry—Are SIMS and FAB the Same? Am. Chem. Soc. Symp. Ser.*, **291**, p. 1.
29. Cooks, R.G. and Busch, K.L. (1983). *Int. J. Mass Spectrom. Ion Phys.*, **53**, 111.
30. Standing, K.G., Chait, B.T., Ens, W., McIntosh, G., and Beavis, R. (1982). *Nucl. Instrum. Methods*, **198**, 33.
31. Briggs, D. and Hearn, M.J. (1986). *Vacuum*, **36**, 1005.

3

SIMS — THE THEORETICAL MODELS

The general outline of the process is widely accepted. When the primary particle of energy E_0 strikes the surface atoms, symmetry will be involved. Some energy may be lost by electronic excitation but most is transferred by a nuclear stopping mechanism during hard-sphere collisions (see Fig. 3.1). Knock-on collision cascades between atoms in the near-surface region are initiated. Some energy will be dissipated into the bulk by displacement cascades. Here the incident ion can remove bulk atoms out of their regular lattice sites. The average number of displaced atoms is given by $E_o/2E_d$ where E_d is the displacement energy, which averages about 25 eV. After displacement, a very efficient process of recombination usually occurs such that the number of defects remaining per ion is $1–10^{-3}$. Some cascades will return to the surface, causing the emission of secondary particles or sputtering. This latter process occurs within 1–2 nm of the surface, whereas displacement or lattice damage can occur down to 10–25 nm.

In seeking to understand the mechanism of secondary ion emission, the major question is whether the two processes of emission and ionization occur simultaneously or consecutively. This is a complex mixture of collisional dynamics and quantum mechanics. We will briefly consider the models of sputtering and ionization separately.

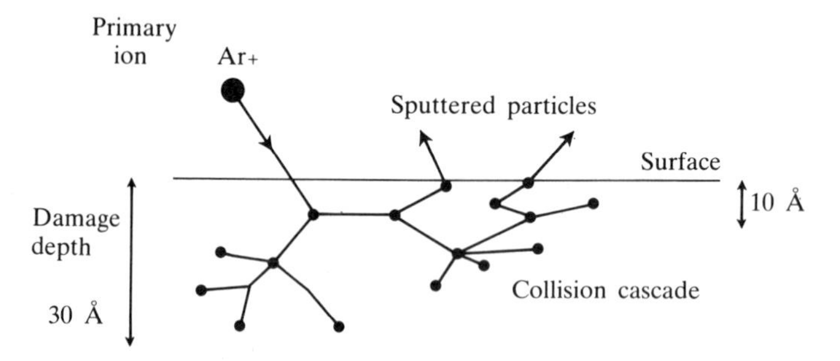

FIG 3.1. Schematic diagram of the main collision processes which occur in the surface layers of a solid bombarded by medium-energy particles.

3.1. Sputtering models

Surfaces erode under particle bombardment and it is this phenomenon which is called sputtering. It is, however, important to remember that other effects also occur under particle bombardment—generation of heat, ionization, etc. Furthermore, not all erosion is caused by sputtering: evaporation, blistering, etc., may be consequences of high bombardment rates. Sigmund in his review of cascade sputtering theory suggests that there are three criteria which define sputtering.[1]

1. It is a class of erosion phenomena observed on a material surface as a consequence of particle bombardment.
2. It is observable in the limit of small incident-particle current. This makes it clear that macroscopic heating and subsequent evaporation by a high-intensity beam is not sputtering.
3. It is observable in the limit of small incident-particle fluence. This ensures that even a single particle can initiate a sputtering event.

Sigmund included a fourth criterion requiring the material to be homogeneous, but this seems unnecessarily restrictive.

The basic sputtering mechanism is best modelled initially at the 'static' limit. Secondary processes such as mixing, and the formation of the altered layer which accompany higher fluences ($> 10^{14}$ ions cm^{-2}) should be tackled thereafter.

As we have seen in Chapter 2, sputter yields vary widely and are dependent on the primary beam and target parameters. The total range of yields per incident particle is about 10^{-5}–10^{3}. Although the elementary event in sputter erosion is an atomic event, the values of the sputter yield show that the erosion effect of one bombarding particle is a statistical variable. The statistical or random nature of sputtering will underlie the proposed models.

There are basically two approaches to modelling the sputtering process. One is to treat the process as a series of hard-sphere collisions in which classical energy transfer equations can be used. No assumptions are made as to structure of the solid and the geometrical distribution of the atoms with respect to each other. Only a so-called surface energy is introduced as a measure of the energy to be overcome if a surface atom is to be emitted. The alternative approach is to use one of the computer simulation methods in which the effect of bombardment on something approaching a realistic three-dimensional structure is attempted.

The first approach is typified by the extensive and successful work of Sigmund.[1] The theory is primarily concerned with elastic collisions or knock-on sputtering.

3.1.1. *Linear collision cascade theory*

When a high-energy particle penetrates a solid matrix there are two major ways in which it can lose energy. The first is by elastic collisions with the nuclei which form the solid; this is termed nuclear stopping. The other, which becomes the major source of energy loss at very high projectile velocities, is via energy transfer to the electrons of the solid. This excitation is a non-collisional inelastic process and is a continuous drain on the energy of the projectile. This is termed electron stopping. Thus for ion bombardment of metallic targets at moderate energies elastic collision processes are most important. For example, a 10 keV argon ion will take about 10^{-13} s to travel 10 nm in vacuum. This is a very long time compared with the relaxation time of conduction electrons (*ca.* 10^{-19} s). Thus although some excitation of electrons will occur, this energy will be immediately dissipated throughout the solid. In insulators the lifetimes of excited electronic states may be long enough to allow this energy to be transferred into atomic motion, perhaps because the excited state is an antibonding state, and thus electronic sputtering may result. Most mechanistic models have restricted themselves to elastic collisions.

Three types of knock-on sputtering can be distinguished. These are all involved to a greater or lesser extent in SIMS:

1. **Single knock-on or prompt sputtering**, which occurs at $t = 10^{-15}$–10^{-14} s after primary particle impact. This is a direct, or almost direct, impact process between the incident particle and the sputtered particle involving only surface atoms.
2. **Slow collisional sputtering**, the linear cascade regime, occurs due to the internal flux of moving target atoms intersecting the surface in the time-scale 10^{-14} s $\leq t \leq 10^{-12}$ s after primary impact. The spatial density of moving atoms is small. It is probably this process with which we are mainly concerned in SIMS (see Fig. 3.2).
3. **Slow thermal sputtering or the spike regime** occurs some 10^{-12} s after particle impact. This is also a consequence of the movement of recoil atoms but the spatial density is high such that the majority of atoms within a certain volume are in motion.

Particle–surface interactions involve forward and backward effects. Forward effects affect the interior of the bombarded target and include the stopping of the incident primary particle and the deposition of energy. Backward effects are those which lead to particle expulsion or sputtering. Clearly, the forward effects are involved in generating sputtering.

3.1.1.1. The basic ideas and equations of collision theory The detailed derivations and arguments of linear cascade theory are long and in places

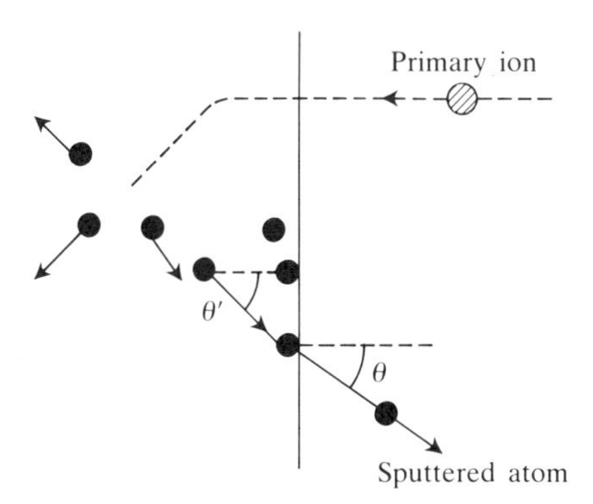

FIG 3.2. Sketch of the primary features of slow collisional sputtering in the linear cascade process. The incident particle creates a random cascade of collisions within the target. The primary impact is only coupled indirectly with the sputtered surface atoms. θ' is the polar angle of the moving target atoms and θ is the polar angle of the sputtered atom. Reproduced with permission from Ref.3.

mathematically complex. It is not possible or desirable to reproduce them here. Some of the basic equations and background argument will be explained to give a physical picture of the model. Readers who wish to pursue the detail further should consult the quoted reviews.

When considering statistically determined processes the idea of the *cross-section* of the process is important. The probability that a particular process occurs during a two-atom collision when a beam of projectile particles collides with a structure of target atoms whose thickness is x and density is N is given by

$$p = Nx\sigma \tag{3.1.1}$$

where σ is the cross-section for the process. Cross-sections for elastic collisions are usually quoted in a differential form, thus the cross-section for the transport of energy in a single collision is given as $\mathrm{d}\sigma(E,T) \equiv (\mathrm{d}\sigma/\mathrm{d}T)\mathrm{d}T$, where E is the initial and T the transferred energy.

The energy transferred in an elastic collision between two atoms is determined by the laws of conservation of energy and momentum.

Thus

$$T_{\mathrm{m}} = \gamma E = 4M_1M_2E/(M_1+M_2)^2 \tag{3.1.2}$$

is the maximum energy which can be transferred from an atom 1 of mass M_1 with initial energy E to atom 2 of mass M_2 whose initial energy was zero when the collision is head-on.

Scattering after high-energy collisions between atoms will obviously be determined by the Coulomb repulsions between the nuclei. It is these that make the colliding atoms somewhat similar to colliding billiard-balls or hard spheres. Bohr has shown that

$$d\sigma(E,T) = \pi(M_1/M_2)Z_1^2Z_2^2e^4dt/ET^2 \tag{3.1.3}$$

where $0 \leq T \leq T_m$ and Z_1e and Z_2e are the nuclear charges. This cross-section strongly prefers collisions with small energy transfers and decreases with increasing E. However, this cross-section is only valid when the energies of collision are high enough that the nuclei do approach very close to each other, closer than the screening radius a, or when $\epsilon >> 1$ where

$$\epsilon = [M_2E/(M_1+M_2)](a/Z_1Z_2e^2) \tag{3.1.4}$$

and

$$a \simeq 0.885a_0(Z_1^{2/3}+Z_2^{2/3})^{-1/2} \tag{3.1.5}$$
$$a_0 = 0.529\ \text{Å}.$$

For eqn (3.1.3) to be valid for a collision between an argon ion and a Cu target, energies of > 100 keV are required. For lower, more realistic (in terms of SIMS) energies, $\epsilon \leq 1$, the nuclei are *not* very hard spheres and the screening of the Coulomb interaction by the surrounding electron cloud has to be included. In this regime the cross-section can be given by

$$d\sigma(E,T) \simeq C_m E^{-m}T^{-1-m}dT; \qquad 0 \leq T \leq T_m \tag{3.1.6}$$

with

$$C_m = \pi/2\lambda_m a_2(M_1/M_2)^m(2Z_1Z_2e^2/a)^{2m}, \tag{3.1.7}$$

where λ_m is a dimensionless function of the parameter m which varies slowly from $m=1$ at high energies (where eqn (3.1.6) becomes eqn (3.1.3) and $\lambda_1 = 1/2$) down to $m \simeq 0$ at very low energies.

An important quantity in understanding the sputtering process is the mean energy lost ΔE in elastic collisions over a travelled path length x into the solid. From eqns (3.1.1) and (3.1.6)

$$\Delta E = Nx\int d\sigma(E,T)\cdot T \equiv NxS_n(E) \tag{3.1.8}$$

where $S_n(E)$ is called the nuclear stopping cross-section and

$$S_n(E) = [1/(1-m)]C_m\gamma^{1-m}E^{1-2m}. \tag{3.1.9}$$

$S_n(E)$ rises approximately proportionally to E at low energies where $m \simeq 0$; it approaches a plateau at intermediate energies ($m = 1/2$) and falls off at higher energies. This can be summarized a little differently:

$$S_n(E) = 4\pi aZ_1Z_2e^2(M_1/M_1+M_2)s_n(\epsilon) \tag{3.1.10}$$

where $s_n(\epsilon)$ is a universal function depending on which expression for Coulomb screening potential is used (see Fig. 3.3). These derivations have

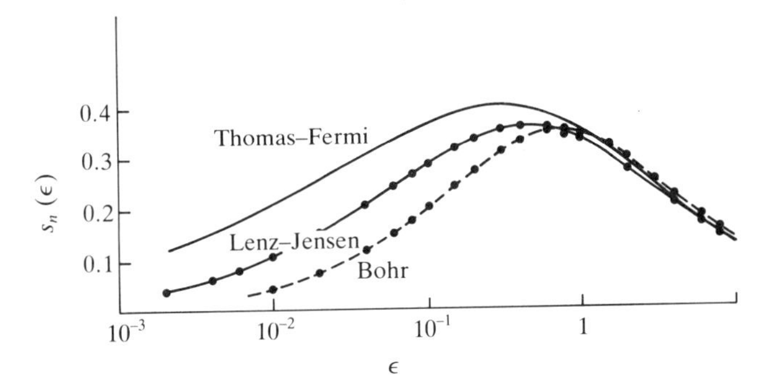

FIG 3.3. Nuclear stopping cross-sections. The three curves refer to three different screening functions for the Coulomb interaction between two colliding atoms. The Thomas-Fermi curve is known to be too high, whilst the Lenz–Jensen curve somewhat underestimates the stopping cross-section. Reproduced with permission from Ref. 1.

been based on the Thomas–Fermi model of atomic interaction. It has been found to be most accurate for weak screening, i.e. at high collision energies where m is close to 1, although it is not so good for heavy screening where $m \simeq 0$ and $\ll 1$.

In the situation which concerns us in SSIMS, where the density of recoil atoms is small, there will be very few collisions. It is possible to determine the unique relationship between the scattering angle and the energy T transferred by an atom 1 of energy E hitting an atom 2 *at rest*. Figure 3.4 shows the scattering geometry. Thus the scattering angle of the projectile atom is given by

$$\cos\phi' = (1 - T/E)^{1/2} + 0.5(1 - M_2/M_1)(T/E)(1 - T/E)^{-1/2} \qquad (3.1.11)$$

and for the target atom 2

$$\cos\phi'' = (T/\gamma E)^{1/2}.$$

Thus recoil atoms are scattered into $0 \leq \phi'' \leq \pi/2$ whilst the projectile atoms are scattered into $0 \leq \phi' \leq \pi$ (for $M_1 < M_2$) or $\pi/2$ (for $M_1 > M_2$).

Thus far, expressions are available to characterize the slowing down of an impinging particle as it enters a solid made up of screened nuclei, and also the scattering of colliding particles. These now have to be combined to provide a description of the loss of energy of a particle as it penetrates through a random medium (see Fig. 3.5). Thus from eqn (3.1.8).

$$dE/dx = -NS(E). \qquad (3.1.12)$$

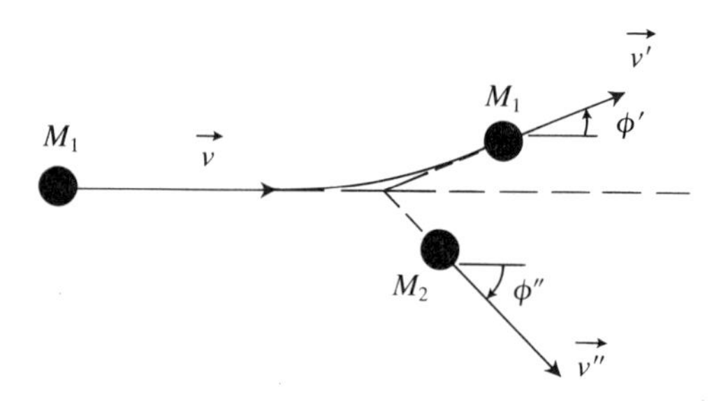

FIG 3.4. The scattering geometry in an elastic collision of an atom of mass M_1 and initial velocity v_1 with atom of mass M_2, initially at rest.

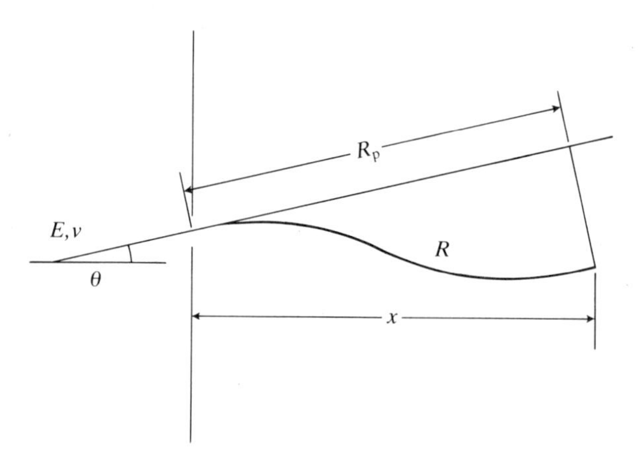

FIG 3.5. The penetration parameters. The initial velocity is v and the energy is E. The angle of incidence to the surface normal is θ. The penetrated path length is R and the projected range is R_p for penetration length of x.

The mean penetrated path length $R(E)$ of a particle before coming to rest can be obtained from

$$R(\epsilon) = \int_0^E dE'/NS(E'). \tag{3.1.13}$$

If all the energy is used in elastic collisions, eqn (3.1.9) yields

$$R(E) \simeq [(1-m)/2m]\gamma^{m-1}(E^{2m}/NC_m). \tag{3.1.14}$$

The mean projected range $R_p(E)$ is normally shorter than $R(E)$ because of the scattering of the projectile (see Fig. 3.5). The path length correction $R_p(E)/R(E)$ is <1 when the mass of the projectile, M_1, is less than the mass of the target atoms, M_2 and collision energies are low ($\epsilon<1$). However, the correction is close to 1 when $M_1>>M_2$ and for high ϵ.

It is interesting that the projected range does not depend explicitly on the beam energy but mainly on M_1/M_2 and m. The **penetration profile** $F_R(x,E,\theta)$, which gives a statistical distribution with respect to x of penetration for a beam incident at angle θ to the surface normal, is characterized by a mean value

$$x(E,\theta) = R_p(E)\cos\theta. \quad (3.1.15)$$

We have considered only elastic collisions; at higher projectile energies, energy loss by electrons dominates the slowing down of the ion. However, for most sputtering situations the ion velocities are far below those where electronic stopping would become dominant. At lower velocities, $v < Z_1^{2/3}e^2/\hbar$, a correction due to Lindhard and Scharff is usually added to the nuclear stopping cross-section.[2]

$$S_e \simeq \xi_e 8\pi e^2 a_0(Z_1Z_2/Z)[v/(e^2/\hbar)] \quad (3.1.16)$$

where $Z = (Z_1^{2/3} + Z_2^{2/3})^{1/2}$ and ξ_e is a function of Z_1. The ranges of operation as a function of ion energy of the nuclear and electron stopping processes are shown in Fig. 3.6.

3.1.1.2. Sputter yields from cascade theory The multiple collision processes of the type shown in Fig. 3.2 are treated by Sigmund using Boltzmann's transport theory. Basically, the treatment of simple collision theory is transferred to a large number of random collisions. There is a

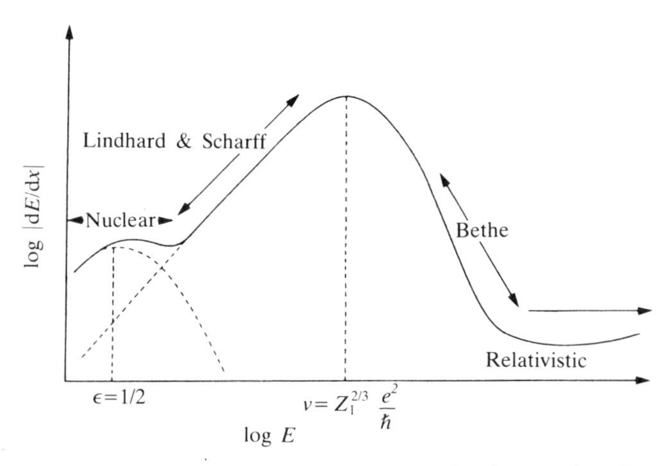

FIG 3.6. Schematic diagram of the stopping power of an ion as a function of energy. At low velocities nuclear stopping dominates for medium and heavy ions. At higher velocities electronic stopping takes over and the projectile is preferably neutral. Beyond the stopping-power maximum the Bethe regime is approached where the projectile is preferably stripped. Reproduced with permission from Ref.1.

central assumption that molecular chaos ensues! After each individual collision the detailed configuration of particles in real and velocity space randomizes. Thus the initial conditions for subsequent collisions are determined by Boltzmann's statistical distribution function and not by the events that occurred previously.

The concepts and physics of dilute collisions, nuclear stopping, nuclear screening and particle scattering which have been introduced are combined with the statistical chaos generated by the Boltzmann equations to model what occurs when a beam of ions collides with a surface to initiate a cascade of recoil atoms in the solid and hence sputtering.

One of the first parameters which is obviously significant is the depth profile of deposited energy. Once this is known it can be combined with an estimate of the internal flux of atoms having a particular range of energies above that required to leave the surface, to yield a sputtered particle flux.

The depth profile of deposited energy follows similar thinking to the derivation of the penetration profile [see eqn (3.1.15)], with the randomization resulting from the application of Boltzmann's equations. Figure 3.7 compares the calculated and experimental estimates. The experimental determination makes use of the proportionality which exists between radiation damage and deposited energy. In this case the bombardment—induced implantation of Bi in silicon also allows the determination of the projected range from a Rutherford back-scattering analysis of the depth distribution of the Bi ions. Thus Fig. 3.7 gives a concentration profile of

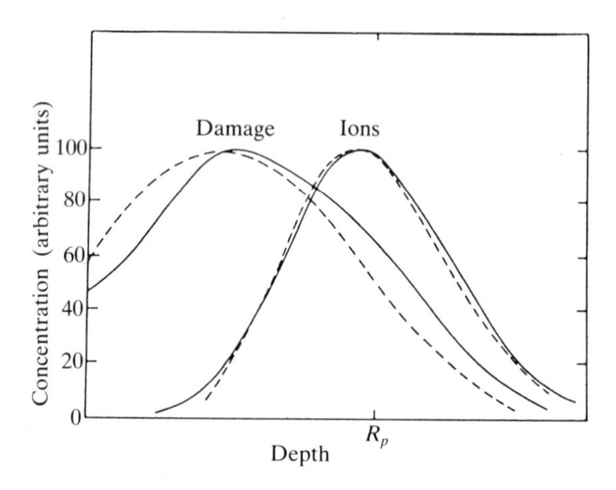

FIG 3.7. Comparison of measured (solid line) and calculated (dashed line) penetration and damage profiles for 50 keV bismuth ions implanted in silicon. Measurements made by Rutherford back-scattering (RBS). Calculations use the Thomas–Fermi interaction and include electronic stopping. Reproduced with permission from Ref.1.

deposited energy. It is almost Gaussian in form. It can be seen that the ion penetrates some 15–40 per cent further than the nuclear damage region. The nuclear stopping power is approximately uniform along the trajectory, whereas the ions themselves tend to stop near the end of the trajectory!

The energy of the incident beam, E, is narrow and well defined, a delta function (see Fig. 3.8(a)). This energy is lost by collisions within the solid. Following the summary presented by Kelly,[3] the total internal movement or flux of target atoms is proportional to Fig. 3.8(b).

$$(E/E_0'^2)\mathrm{d}E_0'\cos\theta'\mathrm{d}\Omega'. \tag{3.1.17}$$

The prime signifies the energy of particles within the solid, thus E_0' is the energy of a moving target atom, θ' is the polar angle of the moving target atom with respect to the surface normal and $\mathrm{d}\Omega = 1/2\sin\theta'\mathrm{d}\theta'$. Equation (3.1.17) assumes that the number of cascade collisions is large. The target behaves as if it is amorphous, $E \gg E_0'$, and the angular velocity distribution of low-energy atoms is isotropic. Figure 3.2 shows a schematic diagram of the slow collisional sputtering process. The total number of sputtered atoms can be obtained if the assumption can be made that equation (3.1.17) is independent of depth, so E can be replaced by $EC_n^{\mathrm{diff}}(x)\mathrm{d}x$, where C_n^{diff} is the differential depth distribution of deposited energy (see Fig. 3.7).

Surface atoms are assumed to be bound with a planar potential energy of magnitude U. This parameter influences sputtering in two ways: first, it provides a barrier which must be surmounted by escaping atoms and hence influences the sputter yield; second, its directional properties will significantly influence the directions taken by ejected atoms. Thus an atom is allowed to leave the surface if its kinetic energy is greater than some value U.

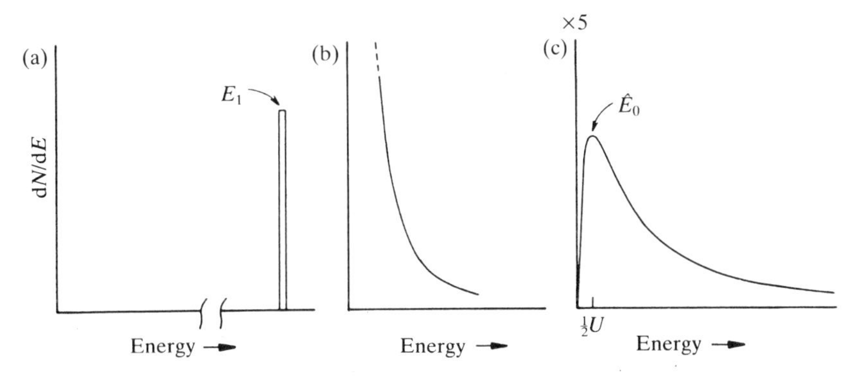

FIG 3.8. (a) Schematic energy distribution of incident ions having energy E_1. (b) Representation of the internal flux of target atoms. (c) External or sputtered flux. The flux peaks at energy $\hat{E}_0 = 1/2U$ and approaches E_0^{-2} behaviour at high energy. Reproduced with permission from Ref. 3.

This is usually related to the sublimation energy which may be corrected for the fact that the bonds binding an atom at the surface are on average about half of those in the bulk. This surface barrier is most easily thought to act in a planar fashion, i.e. it is the energy of the particle normal to the surface, which is important in overcoming the barrier. An alternative is to allow it to operate isotropically.

If the function $H'[E\ C_n^{\text{diff}}(x)\mathrm{d}x, E_0']$ is integrated over E_0' starting at $U/\cos^2\theta'$, and θ' is integrated from 0 to $\pi/2$, substituting $x=0$ and replacing $\mathrm{d}x$ with λ, the mean atomic spacing, the cascade sputter yield is given by

$$Y_{\text{cascade}} = 4.2\lambda^2 E C_n^{\text{diff}}(0)\lambda/U \text{ atoms/ion.} \quad (3.1.18)$$

This can be derived in a slightly simpler approach. If the incident energy is E, then the amount of energy deposited in the outer atomic layer is $EC_n^{\text{diff}}(0)\lambda$. The fraction of this which is directed towards the surface is, assuming an isotropic velocity distribution,

$$\int_0^{\pi/2}(1/2)\cos\theta'\sin\theta'\mathrm{d}\theta = 1/4.$$

Thus the yield is given by

$$Y_{\text{cascade}} = (1/4)EC_n^{\text{diff}}(0)\lambda/U \text{ atoms/ion.} \quad (3.1.19)$$

The two equations (3.1.18) and (3.1.19) are the same when a reasonable interatomic spacing of 0.24 nm is used.

Finally, to obtain the energy distribution of *sputtered* particles, eqn (3.1.17) must be transformed by considering that to be released the internal flux of collisions must overcome the barrier surface energy U. Thus conservation of energy yields

$$E_0' = E_0 + U \quad (3.1.20)$$

and conservation of parallel momentum yields

$$E_0'\sin^2\theta' = E_0\sin^2\theta \quad (3.1.21)$$

Thus the sputtered flux is proportional to

$$EE_0(E_0 + U)^{-3}\mathrm{d}E_0\cos\theta\mathrm{d}\Omega. \quad (3.1.22)$$

Such a flux peaks at $(1/2)U$ and approaches E_0^{-2} at high sputtered particle energies, as is frequently observed experimentally (see Fig. 3.8(c)).

Let us remind ourselves of the assumptions underlying these equations:[1]

1. Linear cascades were assumed, which means that only the small number of target atoms in the cascade volume were set in motion. This is, of course, not true for very high mass ions at intermediate energies.
2. Large recoil cascades were assumed, which may not be true for low energies (<1 keV), and even at moderately low energies, if there are large differences between the ion and target mass.

3. The target surface is assumed not to have an influence on the development of the collision cascades. This will break down in the case of very light ions hitting very heavy targets.
4. Certain models for the elastic and inelastic scattering processes have been assumed.
5. Bulk binding forces and structure have been ignored.

3.1.1.3. Comparison of theoretical and experimental sputter yield data Bearing these considerable assumptions in mind, it is impressive to see how closely many of the experimental sputtering data follow the predicted behaviour. A yield curve as a function of primary ion energy shown in Fig. 3.9 shows that broad agreement is obtained. Equations (3.1.18) and (3.1.19) predict that the yield will be dependent on EC_n^{diff}, which is proportional to the nuclear stopping power. Andersen and Bay have compared all their collected sputter data with the predicted yields according to linear cascade theory.[4] In Chapter 2, Fig. 2.1 is an example of the result for aluminium. The dashed lines are the theoretical curves for Ar^+, and Xe^+ bombardment. There is close agreement between the theoretical and experimental data for sputtering by argon. For xenon, the experimental data are more peaked around the maximum for nuclear-stopping power.

When the data for all target elements are compared there is good

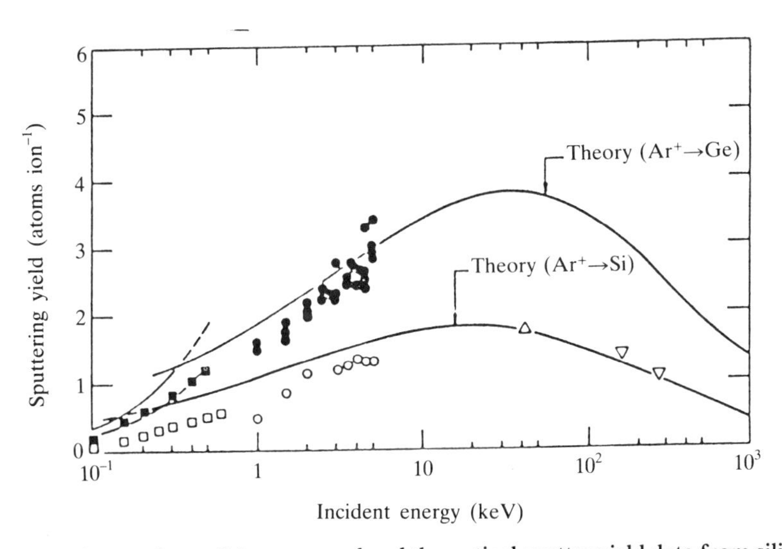

FIG 3.9. Comparison of the measured and theoretical sputter yield data from silicon (□,○,△,▽) and germanium (■, ●) as a function of primary ion incident energy. Agreement is helped by the fact that both materials amorphize under ion bombardment. Reproduced with permission from Ref. 3.

agreement between experiment and theory for high-yield elements (e.g. Cu, Zn, Ag, Au) while the yield for low-yield elements is overestimated by up to a factor three.[4]

The predicted dependence of yield on projectile species is followed very well for light targets, but for heavier targets, for example Au, with heavier projectiles ($Z_1 > 40$) the yield increases faster with Z_1 than predicted by theory (see Fig. 3.10). These are thought to be due to non-linear effects in the collision cascade.

The collision cascade theory demands that the projectile energy be substantially greater than the surface-binding energy if sputtering is to occur. This implies a threshold energy below which no sputtering will occur. It is difficult to measure precisely and there is considerable discussion on what the value is. A value of $E_{th} \simeq 4U$ seems to be reasonable, where U is the surface barrier energy, taken to be the sublimation energy.

Equation (3.1.22) indicates that the energy of the sputtered flux should

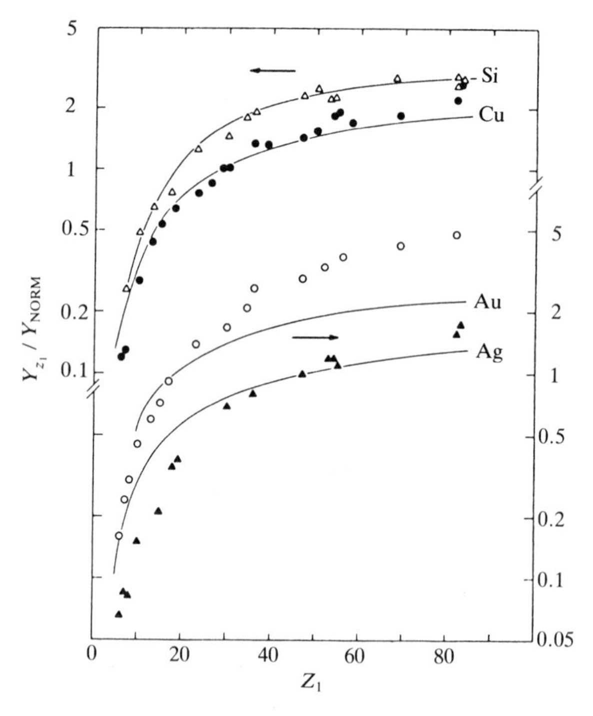

FIG 3.10. The relative sputtering yields for 22 different primary ions at 45 keV for Si, Cu, Ag, and Au targets. △, Si normalized to Y_{Ar}; ●, Cu normalized to Y_{Cu}; ▲, Ag normalized to Y_{Ag}; ○, Au normalized to Y_{Ar}. The solid curves represent the theoretical values. Reproduced with permission from Ref. 4

peak at $1/2U$ and approach E_0^{-2} at high energies. This is indeed borne out by experiment (see Chapter 2, Fig. 2.6).[5] It is this E_0^{-2} dependence which breaks down when spike conditions are reached.

In the light of the experimental data, Andersen and Bay conclude that the region of validity of the linear cascade theory can be summarized by a simple three-dimensional space diagram in projectile atomic number, Z_1; target atomic number, Z_2; and projectile energy E.[3,4] Thus Fig. 3.11(a) shows the region of validity for a 50 keV beam of varying atomic number, Z_1, making impact on targets of varying atomic number, $Z_2 = 50$ bombarded by projectiles of varying energy and Z_1 is shown. It is significant for SIMS to note that in the low-energy region below 10 keV the cascades may not be isotropic such that the model may not adequately reflect observed behaviour.

Secondary effects such as preferential sputtering are more difficult to deal with. Where the masses of the components in an alloy are very similar then a simple extension of eqn (3.1.18) would suggest that the species with the lowest U should be sputtered preferentially. As we have seen in Chapter 2 this is sometimes, but not always, the case.

3.1.2. The spike regime or prompt thermal sputtering

Once the majority of atoms in a given volume have been set in motion, the energy must dissipate in a manner which is quite different from linear collision cascades. The mechanism of energy loss in a spike is uncertain. It

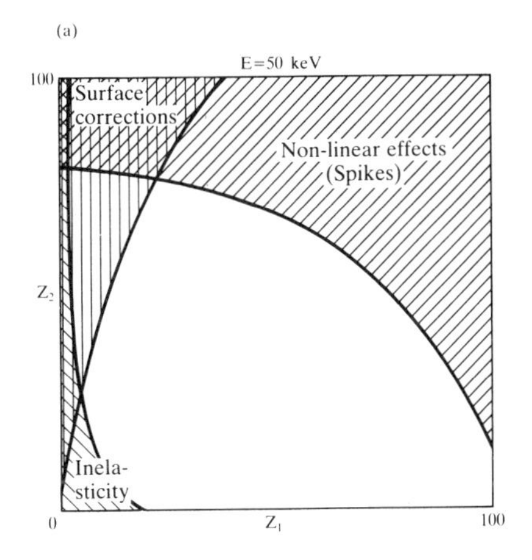

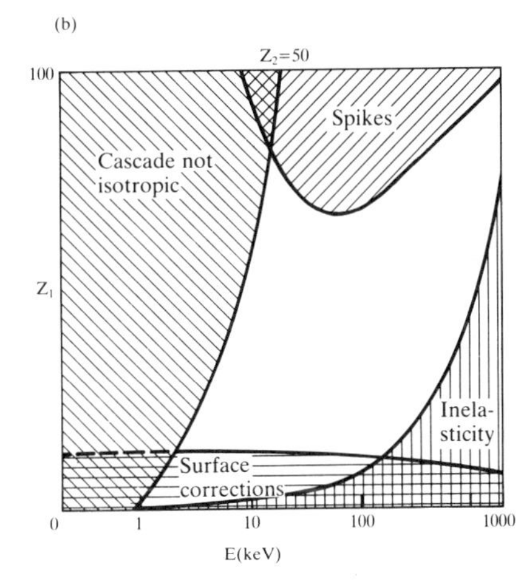

FIG 3.11. The approximate regions of validity of the linear collision cascade sputtering theory illustrated by two planar cuts through the (Z_1, Z_2, E) space. (a) is the (Z_1, Z_2) plane for E 50 keV, while (b) shows the (E_1, Z_1) plane for a material of atomic number, $Z_2 = 50$. Reproduced with permission from Ref. 4.

certainly shows itself in transient vaporization from the surface of the impact region.[6] It is not clear to what extent the loss of material is thermal, due to a shock-wave, or electronic in origin, because thermal and electronic excitation occur together. Sigmund has suggested that the spike region can be approximated to an ideal gas at a local temperature T, where the mean energy per atom is $\theta \simeq 3kT/2$.[1] With typical solid atomic densities, and θ of 1 eV, pressures of 10^5 atm might result. The distinction between thermal and shock-wave induced sputtering might be difficult to determine.

Where the material is a conductor it may well be thermal; however, there are indications that for insulating materials it may be electronic. As indicated above, the E_0^{-2} dependence of the high-energy tail of the sputter yield is lost for spike sputtering. The thermal origin of some spikes is supported by the observation of Maxwellian contributions to the energy spectra of sputtered atoms.

Much cluster ion emission does not obey the E_0^{-2} dependence. As mentioned in Chapter 2, Standing has shown that molecular ion yields from simple amino acids closely follow the amount of energy brought in by the nuclear stopping mechanism; however, for a Li primary ion, where nuclear and electronic stopping both contribute to energy deposition, it was the *total* energy deposited that was the significant factor in determining the molecular ion yield. The two forms of excitation are very different, and it is significant that molecular ion emission is insensitive to these differences.[7]

Michl has suggested that, for insulating materials, electronic excitation may be linked with cluster emission in the thermal spike regime. In his SIMS studies of inorganic frozen solids it is postulated that the simple elemental ions are generated by the collision cascade process, but the cluster ions are formed in the thermal spike regime due to reactions between primary damage centres and hot atoms, molecules and ions in the damage track region. It is argued that because the material is an insulator, charge generated is in the collision track due to collision-induced electronic excitation. This is not easily dissipated and will give rise to further disruption and ionization. Reactions will occur in this region which are analogous to high-energy radiation chemistry and the products will 'explosively expand into the vacuum'.[8]

3.1.3. *Molecular dynamics (MD) models*

It is clear that linear cascade theory cannot easily account for sputtering, which may be influenced by the structure or chemistry of the solid. Attempts have been made to adapt the theory by imposing a structure of atoms on collision theory, but the success has been limited.[9]

Furthermore, almost all of the modelling based on transport theory has been primarily concerned to understand the emission of atomic species. The importance of cluster or molecular ion emission has only been recognized as SSIMS has developed. Oechsner has suggested that there are basically two mechanisms by which clusters are emitted.[10] The atomic combination model (ACM) argues that clusters are formed by the recombination of atomic species after emission. The combining atoms may not be nearest neighbours in the matrix from which they originate. The other mechanism is the direct emission model (DEM), which suggests that the emitted clusters arise by direct fragmentation of the lattice. In both cases, the emission of these species will be structure dependent. Models are required which take account of the full three-dimensional structure of the target crystals. This can, in principle, be accomplished using computer simulation methods.

Two fundamental assumptions are common to all computer simulation models used to study damage production due to the irradiation of solids. First, the constituent particles of solids are assumed to be atoms or atomic ions interacting with one another through conservative pairwise forces. A volume-dependent interaction term may also be included; however, the assumption means that only atomic properties are studied, not electronic properties. Second, model calculations are based on classical mechanics. Clearly, Newtonian mechanics neglects electronic effects and, of course, electronic effects can be very important when collisions between atoms are considered. Electronic rearrangements can occur which will significantly change the interatomic forces. As we have already seen, encounters between energetic atoms can have a significant inelastic component. Corrections may be included but not by adding specifically electronic components to the modelling.

In setting up a model, one has to specify the number of mutually interacting atoms and the boundary conditions which guarantee the stability of the model and specify its interactions with a surrounding matrix. Dynamical models present practical computing problems if high-energy events and sufficiently large numbers of primary particles are to be studied to make the results comparable with experiment. Approximations have to be made.

We will briefly describe two approaches:

(1) displacement cascades in dynamical models,

(2) cascades in the binary collision approximation.

3.1.3.1. The MD model of the sputtering The computational crystallite consists of between 200 and 5000 atoms which interact according to two-body forces derived from a potential $V(r)$. Typical potentials used are the purely repulsive Born–Meyer potential

$$V_{BM}(r) = C_{BM}\exp(-r/a_{BM}) \quad (3.1.23)$$

where $C_{BM} = 52(Z_1Z_2)^{3/4}$ eV and $a_{BM} = 0.219$ Å, and the Morse potential

$$V_{MO}(r) = C_{MO}y(y-2)$$

where

$$y = \exp[-(r-r_0)/a_{MO}] \quad (3.1.24)$$

and C_{MO} and a_{MO} are similar constants. The parameters are obtainable from the literature.[11] The potential function is usually set so that it vanishes smoothly at, say, the second neighbour separation in the crystal. Thus a relatively small number of atoms contribute to a force at a particular point, which is important for reduction in computational effort. If the potential between the atoms is only repulsive the stability of the crystal is maintained by putting special static external forces on the atoms at the crystallite boundaries. Even when the specified interactomic forces are partially attractive, other external boundary forces may be required to define the correct lattice spacing or simulate the cohesive energy or elastic constants. The embedding of the 'crystallite' in its macroscopic medium also has to be specified by a further set of boundary conditions. For example, the atomic displacements in the crystallite may be influenced by the matrix and these could be simulated by a spring force to represent the elastic response of the matrix.

To simulate the impact of an ion on the crystallite the dynamical model selects an atom which is set moving in a particular direction with the desired kinetic energy. The classical equations of motion are integrated for all the particles in the crystallite as long as enough of the original kinetic energy remains. The time steps between integrations are usually 1–3 fs and the whole calculation will require 100–1000 steps.

3.1.3.2. MD studies of atomic sputtering Harrison *et al.* have been active in using the dynamical model to simulate sputtering. For example, in studying the sputtering of Cu,[12] a Born–Mayer potential was used for interatomic separations 1.5 Å and a Morse potential for separations >2.0 Å. A cubic spline interpolates smoothly between them. The model contained 250 atoms arranged in four to six layers depending on the orientation of crystal being studied. The sputtering of Cu(100) surfaces by 5 keV Ar^+ was simulated. Thirty-six incident particles generated 170 sputtered particles, thus Y was 4.7 atoms ion^{-1}, close to the experimental value of 4.1 atoms ion^{-1}.

In an extension of this work, Harrison and co-workers were able to reproduce Wehner-type spots generated as a consequence of particle channelling during Ar^+ bombardment of the (001) face of Ni.[13] The calculations of the angular distributions of emitted particles suggest that it is the high kinetic energy particles (> 10 eV) which are channelled by the surface structure. Figure 3.12(a) shows some experimental data, Fig. 3.12(b) shows the spots generated by all emitted particles, Fig. 3.12(c) shows only particles with energies < 10 eV, and Fig. 3.12(d) shows only particles of energy > 10 eV.

3.1.3.3. MD studies of molecular sputtering Another application of Harrison's approach is the simulation of the sputtering of molecular species.

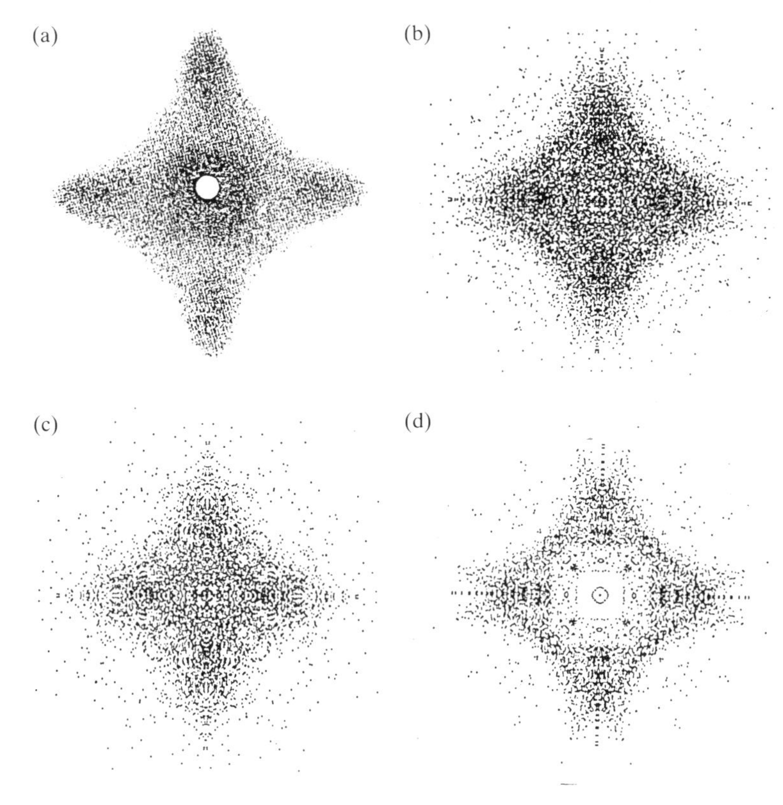

FIG 3.12. (a) Experimental angular distribution of particles ejected due to 1keV Ar^+ bombardment of the (001) face of nickel, so-called Wehner spots. (b) Calculated result, collecting all emitted particles. (c) The calculated result, collecting only particles of energy <10 eV. (d) The calculated result, collecting only particles of energy >10 eV. Reproduced with permission from Ref. 13.

This work started with studies of clean metal surfaces and was extended with the co-operation of Winograd and Garrison to investigate the clusters formed when adsorbed molecules on metal surfaces are sputtered. This has progressed to simulations of the desorption and fragmentation of large organic molecules on metal surfaces. These investigations have obvious relevance to SSIMS and although isolated studies have appeared from other workers, this is really the only group to have carried out significant work. It is appropriate therefore to consider briefly their conclusions.

(a) *Clean metal surfaces*. The early work modelled the formation of clusters from metal surfaces. A study of the sputtering from the three low index planes of Cu exemplifies the approach.[14] The model and pair potentials applied were basically those used earlier by Harrison. The microcrystallite consisted of four layers with about 60 atoms per layer and the bombardment was normal to the crystallite with 600 eV Ar^+. The interaction of Ar^+ with the other species is represented by a purely repulsive term

$$V_{ij} = Ae^{-BR}. \tag{3.1.25}$$

Winograd *et al.* found that the character of the sputter yield was not affected if the area or depth of the microcrystallite was increased. The criterion for the formation of a dimer or multimer is important. It is based on the balance of the total kinetic and potential energy of the apparently close moving dimer or multimer. Thus the relative kinetic energy of the component atoms, T_R, plus the potential energy, V, for all pairs of ejected atoms are first computed. The potential energy between any pair of atoms i and j, V_{ij}, is calculated using a Morse potential

$$V_{ij} = D_e \exp[-\beta(R-R_e)]\{\exp[-\beta(R-R_e)]-2\} \tag{3.1.26}$$

with constants $R_e = 2.22$ Å, $D_e = 2.05$ eV, and $\beta = 1.41$ Å^{-1}. If the total energy of the pair

$$E_{tot} = T_R + V_{ij} \tag{3.1.27}$$

is negative, then the tested dimer is considered bound. Many high-energy impacts are found to yield several bound dimers above the surface. The possibility that these are bound is checked by a similar procedure. The yield and origin of multimers from the three faces was monitored following between 100 and 130 trajectory calculations. The yields of Cu_2^+ and Cu_3^+ relative to Cu^+ are shown to be sensitive to the structure of the surface plane. The Cu_2 data are in reasonable qualitative agreement with the experimental data for the emission of Ni_2 *ions* from Ni surface planes.[15] Of course, direct comparison is difficult because the model only predicts the yields of neutral species; furthermore, the experimental data were obtained by bombardment at 70° to the surface normal whereas the model is for normal incidence. However, more significant is that the authors found that multimers do not emit as intact sections of the solid, they form above the surface and the component atoms do not necessarily arise from atoms which were nearest-neighbours in the surface (see Fig. 3.13). Although on the (110) face most dimers are formed from nearest neighbour atoms, the component atoms can originate from an area of about 70 Å^2. It is difficult to test these conclusions with metal substrates. This mechanism of formation of multimers might appear to reduce the sensitivity of SIMS to surface structure; however, the yield measurements do reflect structural differences.

(b) *Adsorbates on metals*. In Ref. 16 the modelling was extended to the adsorption of oxygen on Cu. Interestingly, the model suggests that frequently a Cu atom will be desorbed from underneath an oxygen without desorption of the oxygen. A large number of clusters were suggested by the

(a)

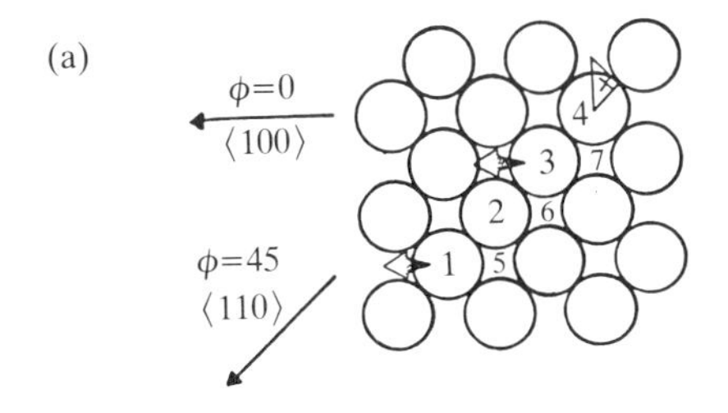

(b)

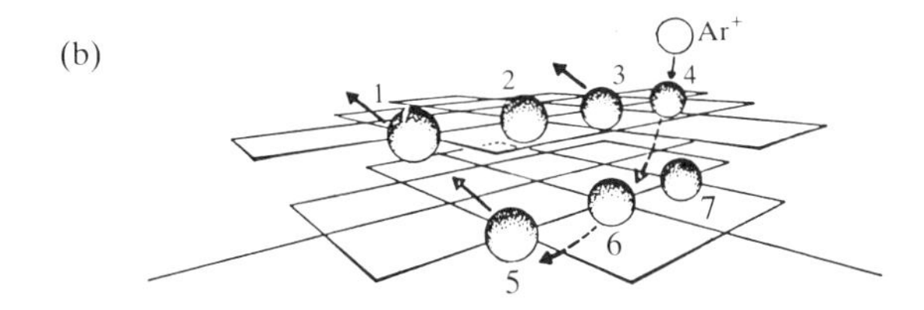

FIG 3.13. Mechanism of formation of the Ni_2 dimer which preferentially ejects in the (100) direction. (a) Ni(001) showing the surface arrangements of the atoms. Atoms 1, 2, 3, 4 are in the top layer, whilst 5, 6, 7 are in the sub-surface layer. × denotes the Ar^+ impact point. Atoms 1 and 3 are ejected as indicated by the arrows and form a dimer moving in the (100) direction. (b) Three dimensional representation of the dimer formation process. Reproduced with permission from Ref. 13.

model—Cu_2, Cu_3, O_2, CuO_3, Cu_2O_3, Cu_3O, Cu_4, Cu_2O_3, Cu_3O_2, Cu_2O_4, Cu_3O_3—and again they are *all* formed above the surface from atoms after sputtering. In many cases, but not all, they were formed from nearest-neighbour atoms from the surface, so a significant relationship to structure is maintained.

Oxidation studies of metals have demonstrated that the type of secondary ions emitted *are* sensitively related to structure (see Chapters 6 and 7). In Chapter 6 it will also be shown that SSIMS is very sensitive to the coverage and surface structure of CO at metal surfaces. Garrison and Winograd have modelled CO adsorption on the smooth (100) and (111) surfaces and the stepped (7,9,11) or Ni(S) – [5(111) × (110)] surface.[17-19] In all these cases they find that the *recombination* mechanism of cluster formation dominates in producing NiCO clusters and indeed the other clusters. Again, a good proportion of the clusters, but by no means all, are formed by combination of atoms and molecules which are nearest neighbours at the surface. Garrison and Winograd conclude that 'there is apparently no direct relationship between these moieties and linear and bridge-bond surface states'. Many of the experimental data would seem to run counter to this rather bald conclusion (see Chapters 6 and 7).

This work has been extended in recent years to model the emission and fragmentation of larger organic molecules from the surface of substrate metals. Experimental data show that molecular or pseudo-molecular ion emission is an important feature of SSIMS or FABMS of organic molecules. It seems remarkable that organic molecules with bond energies of the order of 4–5 eV can be sputtered intact by primary particles with energies of several keV. Figure 3.14 shows the progress of the emission of benzene from the surface of nickel.[20] It illustrates the three basic conclusions which have been drawn.

1. Since benzene has many internal degrees of freedom it can absorb excess energy from an energetic collision and desorb intact, although distortion of the planar molecular geometry may occur. Again, the combination of a substrate Ni atom with the benzene molecule is a consequence of recombination after emission.
2. Desorption is a consequence of the benzene molecule being struck from below by one or more Ni atoms which are at the end of the collision sequence initiated by the Ar^+ bombardment, thus the energy of this desorption collision is rather low.
3. Fragmentation does occur. It is partly a consequence of direct bombardment by the primary particle and partly due to energetic substrate atoms striking the molecule.

In general, MD calculations do not seem to suggest that fragmentation occurs due to unimolecular decay on the way to the detector,[21-23] although there is clear experimental evidence for this.[24]

Reviews of much of the MD modelling of adsorbate systems have appeared in the Refs 25–27.

3.1.3.4. Assessment of the MD approach to sputtering Whilst this modelling is helpful in giving a physical picture of the sputtering process and the mechanism by which secondary atoms and clusters *may* be emitted, it clearly has weaknesses associated with the fact that approximations have to be introduced to make the calculations tractable. The pictorial presentations of the process are easily accepted as representative of the actual process occurring in sputtering and SIMS. This can be dangerous because the validity of the conclusions must be affected by the following factors:

1. There are frequently no boundary conditions to take account of the microcrystallite's location in a larger matrix although edge and corner atoms are not permitted to enter into the sputtering.
2. The same interaction potentials are employed for all atoms of the same type no matter where they are in the solid. Thus, in the study of the stepped surface, the binding potential used in the model between atoms at a step is

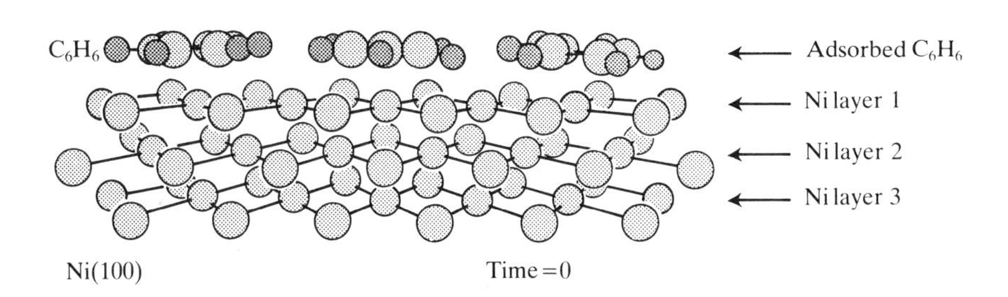

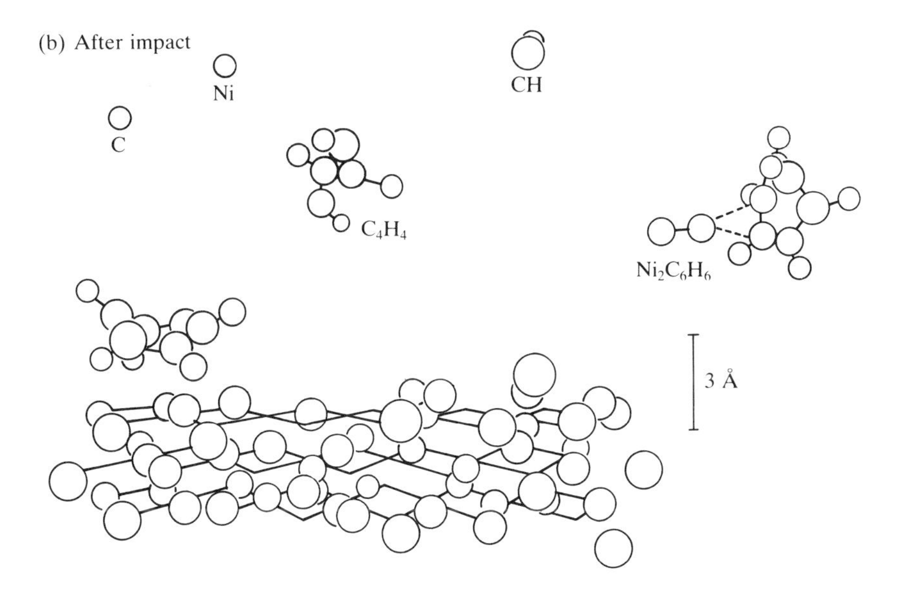

FIG 3.14. Schematic diagram of MD representation of the situation (a) before and (b) 3×10^{-13}s after the impact of one primary particle normal to benzene molecules adsorbed on a Ni (100) surface. Reproduced with permission from Ref. 20.

the same as in the terrace regions. This is obviously not correct. Indeed, even on smooth surfaces it is known that surface atoms are slightly closer and bonding is a little stronger.

3. A similar criticism is that no account is taken of the effect that an adsorbed species may have on the cohesion of the substrate atoms. The bonding of surface atoms to one another and to the bulk is significantly affected by the adsorption of oxygen or other adsorbates. It seems likely that this could profoundly affect the mechanism of cluster emission and formation. One could envisage, for example, that an adsorbed species and the surface atoms to which it is attached may be approximated to a molecular cluster embedded in the surface layers. Thus, just as organic molecules are predicted to desorb intact, such an adsorbate structure may also be removed as a whole unit, perhaps fragmenting after emission. This type of concept has given rise to the intact cluster emission theory for the emission and ionization of clusters from inorganic and organic compounds (see Section 3.2.5).
4. Many experimental studies use low-angle bombardment whereas the models use normal incidence. There are some studies which suggest that lower angles give rise to clusters which are all composed of nearest-neighbour atoms.
5. A general problem with most MD models is that they do not take any account of ionization. The formation and stability of a cluster will be significantly affected once ionized. However, Garrison and Sroubek have joined forces to seek to remove this omission.[28]

Ultimately all such models have to be tested by their adherence to the experimental data.

3.1.4. The binary collision approximation

Molecular dynamics calculations are limited to studies of collision cascades in small numbers at low energy. As we have seen, they are useful for the examination of detailed emission mechanisms. On the other hand, they are usually not capable of dealing effectively with the statistics of cascades, i.e. with the probabilities with which the various mechanisms occur. Furthermore, many practical sputtering situations demand primary energies far higher than can be accommodated by MD calculations.

In the binary collision approximation (BCA), instead of considering the effect of projectile collision on the energy of all the atoms in a defined crystallite, the trajectories of the cascade particles are constructed as a sequence of two-body encounters. There is a very considerable literature on the application of BCA to a wide variety of radiation damage problems; see, for example, Ref. 11. We will briefly examine some of the features of MARLOWE, a computer code based on the BCA.

The basic model assumes that a particle is set in motion if it receives a kinetic energy in excess of a minimum amount, E_m. Projectiles are followed collision by collision, until their kinetic energies fall below an amount, E_c. Only collisions are included which are between projectiles and initially *stationary target atoms*, and they are evaluated only if their encounter parameter is greater than a value, P_c. Since the projectile trajectories are considered to be sequences of two-body encounters, integration of the classical equations of motion reduces to evaluation of the two scattering equations for the scattering angle and time integral (see Fig. 3.1.3). Hence the calculations are very much faster. However, the approximations do seem rather severe. The two-body collisions are isolated from each other, which would not be the case in a crystalline lattice, where the sequence and energy of each subsequent collision would be affected by lattice effects. Second, in two-body collisions it is assumed that projectile and target atom end up far apart from each other after the collision. Finally, no time variable is included to monitor the progress of the collision, as in the MD approach; instead, a scheme is included for ordering the various collisions.

As we saw in Sigmund's cascade theory, there is a requirement to include parameters which take account of the fact that atomic encounters can be significantly inelastic. Sometimes this is a local effect due to the definition of the impact parameter, and the extent to which screening effects or electron stopping are included in each collision. Sometimes it is non-local, intended to represent the influence of distant interactions between energetic ions and target conduction electrons. The theories of Lindhard and Scharff are frequently used to supply the parameters to take account of these factors in MARLOWE.[2]

Further modifications are included in MARLOWE to improve the BCA model. First, the MD models recognize the possibility that a single collision may cause the movement of several atoms. A modification of BCA is to allow a projectile to collide with up to four 'simultaneous' target atoms. Another modification takes account of the binding energy of atoms in a lattice. Thus each new lattice atom added to the cascade has to surmount an energy barrier, E_b. In effect, each lattice atom is bound to its site by energy E_b. Consequently, a target atom receiving energy T in a collision commences to move with kinetic energy $T - E_b$.

An additional surface energy parameter is added to control sputtering. An atom is allowed to leave the surface if its kinetic energy is greater than some value U_s, related to the sublimation energy. This surface barrier is most easily applied as a planar function.

MARLOWE is used to simulate sputtering from polycrystalline and

TABLE 3.1

Model parameters used for MARLOWE investigations of the sputtering of Au (001) surfaces by 700 eV Xe

Lattice constant a	4.0783 Å
Debye temperature θ_D	180 K
Minimum energy for motion E_m	4.5 eV
Particle cut-off energy E_c	4.0 eV
Lattice binding energy E_b	0.5 eV
Maximum impact parameter P_c/a	0.62
Surface binding energy U_s	4.0 eV
Molière parameters a_{12}	
Xe–Au	0.0731 Å
Au–Au	0.0750 Å

amorphous solids by applying a random rotation matrix. A polycrystalline medium is constructed by generating a random rotation matrix at the beginning of each cascade. An amorphous medium is simulated by generating a new random rotation matrix at *every* collision.

Robinson reports an interesting example of the application of MARLOWE to the sputtering of gold by normal incidence 700 eV Xe atoms.[11] Table 3.1 lists the collisional and the Molière potential parameters used. Xe bombardment was distributed over all possible geometries of the unit mesh. Figure 3.15 shows the dependence of sputter yields on target thickness using the planar binding model. The ordinary reflected sputter yield reaches a limiting value after a thickness of $2a$, or about five atomic layers. Table 3.2(a) shows the calculated depth of origin of the sputtered Au atoms. Less than 1 per cent of the ejected atoms originate from *beyond* the second layer. Table 3.2(b) shows that over 66 per cent of ejected particles arise from a single collision. Linear cascades do occur but not more than 25 per cent of ejęcted atoms arise from them. The calculated sputter yields are compared with experiment in Table 3.2(c), and good agreement is obtained when the planar binding model is used.

3.1.5. *Other cluster emission models*

To limit the desorption process to primarily collision mechanisms has been shown to be unsatisfactory, particularly in seeking an understanding of cluster emission. MD calculations still consider classical collision dyanamics. However, the contribution of other processes is evident in insulating organic materials.

Macfarlane has pointed out that whilst the primary excitation process both in the keV and MeV regime is clear, how this energy is transformed into the

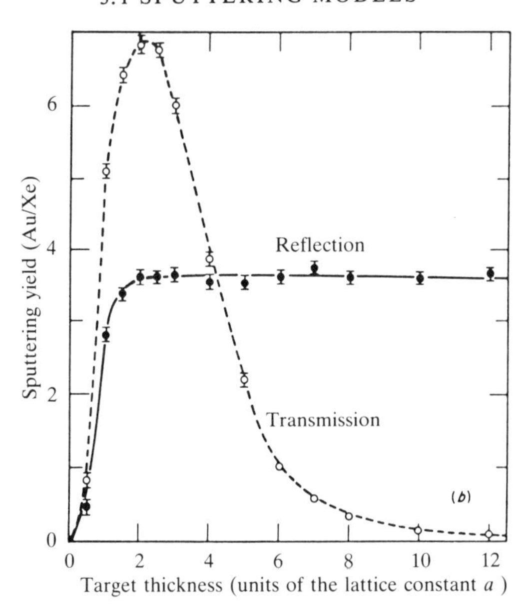

FIG 3.15. Sputtering yields calculated for 700 eV Xe atoms normally incident on a Au (001) surface using the BCA program MARLOWE. Reproduced with permission from Ref. 11.

form required for the ejection of secondary molecular species is still not understood.[29] If the material is crystalline, the volume which has been excited is melted and freezes into an amorphous state. It is evident that in the energy dispersion process lattice vibrations are excited to a level where intermolecular bonds are broken. It can be postulated that if the material is an insulator the primary deposited energy is transferred away by excitons (quanta of electronic excitation) which can propagate through the solid.[30] These radiationless transitions deposit their energy into vibrational excitation within the matrix. Macfarlane points out that the attractive feature of the exciton model is that exciton states can be made either by electronic excitation (so-called transverse excitons) or by collisional processes (longitudinal excitons) and *both* could decay into the same pattern of electronic and vibrational excitations. This approach provides an explanation which could account for the similarities between the mass spectra from SIMS and ^{252}Cf PDMS. It would also explain why the emission of secondary cluster ions is sensitive to the matrix, because the energy dispersion in an exciton–phonon process will be influenced by the physical features of the solid.

Thus desorption can be described as vibrational and, if the energy reaching the molecule is low, intermolecular bonds will be broken and the desorption

TABLE 3.2

Some properties of atoms ejected from Au(001) surfaces by normally incident 700 eV Xe atoms

(a) Depths of origin of sputtered atoms

Atom layer	1	2	3	4	5	6
Yield Au/Xe	3.291	0.285	0.018	0.005	0.006	0.001
Fraction of emissions	0.913	0.079	0.005	0.0014	0.0018	0.0001

(b) Ejection of atoms by linear collision sequences

Sequence length	0	1	2	3	4	5
Yield Au/Xe	2.459	0.837	0.219	0.054	0.026	0.006
Fraction of emissions	0.682	0.232	0.061	0.015	0.007	0.0016

(c) Sputtering yields for 700 eV Xe normally incident on Au targets

	Inelastic		Calculated sputter yield Au/Xe	
Target	loss	Temperature	Isotropic binding	Planar binding
Polycrystalline	Yes	Static	5.161	3.221
(001)	Yes	Static	4.809	3.606
(001)	Yes	120 K	4.845	3.679
(001)	No	120 K	5.093	3.661

Experimental yields for Xe^+ ions on polycrystalline Au

Ion energy (eV)	400	500	600	700
Observed yield (Au/Xe)	2.5	3.2	3.4	3.7

of an intact molecule will result. Where the energy arriving at the molecule is high, fragmentation will result.

Taking this view as a starting point, King *et al.* used RRKM (Rice, Ramsperger, and Kassel Model) unimolecular decomposition theory to predict the mass distribution of ejected molecules and clusters when insulin was sputtered in ^{252}Cf PDMS.[31] Basically, they regard the primary energy as having been deposited into vibrational modes, whether from a nuclear or electronic stopping process. Thus particle ejection arises from a chemical process in which an energized species M* decomposes unimolecularly via an activated or transition state $M^{\neq}$ according to the reaction scheme:

$$M^* \rightleftharpoons M^{\neq} \rightarrow \text{products.}$$

In RRKM theory a molecule in the target is regarded as a system of loosely coupled oscillators, which are conveniently regarded as the normal modes of vibration of the molecule. The rate constant, k, for ejection of the active molecule is regarded as increasing with the energy possessed by the molecule in its various degrees of freedom. The calculation involves distributing j quanta of energy between n degrees of vibrational freedom using statistical mechanics. The process of ejection involves depositing sufficient energy, $E^{\neq}$, into M* for it to attain the transition state. The authors wished to find the parameters which reproduced the observed the relative yield of the insulin molecular ion, $m/z = 5800$, and the two fragment chains at $m/z = 2400$ and 3400. Thus there were three linked reactions to consider. The calculation was able to reproduce the observed data with apparently reasonable activation energies of 8.45 eV for the ejection of the positive molecular ion, 9.77 eV for the $m/z = 3400$ fragment, and 8.24 eV for the $m/z = 2400$ fragment. Approximately 450 normal modes were involved in the ejection process and the total energy deposited into each molecule was 780 eV.

In avoiding the complexities of the initial energy deposition regime, this approach gives a feel for the processes occurring in the ejecting molecule and the energies required.

3.1.6. *Electronic sputtering*

The possibility of an electronic contribution to sputtering has only fairly recently been recognized as being a significant process. Most of the examples where it has been identified are concerned with charged particle bombardment of poorly conducting targets. This process may be slow or fast, thus $t >$ or $< 10^{-11}$–10^{-10} s. This process may be of some importance in static SIMS of poor conductors. Three or four possible mechanisms have been suggested.

3.1.6.1 Direct interaction of the impacting ion with the lattice will deposit a significant amount of electronic energy. For example, the electronic stopping power of 10 keV Ar^+ is ~ 90 eV nm^{-1}.[32] This is sufficient to cause ionization in the surface layers of halides or oxides, generating antibonding states with consequential loss of bonding.

3.1.6.2. Interatomic Auger decay may occur if ion bombardment generates an electron vacancy in a metal ion which can *only* be filled by an *interatomic* Auger process, say from an O^{2-} ion. As a consequence, two or even three electrons may be lost by the oxygen, leading to loss of bonding and desorption of oxygen.[33]

3.1.6.3. Diffusing defects generated as a consequence of ion bombardment can also give rise to sputtering when they arrive back at the surface.[34]

3.1.6.4. Ion-explosion sputtering has been proposed for sputtering of insulating materials by MeV ions and is thought to be particularly important in ^{252}Cf PDMS. The basic assumption is that a high density of ionization occurs along the track of the ion. The electrons generated travel atomically great distances away so that localized Coulombic explosion occurs, with resultant sputtering.[35]

3.2. Ionization

Although surface annealing is thought to occur within 10^{-10} s, in the region of the primary particle impact zone the crystalline structure of the surface will be considerably disrupted. It does not seem appropriate, therefore, to consider the electronic structure to be that of the perfect solid. In other words, models which rely on a normal solid band structure are not likely to be valid. It is probably more satisfactory to consider that the area of emission is amorphous with a continuum of energy states. Many treatments consider that the ionization probability of a departing atom is determined by an atom–surface interaction within a few tenths of nanometres from the surface. Various mechanisms of the interaction have been analysed and are reviewed in Ref. 36

Four examples will be briefly described.

3.2.1. The perturbation model[37,38]

This approach suggests that the ionization probability is a maximum closest to the surface, where atom–surface coupling varies most strongly. Of course, this coupling also produces efficient neutralization. As the distance from the surface increases, the coupling becomes weaker and the ionization probability is less but the *ion* escape probability begins to be finite. Thus the probability of creating and observing an ion reaches a maximum in a region a few Ångstroms from the surface. Thus the ionization probability is given by

$$\alpha^+ \propto \exp\left[-(I-\psi)/\epsilon_0\right] \qquad (3.2.1)$$

where I is the ionization potential, ψ is the work function and $\epsilon_0 = hav/C\pi$ in which h is Planck's constant, a is a range factor approximately equal to 2 $Å^{-1}$, and v is the velocity of the departing atom. For metals, this model rationalizes the well-known dependence of yield on ionization potential, and on adsorption or matrix effects which give rise to work function changes. However, for non-metals, where the valence levels of the departing atom correspond to an energy in the band-gap, this model does not predict surface–atom coupling.

3.2.2. *The surface excitation model*[36]

This model is a development of the above approach. Resonant electron transfer between the surface and the departing atom occurs up to a distance of 5–10 Å. For ionization to occur and the ion to escape it is necessary that a valence level in the departing atom is isoenergetic with a *vacant* level in the surface *below* the photo-threshold (Fermi level in a metal or valence band edge in a semiconductor or insulator). The surface excitation resulting from the primary ion impact greatly increases the probability (P_e) of this occurring. The surface disruption will result in a transient continuum of energy states. Thus the ionization probability will be given by

$$\alpha^+ = P_e/(1-P_e). \qquad (3.2.2)$$

This model unifies discussion of metal and insulator surface by focusing attention on the local transient disorder at the sputtering site and thus relating ionization probabilities to the local photo-threshold. Local changes in the surface matrix as a consequence of, for example, oxygen adsorption and incorporation, which have been reported to enhance *both* positive and negative ion emission, can be explained by this model.

3.2.3. *The bond-breaking model*[39]

Here we are specificially concerned with ion emission from ionic solids. The key assumption is that the ion state is the ground state, and the probability of observing a secondary ion depends simply on whether the ion state remains energetically favoured over the neutral state as the ion moves away from the surface. This depends on the distance at which the ion and neutral potential energy surfaces cross. If the atom–surface separation is <10 Å, the ion and neutral states will be strongly mixed. The features which influence the crossing distance are:

1) the ground state energy of the associated species;
2) the difference between the ionization potential and electron affinity of the positive and negative ions; and
3) the range of the electrostatic interaction.

Sometimes gas-phase ion-pair interactions have been used in an attempt to understand the effects. These are really not appropriate. For example, level crossing occurs in the gas phase for the Si – O system whereas it does not from solid SiO_2. The effect of oxidation on ion yield can then be thought of in terms of moving the curve-crossing further from the surface.

3.2.4. The molecular model[40]

The main contention of this model is that because electronic transition rates are so rapid (10^{14}–10^{16} s^{-1}), it should really be expected that no excited or ionized species formed at the surface could escape without de-excitation in the 10^{-13} s required to move outside the range of surface electronic interaction, so ionized species can survive only if they are formed some distance away from the surface. Thomas suggests that this can happen via level-crossing processes in the dissociating sputtered molecular species (see Fig. 3.16).

This model has been developed further to the nascent ion molecule model due to Gerhard and Plog.[41] They also suggest that the rapid electronic transitions which occur in the surface region will neutralize any ions before they can escape. Secondary ions are thought to result as a consequence of dissociation of sputtered neutral molecular species some distance from the surface. Energy transfer in the collision cascade in the solid gives rise to collision energy transfer between the atoms of the molecule leaving the surface; consequently, dissociative ion formation may occur with charge exchange during dissociation of the molecule. In the terminology of the model, ions are formed by the non-adiabatic dissociation of nascent ion molecules (neutral molecules).

The conditions for neutral emission were found to depend on the mass ratio of the constituents of the sputtered molecule. Most of the neutral molecules originate from direct emission of ion pairs such as MeO and keep their molecular character after leaving the surface (see Fig. 3.17). Only a few molecules have enough energy to dissociate into their constituents. The dissociation takes place at a distance from the surface, where the electronic influence of the surface will be much smaller than in the case of bond-breaking emission. Bond-breaking considers the system solid–Me^+, whereas the nascent ion molecule considers the system MeO^0.

The important feature of this model is its illustration of the mass dependence of secondary Me^+ yields emitted from oxide specimens under Ar^+ bombardment. The model contains three steps of mass dependence:

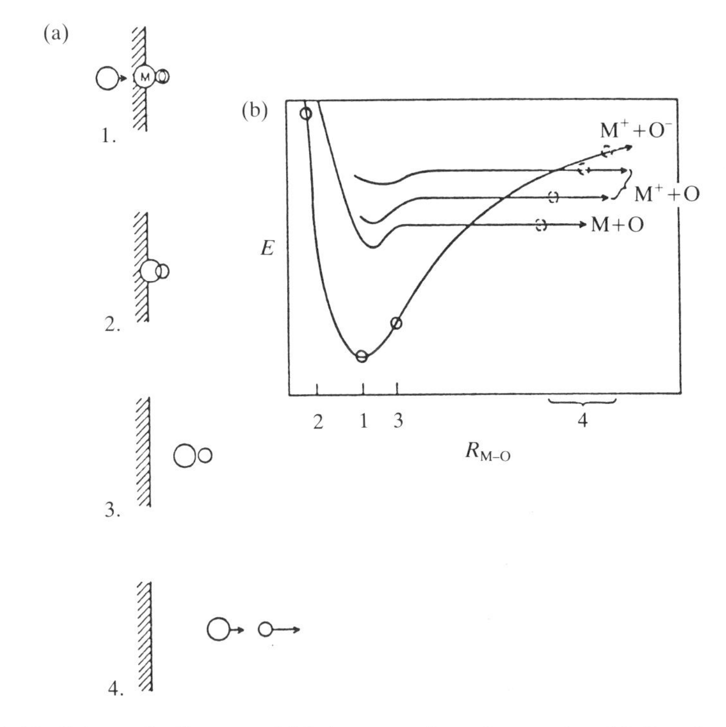

FIG 3.16. Schematic diagram of (a) the sputtering of a transient M-O molecule and (b) the possible level-crossing processes during dissociation of this entity. Reproduced with permission from Ref. 34.

1) energy in the collision cascade of a matrix which is inhomogeneous in mass;
2) collision energy transfer between atoms of the nascent ion molecule leaving the surface;
3) charge exchange during dynamical dissociation of the ionically bonded atoms of the nascent ion molecule.

The treatment leads to a mass dependence of Me^+ yield from oxides and chlorides which mirrors that derived from the S^+_{max} values derived from Benninghoven's so-called valence model[42] (see Section 7.5.3).

$$S^+_{max}(m_{metal}) = \text{const}.m_{metal}{}^{-2.4} \tag{3.2.3}$$

The concept of the molecular model is a helpful one, especially in understanding the emission of cluster or molecular ions. The same parameters are involved, namely the dynamic interaction of atoms in the surface region or selvedge, and the ionization mechanism which operates for those species which leave the sample surface. We have seen above that MD and other

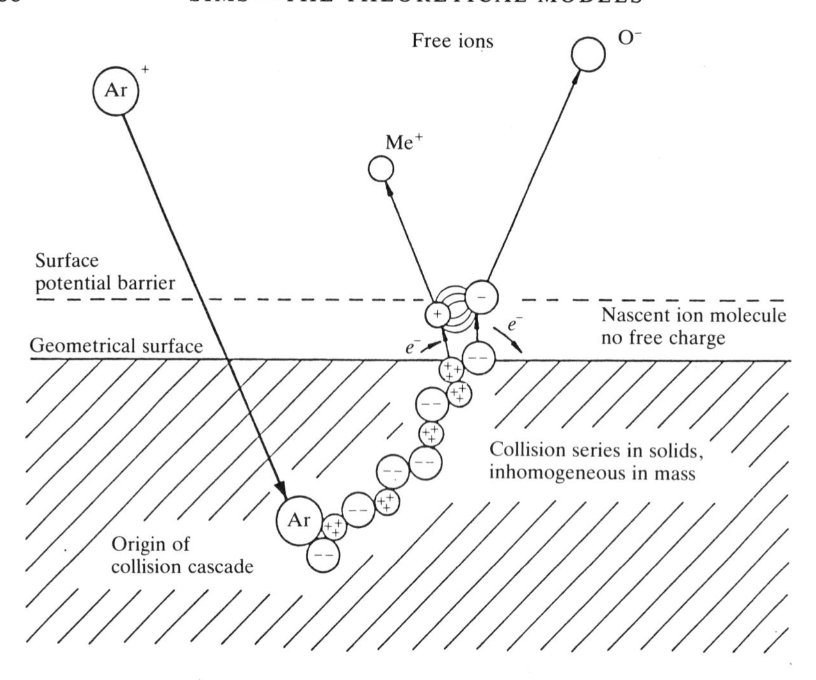

FIG 3.17. The basic processes occurring during the emission of clusters and ions according to the nascent ion molecule model. Secondary ions are formed by the dissociation of the neutral molecular species emitted from the surface. Reproduced with permission from Ref. 41.

collisional models have contributed to the first parameter. Plog and Gerhard suggest on the basis of their nascent ion molecule model that MeO_m^+ ions originate from MeO^0 and other *mono*-metal cluster ions arise from the dissociation of $Me_nO_m^0$ species. It is thought that di- and tri-metal clusters would not survive the surface collision as one molecule. To overcome this problem, Gerhard has suggested a statistical molecule emission process in which ions sputtered from *adjacent* sites in the selvedge are coupled to form clusters.

The two models of cluster generation suggested by Oechsner—the atomic combination model (ACM) and the direct emission model (DEM)—are clearly relevant here. In studies of oxidized Ti, Nb, and V, Yu found that at low oxygen exposures the emission of MeO_m^+ ($m = 1,2,3$) was consistent with the recombination of non-adjacent Me^+ with O^0.[43] However, higher oxygen coverages seemed to promote the formation of MeO_m^+ by lattice fragmentation. The negative ions MeO_m^-($m = 1,2,3$) showed strong evidence of formation by this mechanism even when their positive counterparts showed evidence of exclusive formation by the recombination model. The data were interpreted in terms of the potential energy curves of the neutral molecule as compared with the negative ion, an analogous approach to the models above. Figure 3.18 depicts three possible relations between the potential energy curves of $A^0 + B^-$ and $A^0 + B^0 + e^-$. Stable AB^- formation by lattice fragmentation is likely to occur when the potential minimum of the neutral molecule is within the potential well of the negative ion (see Fig. 3.18(a)) or when the minimum is outside the well of the negative molecular ion (see Fig. 3.18(b)) *if* the initial internuclear distance is restricted to the order of a lattice spacing, i.e. between x_1 and x_2. Where the electron affinity, EA, of the molecule is greater than the dissociation energy, D, of the molecular ion,

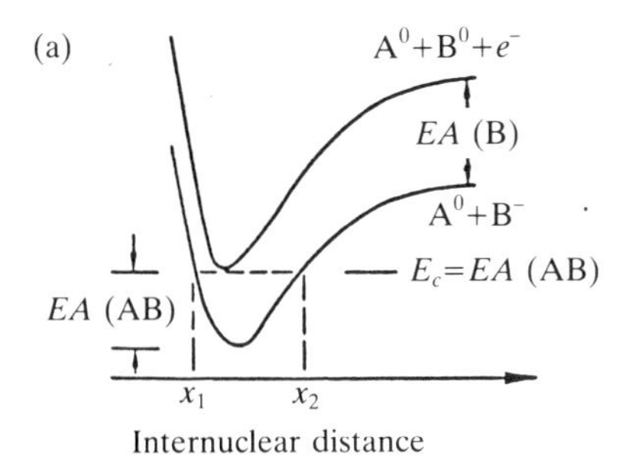

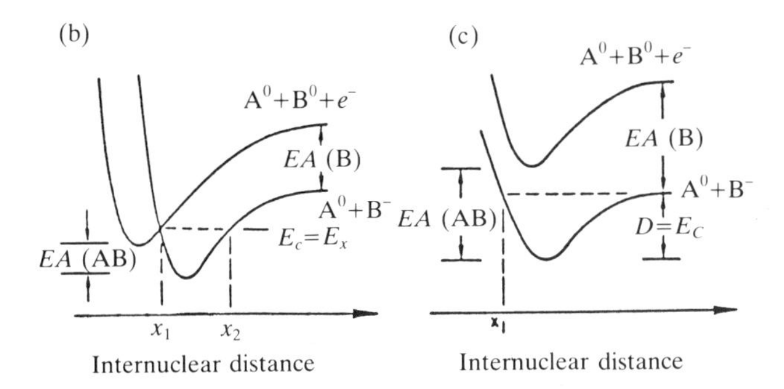

FIG 3.18. The formation of the cluster ion AB^-. Schematic diagram of the three possible relations between the potential energy curves of the negative ion A^0+B^- (AB^-) and the neutral molecule, $A^0+B^0+e^-$. (a)$E_c=EA$(AB); (b) E_cE_x; (c)$E_c=D$(AB). Stable AB^- formation is likely by lattice fragmentation if the potential minimum of the neutral molecule is within the potential well of the negative ion. See text. D is the dissociation energy of AB^-; EA is the electron affinity; E_x is the potential energy at curve crossing; E_c is the maximum allowable excess energy; and X_1-X_2 is the region of internuclear separation for stable AB^- formation. Reproduced with permission from Ref. 43.

formation by recombination of non-adjacent atoms and ions is possible (see Fig. 3.18(c)). Yu's results suggest that positive molecular ions could be formed by either mechanism but that negative polyatomic ions were only stable when formed by DEM.

3.2.5. Desorption ionization

To try to understand molecular and cluster ion emission, particularly from organic materials, where it has been suggested that vibrational excitation (see Section 3.1.5) may be important, Cooks and Busch have advocated the **desorption ionization** (DI) model.[44] This model emphasizes that the processes of desorption and ionization can be considered separately and it seeks to rationalize the relative abundances of the ions generally observed in the SIMS spectra. Whatever the form of the initial excitation process, the energy is transformed into thermal/vibrational motion as far as the molecules are concerned. Following the desorption of *preformed* ions C^+ or A^-, note that no ionization step occurs, and the majority of these ions traverse the uppermost layers (termed the selvedge) unperturbed and are observed in high abundance in the spectrum. Most ions are believed to be desorbed with very low internal energy, or are unlikely to be energized by subsequent collision, so fragmentation is minimized. Neutral molecules are desorbed in high yield but to be detected must undergo an ionization process such as cationization.

To generate other ions the DI model suggests that desorption is followed by chemical reaction which may be one of two forms:

(1) in the selvedge fast ion/molecule reactions or electron ionization can occur;

(2) in free vacuum, unimolecular dissociations may occur, governed by the internal energy of the parent ion giving rise to fragment ions (see Fig. 3.19 and Chapter 2, Table 2.5).

Thus, according to this model, the desorption event is of relatively low energy. There is *no* net creation of ions during desorption. The analyte if charged may be desorbed as an ion, or if desorbed as a neutral, it may be ionized by reaction with a desorbed ion, or an electron. The model is consistent with the fact that in SSIMS little low-mass fragmentation occurs—it is mainly molecular and large cluster ions which are observed. However, it does not seem wholly reasonable to suggest that the initial collisional process does not produce significant numbers of additional ions relative to those existing preformed in the material. Many molecular materials do not of themselves contain ions and presumably rely on the initial collisional process to generate ions. This certainly occurs in the frozen inorganic solids studied by Michl (see Section 2.5.1). The model also implies,

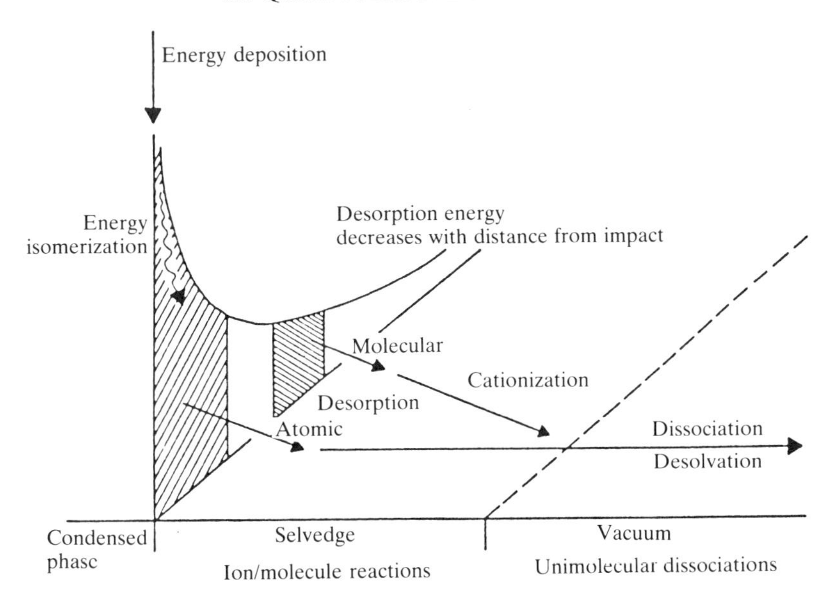

FIG 3.19. Summary of processes thought to occur during desorption ionization. Reproduced with permission from Ref. 44.

in agreement with Michl's ideas, that there is a delay between prompt collisional sputtering and molecular ion emission. Such a delay has been observed in laser-induced molecular ion emission. This qualitative description is analogous to other forms of mass spectrometry and makes a good deal of sense in unravelling the SSIMS spectra of organic materials.

3.3. Quantitative analysis

Several computer programs have been developed to enable the analyst to derive quantitative concentration data from the detected (mainly dynamic) SIMS signals. These entail bringing together the sputtering and ionization models. Clearly, there are a variety of ways in which this can be done. For example, Gries and Rüdenauer[45] brought together the elements of Sigmund's collision cascade model with the surface excitation ionization theory.

3.3.1. Local thermal equilibrium theory

Another approach of long standing is the local thermal equilibrium (LTE) model of Andersen and Hinthorne,[46] which assumes that a plasma in complete thermodynamic equilibrium is generated locally in the solid by ion bombardment. Neutralization processes are ignored and it is assumed that positive ions are generated by

$$X \leftrightharpoons X^{+} + e.$$

A similar process holds for negative ions. Assuming equilibrium, the law of mass action can be applied to find the ratio of ions, neutrals and electrons:

$$n_e n_X^{+}/n_X = K.$$

From thermodynamics or statistical mechanics the value of K can be calculated to yield the Saha–Eggert equation

$$n_e n_X^{+}/n_X = \{2Z^{+}(T)/Z^{0}(T)\}[(2\pi mkT)^{3/2}/h^{3}]\exp[-(E_i - \Delta E_i)/kT] \qquad (3.3.1)$$

where Z^{+}, Z^{0}, and 2 are the partition functions of ions, neutrals, and electrons respectively, E_i is the ionization energy, ΔE_i is the depression of ionization energy in the plasma, and T the temperature which acts as a parameter correcting for matric effects. If we take $n_X{}^{+} = I_s$ (secondary ion current of X^{+}), $n_X \propto$ concentration c_X, and plotting log $(2n_X + Z^0/n^0Z^{+})$ against the ionization energy, where I_s are the ion currents obtained from elements with known concentrations, a straight line is obtained if eqn (3.3.1) is valid. This relationship was demonstrated by Andersen and Hinthorne, and others. Thus the concentration of an element can be calculated from eqn (3.3.1) if the electron density, n_e, and the temperature, T, have been determined for the matrix being analysed by measuring the ion currents from at least two elements of known composition beforehand. Various computer programs have been developed to make the method general in its application. However, the use of two adjustable parameters is involved whose values can vary widely depending on the elements used as internal standard.

The LTE model does not specify any mechanism of sputtering. Morgan and Werner have sought to improve matters by incorporating the collision cascade model of sputtering as the initiator of the plasma. This leads to a more satisfactory one-parameter equation. This and other quantification models are reviewed in Ref. 47.

3.4. Conclusions

This brief survey demonstrates the complexity of the secondary ion emission process.

Atomic sputtering theory has attained a high degree of sophistication and provides a good physical model which enables us to understand the process rather well. In contrast, cluster emission is still an area of considerable uncertainty. Much work has still to be done before a fully satisfactory model is available. Since cluster emission is of particular interest for the analysis of complex inorganic and organic materials it is likely that particular models will have to be developed applicable to the physical properties of the different materials.

It has been realized that this approach is necessary when modelling the ionization process. The ionization theories add a further considerable complication to modelling the overall secondary ion emission process. However, although a fully satisfactory quantitative theory has yet to emerge, the qualitative ideas do help considerably in understanding the parameters important in influencing secondary ion yield.

The following chapters will show that the lack of a fully comprehensive theory does not in fact greatly inhibit the successful empirical application of SIMS to many problems of great scientific and technological importance.

References

1. Sigmund, P. (1981). In *Sputtering by Particle Bombardment I*, Springer Series Topics in Applied Physics, Vol 47 (ed. R. Behrisch), p. 9. Springer Verlag, Berlin.
2. Lindhard, J. and Scharff, M. (1961) *Phys. Rev.*, **124**, 128.
3. Kelly, R. (1984) In *Ion Bombardment Modification of Surfaces: Fundamentals and Applications* (ed. O. Auciello and R. Kelly), p. 27. Elsevier, Amsterdam.
4. Andersen, H. H. and Bay, H. L. (1981). In *Sputtering by Particle Bombardment I*, Springer Series Topics in Applied Physics, Vol 47 (ed. R. Behrisch), p. 143. Springer Verlag, Berlin.
5. Hucks, P., Stocklin, G., Vietzke, E., and Vogelbruch, K. (1978). *J. Nucl. Mat.*, **76/77**, 136.
6. Kelly, R. (1979). *Surf. Sci.*, **90**, 280; (1977). *Radiat Effects*. **32**, 91.
7. Standing, K. G., Chait, B. T., Ens, W., McIntosh, G., and Beavis, R. (1982). *Nucl. Instrum. Methods*, **198**, 33.
8. Michl, J. (1983). *Int. J. Mass Spectrom. Ion Phys.*, **53**, 255.
9. Lehmann, C. and Sigmund, P. (1966). *Phys. Stat. Sol.*, **16**, 507.
10. Oechsner, H. (1982) *Secondary Ion Mass Spectrometry SIMS III*, Springer Series in Chemical Physics, Vol 9 (ed. A. Benninghoven, J. Giber, J. László, M. Riedel, and H. W. Werner), p. 106. Springer Verlag, Berlin.
11. Robinson, M. T. (1981). In *Sputtering by Particle Bombardment I*, Springer Series Topics in Applied Physics, Vol 47 (ed. R. Behrisch), p. 73. Springer Verlag, Berlin.
12. Harrison, D. E., Moore, W. L., and Holcombe, H. T. (1973) *Radiat. Effects*, **17**, 167.
13. Winograd, N., Garrison, B. J., and Harrison, D. E. (1978). *Phys. Rev. Lett.*, **41**, 1120.
14. Winograd, N., Harrison, D. E., and Garrison, B. J. *Surf. Sci.*, 78, 467 (1978).
15. Barber, M., Bordoli, R.S., Wolstenholme, J., and Vickerman, J.C. (1977). *Proc. 7th Int. Conf. Vac. Sci. and 3rd Int. Conf. on Solid Surf.* (Vienna, 1977), p. 983.

16. Garrison, B. J., Winograd, N., and Harrison, D.E. (1978). *Phys. Rev.*, **B18**, 6000.
17. Winograd, N., Garrison, B. J., and Harrison, D. E. (1980). *J. Chem. Phys.*, **73**, 3473.
18. Gibbs, R. A., Holland, S. P., Foley, K. E., Garrison, B. J., and Winograd, N. (1981). *Phys. Rev.*, **B24**, 6178; (1982). *J. Chem. Phys.*, **76**, 684.
19. Foley, K.E., Winograd, N., and Garrison, B. J. (1984). *J. Chem. Phys.*, **80**, 5254.
20. Garrison, B. J. (1980). *J. Am. Chem. Soc.*, **102**, 6553.
21. Garrison, B. J. (1982). *J. Am. Chem. Soc.*, **104**, 6211.
22. Winograd, N., Garrison, B. J., and Harrison, D. E. (1986). *J. Chem. Phys.*, **73**, 3476.
23. Moon, D. W. and Winograd, N. (1983). *Int. J. Mass Spectrom. Ion Phys.*, **51**, 217.
24. Ens, W., Beavis, R., and Standing, K. G. (1983). *Phys. Rev. Lett.*, **50**, 27.
25. Winograd, N. (1981). *Prog. Solid State Chem.*, **13**, 285.
26. Garrison, B. J. and Winograd, N. (1982). *Science*, **216**, 805.
27. Garrison, B. J. (1983). *Int. J. Mass Spectrom. Ion Phys.*, **53**, 243; (1986). In *Secondary Ion Mass Spectrometry, SIMS V*, Springer Series in Chemical Physics, Vol. 44 (ed. A. Benninghoven, R. J. Colton, D. S. Simmons, and H. W. Werner), p. 462. Springer Verlag, Berlin.
28. Garrison, B. J., Diebold, A. C., Lin, J. H., and Sroubek, Z. (1983). *Surf. Sci.*, **124**, 461.
29. Macfarlane, R. D. (1985). In *Desorption Mass Spectrometry, Am. Chem. Soc. Symp. Ser.*, **291**, 56.
30. Forster, T. (1965). In *Modern Quantum Chemistry*, Vol III, p. 93. Academic Press, New York.
31. King, B. V., Ziv, A. R., Lin, S. H., and Tsong, I. S. (1985) *J. Chem. Phys.*, **82**, 3641.
32. Overeijnder, H., Haring, A., and de Vries, A. E. (1978). *Radiat. Effects*, **37**, 205.
33. Knotek, M. L. and Freibelman, P. J. (1978). *Phys. Rev Lett.*, **40**, 964.
34. Kelly, R. (1979). *Surf. Sci.*, **90**, 280; (1977). *Radiat. Effects*, **32**, 91.
35. Fleischer, R. L. (1981). *Prog Mat. Sci.*, 97.
36. Williams, P. (1979). *Surf. Sci.*, **90**, 588; (1982). *Appl. Surf. Sci.*, **13**, 241.
37. Blaise, G. and Nourtier, A. (1979). *Surf. Sci.*, **90**, 495.
38. Norskov, J. K. and Lundqvist, B. I. (1979). *Phys. Rev.*, **B19**, 5661.
39. Wittmaack, K. (1977). In *Inelastic Ion-Surface Collisions* (ed. N. H. Tolk, J. C. Tully, W. Heiland, and C. W. White), p. 153. Academic Press, New York.
40. Thomas, G. E. (1977). *Radiat. Effects*, **31**, 185.
41. Gerhard, W. and Plog, C. (1983). *Z Phys. B, Cond. Matt.*, **54**, 59 and 71.
42. Plog, C., Wiedermann, L., and Benninghoven, A. (1977). *Surf. Sci.*, **67**, 565.
43. Yu, M. L. (1982). *Appl. Surf. Sci.*, **11/12**, 196.
44. Cooks, R. G. and Busch, K. L. (1983). *Int. J. Mass Spectrom. Ion Phys.*, **53**, 111.
45. Gries, W. H. and Rüdenauer, F. G. (1975). *Int. J. Mass Spectrom. Ion Phys.*, **18**, 111.
46. Andersen, C. A. and Hinthorne, J. R. (1973). *Anal. Chem.*, **45**, 1421.
47. Werner, H. W. (1980). *Surf. Interface Anal.*, **2**, 56.

4
SIMS INSTRUMENTATION

4.1. Introduction

The basic requirements for the SIMS experiment were indicated in Chapters 1 and 2. In this chapter the necessary instrumentation will be described in more detail but since instrumentation is developing rapidly, no reference will be made to individual manufacturers' systems. Although the principles of ion-optical design are relevant, since these are comprehensively dealt with in detailed texts,[1,2] they will not be explained here. Rather, the aim is to indicate the particular advantages of the different types of ion source and mass spectrometer available, and how these relate to the uses of the systems in which they are found.

The reader is also referred to the proceedings of the bi-annual SIMS conference held since 1976, especially the more recent.[3-5] As well as describing the instrumental techniques in greater practical details, the presented papers also give feeling for the information (and problems!) produced by SIMS analysis using the equipment described here.

Static, dynamic, and imaging SIMS experiments have very different aims, but they may be conveniently classified for instrumentation purposes in terms of the profile and current of the primary beam. The types of ion gun used to form the required beam, and the choice of mass spectrometer best suited to the experiment, are described in the following sections.

4.2. Vacuum systems

SIMS experiments are performed in high vacuum for two separate reasons: first, to avoid scattering of the primary and secondary beams; second, to prevent interfering adsorption of gases on the surface under investigation. The first of these requirements is the less stringent, a pressure better than 10^{-5} mbar being sufficient to ensure a mean free path that is long compared with the beam path. As a consequence, system components are normally made with insulation clearances, e.g. filaments, etc., designed for operation at these pressures, and a high-pressure surge ($>10^{-3}$ mbar) during operation could therefore cause extensive damage. Some form of safety trip is usually incorporated to prevent this.

The requirements of surface cleanliness depend on the type of SIMS analysis to be performed. It should be borne in mind that the sticking probability of the gas molecule when it strikes the surface is important, so that a relatively high partial pressure of an inert gas such as argon can be tolerated, whereas a reactive gas such as oxygen leads to contamination problems at much lower concentrations. As a guide, one monolayer of gas will form in 1 s at a pressure of 10^{-6} mbar. Thus for SSIMS analyses of single crystals, a pressure of $\simeq 10^{-10}$ mbar is needed to allow adequate time to complete the experiment. Conversely, in a dynamic SIMS experiment, where the sample etch rate may be several monolayers s^{-1}, higher pressures are permissible. The exception to this is if the profile is intended to detect low concentrations of an element that is also present in the residual gases; hydrogen in silicon is the classic example.

A full discussion of the techniques of producing ultra-high vacuum (UHV) is presented in Ref. 6. Suffice it to say that the vacuum chamber is normally stainless steel (ion beams are fortunately insensitive to magnetic fields so that mu-metal shielding is not required), and pumped by a variety of turbomolecular pumps, ion pumps, and diffusion pumps (which require liquid nitrogen cold traps to condense organic vapours). Vibration from rotary and turbomolecular pumps is a problem when imaging with μm resolution, requiring the use of vibration isolators, or else reservoir volumes so that backing pumps can be turned off for short periods. Vibration and electrical interference from external sources should also be minimized. A design incorporating several connecting chambers allows rapid interchange or pretreatment of samples whilst maintaining UHV in the analysis region.

The presence of a mass spectrometer for SIMS suggests that mass analysis of the residual gases in the system is possible. The background gases may contain nitrogen–oxygen (indicating an (air) leak), water (indicating the system requires baking), or specific hydrocarbon fragments (indicating contamination from pump oils). By probing the outside of the system with a search gas (typically helium) and watching for its appearance in the spectrum, the integrity of the vacuum vessel can be checked and any leaks pinpointed. This residual gas analysis (RGA) thus contains much useful diagnostic information. Ionization is usually effected by electron bombardment, which involves building a rudimentary hot filament electron source in the mass spectrometer entrance optics.

Equally, the SIMS ion gun may also find auxiliary applications in sputter cleaning of samples, or controlled etching to specific depths for analysis by other techniques.

4.3. Ion guns

The primary ion beam in a SIMS experiment must satisfy a number of criteria, depending on the exact usage.

4.3.1. *General principles*

4.3.1.1. Static SIMS Generally a broad beam source is used, unless analysis of specific features is required. A beam diameter of 1 mm–1 cm is common (larger areas present problems in secondary ion collection), and uniform current density is desirable. Beam currents of 10^{-10}–10^{-8} A are used (depending on the analysis area), giving monolayer lifetimes >100 s. The beam energy lies between 500 eV and 5 keV. The lower values simplify construction of the gun and power supplies and cause less disruption in the surface and sub-surface of the sample. Less disruption gives more confidence that the analysis is representative of the original surface, but also means a low sputter rate. High beam accelerating voltages give higher sputter rates, and also give higher beam currents from a given ion source. Higher values of beam energy therefore contribute to a more rapid analysis. The beam species must also be sufficiently heavy ($m > 30$), to produce efficient sputtering, and the inert gases argon and xenon are often chosen to preclude chemical modification of the surface.

Unfortunately, ion beams usually have non-uniform, approximately Gaussian cross-sectional profiles, with decreasing current density away from the optical axis. The simplest solution is to defocus the beam until its diameter is much larger than the region of interest, and then use only the approximately uniform centre of the beam by inserting a limiting aperture; this approach is inherently wasteful of beam current. Alternatively, the beam can be focused until its diameter is much smaller than the region of interest, and then the point beam is rastered evenly over the analysis area.

4.3.1.2. Dynamic SIMS Uniform current density and hence etch rate is of utmost importance in dynamic SIMS. An analysis area smaller than 1 mm^2 must be used to give acceptable etch rates (>1 monolayer s^{-1}) at realistic beam currents (10^{-6} A).

Uniform irradiation coupled with higher beam currents results in uniform erosion of the surface. However, the imperfections of practical systems mean that the etched crater walls will always be sloping, and are thus a source of secondary ions in addition to the crater bottom. To eliminate these edge effects that otherwise limit dynamic range requires that ions from only the centre of the crater contribute to the analysis. This implies the use of a system that has some form of spatial resolution to distinguish between secondary ions from different points on the surface. From this, it is a small conceptual leap to a system which produces two-dimensional chemical images by precisely mapping the variations in intensity of emission of any particular secondary ion.

This type of system can be designed in two very distinct ways: the Imaging approach (also known as an ion microscope) or the Scanning approach (also known as an ion microprobe). If the Scanning approach is used, the primary beam is rastered over the sample (see Fig. 4.2) and beam diameters $<50\ \mu m$ are necessary to produce square profile craters. To eliminate crater edge effects the detection system is blanked except when the beam is over the centre of the crater (**electronic gating**). Imaging systems (see Fig. 4.1) use a broad primary beam that continuously irradiates the area of interest. For dynamic SIMS, it is then sufficient to design a set of collection optics which only accepts ions from a limited region in the centre of this area (**optical gating**). Uniform irradiation is often also achieved by beam rastering, but the relaxation on acceptable crater shape (since secondary ions from the sloping walls are no longer detected) allows spot sizes $\simeq 100\ \mu m$ to be used. High beam energies (>5 keV) are generally used to give high sputter rates and are needed to produce the highly focused beams. Oxygen beams to increase positive secondary ion emission are the most commonly found, since the detection of low-concentration metallic species is frequently required. Similarly, caesium beams give enhanced yields of electronegative species (e.g. sulphur) and are again in widespread use.

4.3.1.3. SIMS imaging As indicated in the above section, SIMS images can be obtained using either an Imaging SIMS system, or a Scanning system. There is clearly a problem of nomenclature here; for clarity, we will use the terms microscope and microprobe.

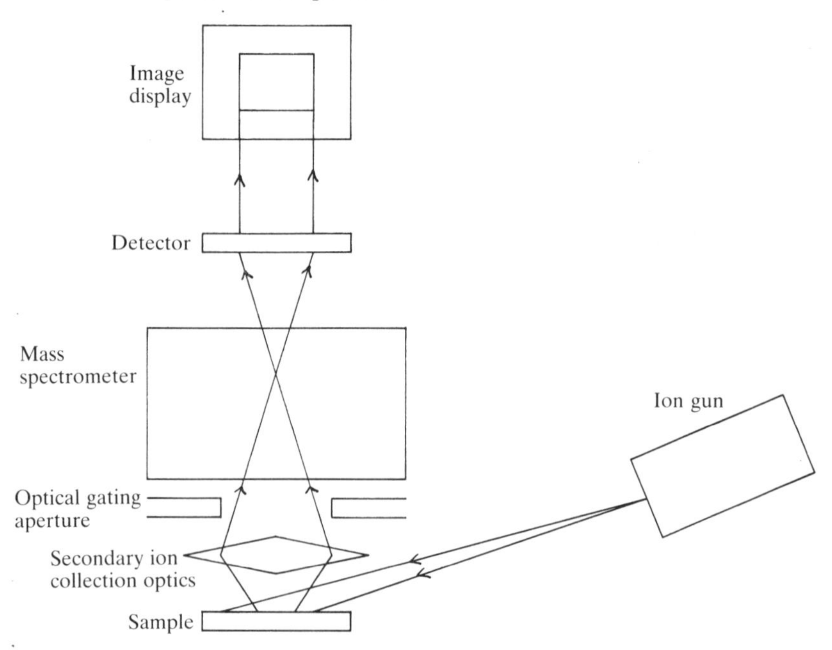

FIG 4.1. Schematic diagram of a secondary ion microscope, based on a focusing mass spectrometer–secondary ion optics systems and a broad primary ion beam.

In an **Ion Microscope** (Fig 4.1) the collection and transport optics, and the mass analyser (a sector magnet), are designed so that the spatial distribution of the ions is retained throughout. A complete image of the surface is therefore formed simultaneously at the detector. The image is viewed using either a fluorescent screen or a position-sensitive detector linked to a computerized data system.

The **Imaging Microprobe** (Fig. 4.2) uses a focused primary ion beam rastered over the sample. In this way, each point on the target is individually bombarded in turn, so that secondary ion emission is localized. For SIMS images, the intensity of a particular secondary ion is monitored for each position of the primary beam, and the result shown at the corresponding point of a synchronized oscilloscope display or computerized data system. A complete picture is thus built up in stepwise fashion.

Several variations on this theme are possible. A trivial modification would be to keep the ion beam stationary while the target is scanned under it in a raster pattern. However, the required mechanical complexity of the sample stage would be expensive and ill suited for use in a high vacuum system, and hence this approach is not found in practice. Alternatively, the entire area of interest can be continually irradiated by a broad primary beam and the

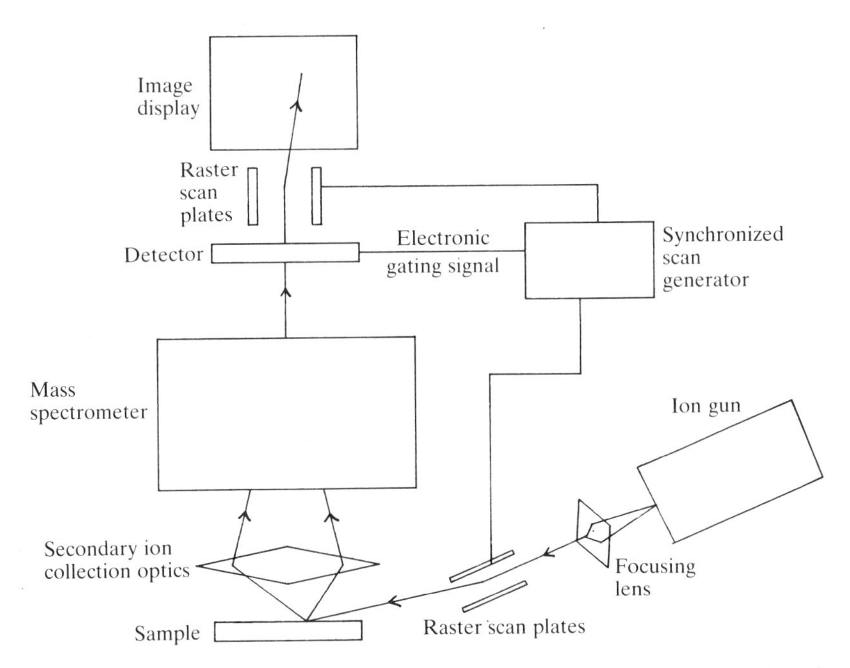

FIG 4.2. Schematic diagram of a secondary ion microprobe, based on a non-focusing mass spectrometer–secondary ion optics system and a scanning focused primary ion beam.

acceptance area of the secondary ion collection options scanned over the sample, again building up the image one point at a time. This approach is inherently inefficient, since at any one time only a small fraction of the generated ions is being used to form the image. However, by scanning both primary beam and secondary optics in tandem, ion collection can be optimized at each point (**dynamic emittance matching**), and this approach is sometimes found in practice.[7]

Lateral resolution is clearly of considerable importance for SIMS image acquisition. When the primary ion strikes the sample, it penetrates a short distance before coming to rest. The kinetic energy of the ion is dissipated amongst the nearby atoms of the solid, so that they are set in motion. Some of these collision cascades return to the surface, where the energy given to the sample atoms allows them to overcome the forces binding them in the solid. Secondary ions thus originate at points remote from the actual impact position. The diameter of the disturbed region is thought to be of the order of 10 nm and this represents the fundamental limit for the SIMS technique.[8] However, instrumental imperfections currently limit lateral resolution above this value.

In the ion microscope the problem stems from the fact that secondary ions are formed with a significant range of energies (many eV). This effect can be minimized by rapidly accelerating the ions to high energy (keV) over a short distance (mm), but the practical difficulties of producing high extraction fields mean that this effective reduction of percentage energy spread can only be taken so far. Chromatic (energy-dependent) aberrations of the secondary ion transport optics are thus important, and start to limit performance at 10 μm lateral resolution. Better lateral resolution can be obtained by restricting the energy range of the secondary ions so that only some of them are used in the analysis, but the corresponding reduction in sensitivity becomes unacceptable when resolution <1 μm is required.

In the ion microprobe, lateral resolution is set by the diameter of the primary beam. In many cases, this is limited by the quality of the ion source, in particular by a parameter known as brightness. This is analogous to the term as used for light sources, and is a measure of the current density available from the source. Bearing in mind that the ion gun optical column cannot actually improve the quality of the beam that leaves the source, a reasonably high current can only be focused into a small spot if the ion source has a high current capability for its size, i.e. high brightness. Equally, it follows that as the focused spot size is decreased, so the ion current must be lowered. When a high-brightness source is used, then the aberrations of the optical column also start to become significant. These factors are discussed in more detail in a later section. The best ion guns currently available incorporate a liquid metal source running at a high accelerating voltage, and spot sizes below 50 nm have been reported.[9]

From the above arguments, it should be apparent that the ion microscope allows the use of high beam currents and offers lateral resolution of $\simeq 10\ \mu m$. To attain high spatial resolution, the transmission of the secondary ion transport system has to be reduced by aperturing. The imaging ion microprobe effectively decouples instrument transmission from spatial resolution and the secondary ion transport system operates at maximum transmission at all times. The former is thus well suited to dynamic SIMS systems (or rapid, low-resolution imaging); the latter is more often found in high-resolution, high-sensitivity imaging applications.

4.3.2. Ion gun components

All ion guns (see Fig. 4.3) feature a source of ions, and some form of lens to extract the ions from the source chamber. To force the ions to move from there to the sample, a potential difference must be applied between these two points, and for convenience this is generally achieved by leaving the target at earth and floating the source region and power supplies to the accelerating voltage (0.5–30 kV); this also sets the final energy of the ions and hence the beam energy. In some ion guns, extraction and acceleration are combined in one voltage step, and in all cases the acceleration point is as near to the source as practical. This is done because a fast-moving ion beam has less time to expand either under the mutual repulsion of the ions in the beam or because of the original transverse component of velocity (so that the energy spread inherent in any ion source becomes relatively less important). These factors mean that early acceleration results in less degradation of the beam during its passage to the target, and equally that optimum beam quality requires high accelerating voltages; hence the keV beam energies found in focused beam systems.

In the simplest ion guns, the source and extractor–accelerator lens combination is all that is required. However, a second lens is usually incorporated to give some degree of focusing at the target. *xy* deflection plates are also included, either for simple positioning of the beam or for raster scanning. A quadrupole or octupole stigmator is also necessary to ensure equal resolution in x and y planes when the beam is to be focused below 10 μm. A mass–energy filter may be included to remove contaminant ions either from impurities in the source feed material or from sputtering of the source walls, etc. Such a filter will also remove multiply charged and cluster ions of the main beam species (e.g. Ar^{2+} and Ar_2^+), which is necessary prior to focusing of the beam. The most common form of mass separator is a Wien filter (see Fig. 4.3(b)). In this device, crossed electrostatic and magnetic fields act in opposition so that the forces cancel for ions of a particular m/z value. Ions of the required mass and energy thus pass straight through, whilst unwanted species are deflected out of the beam.

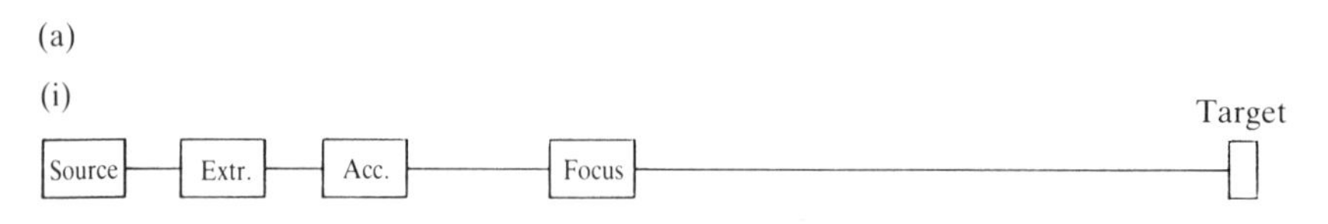

(ii)
Source Extr. Acc. M. Filt Stig. Defl. Target Focus Scan

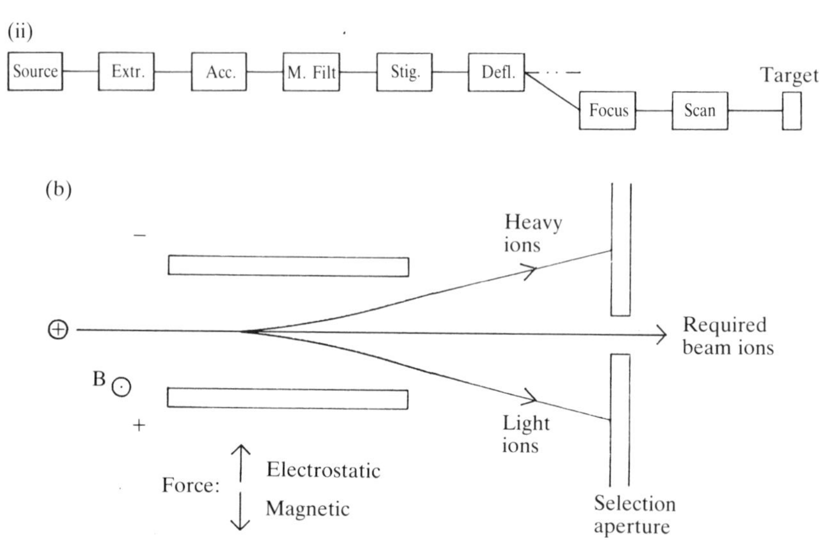

FIG 4.3. (a) Major components of a (i) broad beam and (ii) scanning, focused ion gun. Source = ion source; Extr. = extractor; Acc. = accelerator; M. Filt. = mass/energy filter; Stig. = stigmator; Defl. = neutral elimination bend; Focus = focusing lens; Scan = *xy* raster. (b) Operation of a Wien filter for a mass selected ion beam. Balance condition is $m/z = 2V(B/E)^2$.

Unfortunately, any neutral particles are unaffected by the fields and are also transmitted. These must also be removed from the beam prior to focusing, otherwise an unfocused halo will be present around the final beam. The ion beam is thus deflected through a few degrees in the later part of the optical column, so that the undeviated neutral particles are separated out. Finally, differential pumping of the ion column may be required to maintain acceptable pressures (10^{-6} mbar) in the column. This especially applies if the source produces a large gas load either from a gaseous feed or from outgassing due to operation at high temperatures. High pressures would result in scattering of the beam and loss of beam current through charge exchange processes.

The performance of an ion gun can be seriously degraded by poor mechanical (e.g. misalignment) or electrical (e.g. voltage instabilities) construction. The necessity of high source brightness and low energy spread to give a high-quality input to the ion optical column has already been explained. Aberrations are also introduced by the ion optical column itself. The two major forms of aberration are known as spherical and chromatic aberration. Spherical aberration occurs because of the finite width of a real beam, which means that some ions pass closer the lens electrodes than others. Chromatic aberration occurs because of the finite energy spread of a real beam, which means that the voltages applied to the lens electrodes are relatively different for different ions. Both of these effects alter the point at which ions are brought to a focus, and hence the beam focus as a whole is smeared out, thus increasing the spot size at any given point. If a low-brightness source is used, the area of the source must be large in order to supply the required total current, and hence the action of the final lens must be strongly demagnifying for a microfocused beam. Aberrations introduced in the earlier part of the column are therefore also demagnified, so in this case it is most important to minimize the aberrations produced by the final lens itself. However, for high-brightness, small-area sources the magnification factor is approximately unity, and so the aberrations of the whole column become significant. A weighted average aberration coefficient can be defined for the column, in which the aberration coefficients for the isolated lenses are multiplied by a magnification-related factor, to take account of this effect.

Optimum performance in the submicron spot regime is generally achieved at a lens–target working distance of 10–20 mm. However, until the recent advent of liquid metal sources, the relatively low values of source brightness were the limiting factor in focusing, and hence relatively little effort has been expended in lens design. The most common form of lens is the einzel (unipotential) lens; this consists of three electrodes, the first and last of which are earthed. The lens is rotationally symmetric, and often longitudinally symmetric as well. The design offers simplicity in use and in the power supplies required, and the beam energy is unaltered on exit from the lens.

4.3.3. Types of ion source

Ions can be formed either by transfer of energy to the electron population of the atom, or by distortion of the electron energy levels; either process allows movement of an electron between the atom and the outside world. The effect can be achieved in many ways, three of which are of interest for use in a SIMS primary ion gun. Each will be seen to have strengths in certain areas of SIMS analyses, and instruments are increasingly fitted with more than one source to take advantage of this fact.

4.3.3.1. Electron bombardment plasma sources When an energetic electron interacts with an atom, some of its kinetic energy is transformed into potential energy of the electrons of the atom (see Fig. 4.4). If the energy of

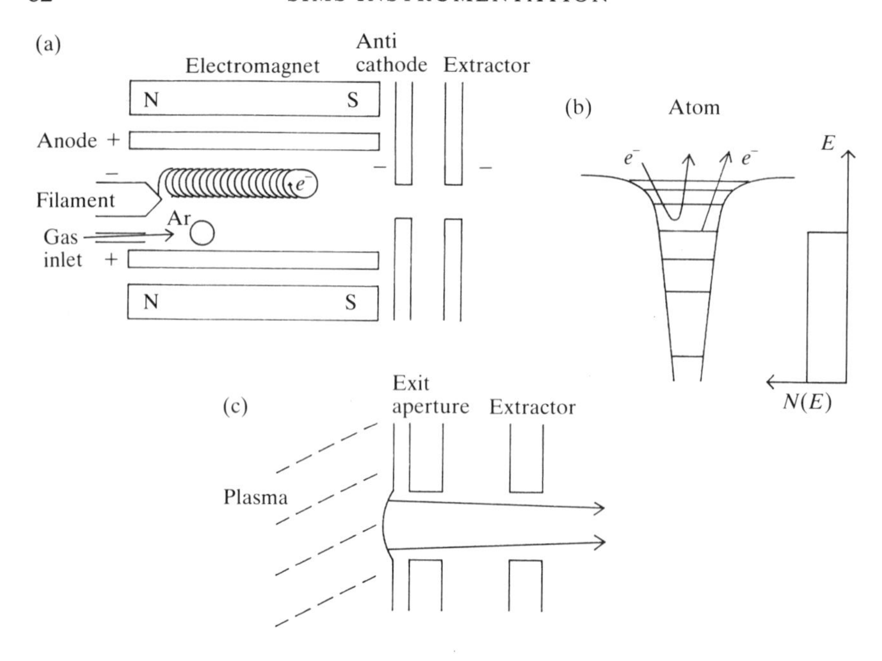

FIG 4.4. Principle of an oscillating electron bombardment discharge ion source: (a) schematic; (b) energy diagram of ionization process; (c) close-up of extraction region.

the bombarding electrons is chosen to be roughly two or three times the ionization energy of the atom, then efficient ionization is found to occur. The most common source of electrons is a filament (cathode) heated by the passage of current to the point of thermionic emission; the electrons are then accelerated towards an anode to give them the required energy. The path length of the electrons can be increased by the use of electrostatic and magnetic fields that force the electrons to move in oscillating or spiral trajectories during their flight to the anode. There is then a significant chance of an ionizing interaction occurring even at relatively low gas/vapour pressures (10^{-3} mbar). Alternatively, sufficient electrons may be produced by positive ion bombardment of a cathode to sustain the discharge. The removal of the cathode heating requirement allows the use of reactive gases and of more massive cathode structures, and hence the need for routine filament replacement disappears.

When the number of ions, electrons, and neutrals in the source exceeds a critical value, a plasma is formed. All sources must include an exit aperture through which ions are extracted into the optical column, and the fields in this region form a boundary to the plasma across the orifice. This plasma boundary is an equipotential surface, and it is from here that the ions are extracted. The ions are therefore effectively drawn from the same point with the same energy, giving a high-quality ion beam. The brightness of such discharge sources is $\simeq 10^4$–10^6 A m^{-2} Sr^{-1} and the energy spread is <5 eV.

A similar type of source is the plasmatron, but here the ions are extracted through an aperture in a planar anode and an additional electrode in this exit region physically compresses the plasma. A magnetic field may also be present to concentrate the discharge further, and the source is then known as a duoplasmatron. This dense plasma means increased sputtering of the electrodes and increased heat to be dissipated, making the sources less reliable in operation. However, the denser plasma at the extraction point gives the significant advantage of higher brightness ($10^6 - 10^7$ A m^{-2} Sr^{-1}). Energy spread is generally <10 eV.

These gas feed sources are well suited to the production of inert gas beams (commonly Ar^+ or Xe^+) which allow the chemistry of a surface to be studied by SIMS without modification by the bombarding species. The cold cathode versions with no heated filament also allow the use of oxygen (O_2^+) beams for enhanced positive secondary ion emission, and such beams are the usual choice for depth profiling experiments. The achievable values of source brightness allow μA currents into spot diameters $\simeq 50$ μm (for dynamic SIMS) or nA currents into spot diameters <5 μm (for imaging SIMS). However, for smaller-diameter beams (and hence improved spatial resolution) a different, higher-brightness source must be used.

4.3.3.2. Field ionization and liquid metal sources When a gas atom is close to a surface of high electric field, the initially symmetric potential well of the atom is distorted. This lowers and narrows one side of the barrier, allowing an electron to tunnel from the atom to the surface, leaving the atom ionized as required. The surface must be a metal of high work function, and must be formed into a sharp tip of radius <1 μm to allow generation of the high electric field; iridium or tungsten are the most common choices. The small source size means a high-brightness source (10^{10} A m^{-2} Sr^{-1}), but so far the achievable beam currents have been too low for practical use in SIMS. However, the source does have the potential for submicron beams of inert gases, and hence research is continuing in this area.

A variation on this design is the liquid metal source in which the source feed material is supplied not as a gas or vapour but as a liquid film flowing over the needle shank to the tip[10] (see Fig. 4.5). The tip is now relatively blunt (10 μm radius) and acts only as an anchoring point. If the feed material is a metal, the opposing electrostatic and surface tension forces at the tip form the liquid into a sharp point known as the Taylor cone. Whether ionization then occurs in the liquid or in the vapour phase is unclear, but in practice the sources have high brightness (10^{10} A m^{-2} Sr^{-1}) and can deliver currents of over 100 μA. The energy spread is unfortunately still quite high (10 eV), especially at high beam currents.

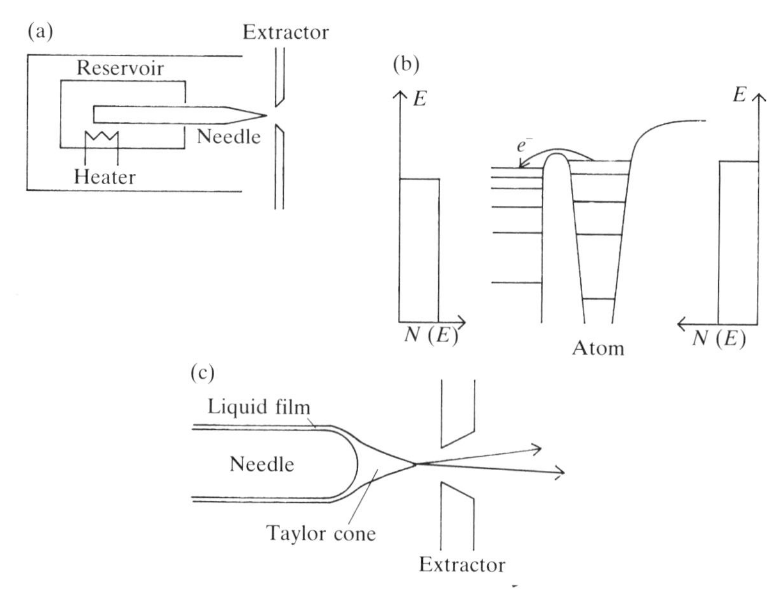

FIG 4.5. Principle of a liquid metal ion source: (a) source schematic; (b) energy diagram; (c) close-up of extraction region.

The source feed must be a metal or alloy of low melting point; gallium and indium are the two most commonly found (this type of source can also be used for caesium beam production). nA currents may be focused into spot sizes from 0.1 to 0.02 μm for beam energies from 10 to 50 keV (basically chromatic aberration limited owing to the high source energy spread). The beam currents are, however, too small to allow depth profiling of large areas (100 μm × 100 μm), though small area dynamic SIMS is possible. Note that the beam species and energy are here dictated by the source requirements, not the SIMS experiment (liquid metal sources were originally investigated as a possible propulsion system for space travel). This gives an idea of the importance attached to submicron SIMS imaging, which is currently only achievable using such sources.

The relative performances of the electron bombardment, duoplasmatron and liquid metal sources are summarized in Fig. 4.6.

4.3.3.3. Surface ionization sources When an low ionization potential atom is adsorbed on a high work function metal, the electronic distributions of the atom and surface broaden and overlap, allowing movement of electrons between the two parts of the system (see Fig. 4.7). If the temperature is then raised so that the rate of desorption exceeds the rate of arrival of the atoms at the surface, then the composite work function of the metal–adsorbate will

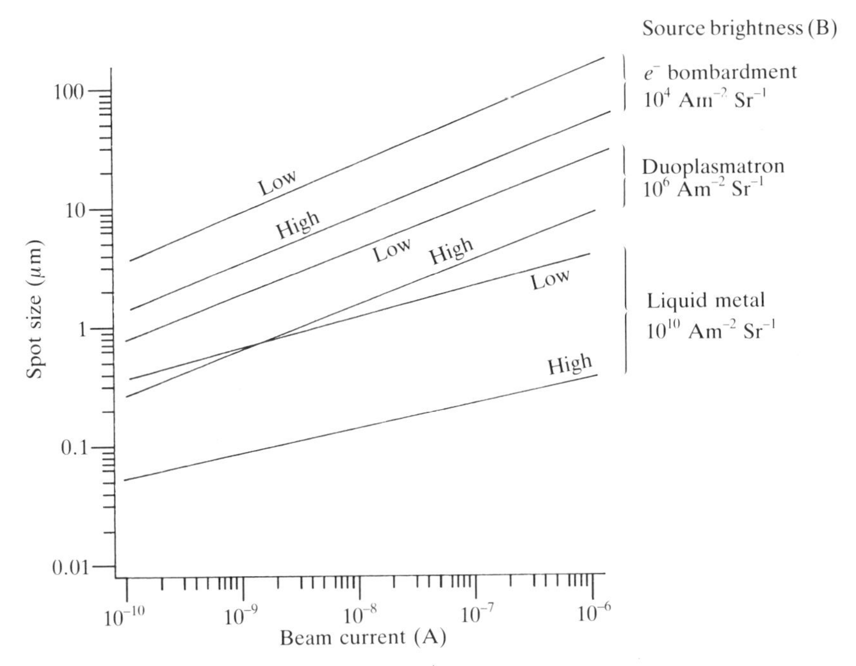

FIG 4.6. Theoretical variation of spot size (d) with beam current (I) for three types of source, showing the importance of both source brightness (B) and lens quality (high or low). Values assume final beam convergence half-angle (i.e. lens working distance) is optimized at each beam current which is not possible in practice. (After Smith, K. C. A. (1957). PhD thesis, University of Cambridge. See also Drummond, I. W. (1981). *Vacuum*, **31**, 579.) Experimental beam diameters are typically $\times 5$ larger than the graph suggests. The weighted average chromatic aberration coefficient $\bar{C}_c$ limits focusing performance for liquid metal source (high energy spread, $\Delta E/E = 10^{-3}$); for other sources it is the spherical aberration coefficient $\bar{C}_s$ which limits performances. Ion optical column quality: high when $\bar{C}_s = \bar{C}_c = 1$ cm; low when $\bar{C}_s = \bar{C}_c = 100$ cm.

increase, resulting in an exponential increase in the ratio of adsorbed ions to atoms as electrons move down the increased potential slope to the metal surface. The element is then desorbed as ions, and almost 100 per cent efficiency can be achieved in practice. Source brightness depends on the size of the emitting area but values $>10^6$ A m^{-2} Sr^{-1} have been achieved. The energy spread of such an ion source is very low since it is set by thermal energy variations (<1 eV).

The materials that can be ionized in such a source are limited—either the alkali metals from, e.g. iridium, or, by operating the above principle in

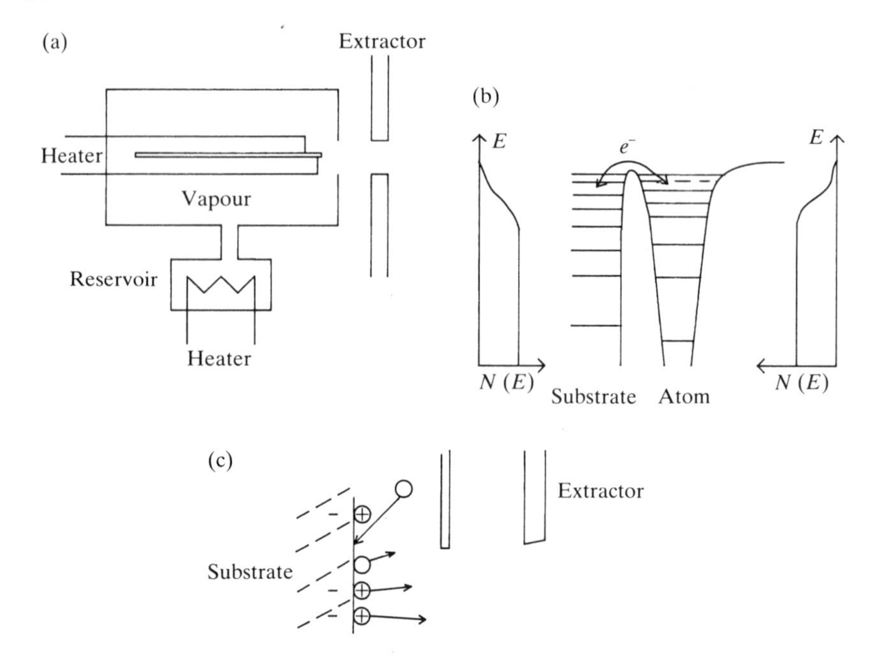

FIG 4.7. Principle of a surface ionization source; (a) source schematic (b) energy diagram, (c) close-up of extraction region.

reverse, negative halogen beams can be generated at a low work function surface, e.g. thorium. These materials are not particularly suitable for SIMS analyses, with the very important exception of caesium, which is often used in SIMS to enhance the yield of negative secondary ions. Spot size–current characteristics lie between liquid metal and plasma discharge sources.

The performance of the different ion beam systems is summarized in Table 4.1.

4.3.3.4. Lasers Laser beams are also used for sputtering. A laser pulse generates a short but intense packet of secondary ions. The mechanism of ion production is unclear, but it is established that the information depth is much greater than with a conventional ion beam, so that true surface information is not obtained and the technique should be regarded as being significantly different from ion beam induced SIMS. Cluster ion production also seems to be dependent on laser power and can be accidentally suppressed. The advantage is speed of analysis, with one pulse often being sufficient to record a spectrum, and reduced sample charging problems by using an uncharged incident beam. Laser ionization is explored further in Chapter 9.

TABLE 4.1

Ion gun comparison

	Species	Brightness ($Am^{-2}Sr^{-1}$)	Energy spread (eV)	Maximum current (μA)	Typical minimum spot size (μm)
Oscillating electron discharge	Ar^+, Xe^+	10^5	<5	10^2	5
Duoplasmatron	Ar^+, Xe^+, O_2^+	10^6	$\simeq 10$	10^3	1
Liquid metal	Ga^+, In^+, Cs^+	10^{10}	>10	10^2	0.05
Surface ionization	Cs^+	10^7	<1	10^2	0.1

4.3.4. Primary beams for insulating samples

In common with other surface analysis techniques, problems occur in SIMS analyses of insulators. The arrival of primary ions (usually positive) and the ejection of secondary electrons and ions leads to a charge imbalance at the surface which cannot be dissipated by the low-conductivity material. In theory, charging to the primary beam potential (several keV) could occur; in practice, electrical breakdown usually occurs before this point is reached (see Fig. 2.15). In severe cases, charging of the sample completely inhibits the emission of secondary ions of the opposite polarity. Otherwise, a constantly changing sample potential leads to a wide range of secondary ion energies which degrades mass resolution and makes optimization of energy filters (especially on quadrupole systems) difficult or impossible. Other unwanted effects of charging include migration of small ions (e.g. Na^+) in the host matrix, deflection of primary and secondary beams, giving distorted images, and even sample movement in fine powders. All of these factors mean that a representative SIMS spectrum is not obtained.

Charging can obviously be minimized by using thin samples and low beam current densities but these are not always practical solutions especially for imaging or dynamic SIMS. Some success has been achieved using negative primary beams to balance the secondary electron egress, or by the use of metal grids over the sample to act as a source or sink of electrons. However, there are two main techniques in widespread use at the present.

4.3.4.1. Auxiliary electron bombardment In a SSIMS analysis, the use of a low-energy (10 eV) electron beam to flood the surface compensates for the effects of positive ions arriving and negative electrons leaving the sample. As localized charging begins to occur, secondary electron emission is suppressed and the auxiliary electrons are attracted to the region. The system is thus self-stabilizing, and the surface potential remains steady at a few volts above ground.[11]

For the higher current densities encountered in imaging and dynamic SIMS, a higher current electron beam is required. This can only be transported to the target by using higher accelerating energies (>500 eV).[12] This prevents the self-stabilizing mechanism from operating effectively, so that careful balancing of ion and electron fluxes is required. The technique is thus not well suited to samples of non-uniform depth or surface composition. High-energy electron beams can also cause changes in the sample, including emission of secondary ions. Problems can also occur in positioning an electron beam with low energy at the surface if a high extraction field is present, e.g. in magnetic sector or time-of-flight instruments. Two recent attempts to circumvent this problem are worth mentioning. The first concerns negative secondary ion analysis in a magnetic sector system.[13] High-energy electrons are injected back down the secondary ion column using magnetic prisms (which do not significantly affect the secondary ion trajectories). The high ion extraction field at the sample repels the incoming electrons so that they arrive at the sample surface with virtually zero energy as is required. Since the electrons are approaching along the surface normal, there is no deflection of the beam during the deceleration stage. It is this geometrical arrangement, i.e. with electron flood and secondary ion extraction both optimized at 90° to the sample surface, which is so important. An additional advantage is that secondary electrons, which would normally be recorded as spectrum background noise, are deflected out of the analyser by the magnetic fields in the ion optical column.

The second technique is applicable to time-of-flight instruments where the primary ion beam is pulsed (see later section).[14] The relatively long time (10^{-4} s) between pulses allows the extraction field to be turned off, and the electron beam pulsed onto the earthed sample. The technique is not easy to implement as it requires rapid switching of kV voltages. A possible problem is that the sample charges during the ion pulse (< 10 ns) so that neutralization after the event is ineffective.

With an awareness of the various problems, the technique of auxiliary electron bombardment is very effective and widely used. Note that the electron beam currents used are frequently orders of magnitude higher than the primary ion beam current, so that electron guns with current capabilities $> 10\ \mu$A are needed.

An electron flood gun for SIMS is rarely a sophisticated piece of equipment. Electron sources such as a tungsten hairpin filament have high brightness (10^8 A m^{-2} Sr^{-1}) and can thus supply the $> 10^{-5}$ A beam currents required. An acceleration voltage of 10–20 V makes electronic and mechanical design straightforward, as high-voltage insulation is not required. The gun is completed by a single-element focusing lens to give spot sizes of $\simeq 10^{-3}$ m, and electrostatic *xy* deflection plates may be included—otherwise the gun may be mounted on adjustable bellows to allow positioning of the beam.

For depth profiling or imaging involving higher ion current densities and small area analysis, more precise focusing ($< 10^{-3}$ m) and positioning are required. The appropriate gun for this type of analysis may incorporate a higher-brightness LaB_6 filament (10^{10} A m^{-2} Sr^{-1}) and will operate at higher voltages ($10^2 - 10^3$ V) to give higher electron current densities in the focused beam as already indicated; *xy* deflection is now mandatory.

4.3.4.2. Fast atom bombardment The alternative solution is to use neutral particles as the primary beam so that charging is due only to secondary electron emission.[15] As already stated, this is suppressed once a small positive sample potential is reached, allowing SIMS spectra to be obtained directly in many cases.[16] An auxiliary low-energy electron beam, now of much-reduced intensity may, however, also be necessary for optimum results, particularly for the detection of negative ions. The technique can be applied to a wider range of non-uniform and electron-sensitive samples than electron compensation alone.

The usual method of producing a neutral beam is to generate a beam of argon or oxygen in the normal way, and then pass it through a cell containing the same gas at high pressure (10^{-3} mbar), see Fig. 4.8. A phenomenon termed **charge exchange** then occurs, in which an electron is transferred from the effectively stationary gas atom to the fast-moving ion. The process involves little change in momentum, so that the resulting neutral beam exhibits almost the same energy and spatial characteristics as the precursor ion beam: even scanning, focused neutral beams have been produced in this way. An electrostatic plate is usually incorporated at the end of the high-pressure region to deflect the unneutralized ions (50–90 per cent) out of the beam. Figure 4.9 shows the detailed arrangement for a scanning neutral beam source capable of producing a microfocused beam (full width at half maximum (FWHM) = 5 μm) of argon atoms at 10 keV. Other methods involve the efficient neutralization of ions on reflection from surfaces at glancing incidence, but this generally results in a less well defined beam.

The drawbacks in using neutral beams are a reduced effective beam current from the inefficient neutralization process, a limitation on possible beam species by the need for a gas–vapour of the element in the charge exchange process, and a high gas load on the system. Neutral beams do, however, have the additional advantage that since they are unaffected by high electric fields, they can be used in high extraction voltage systems (e.g. magnetic sector or time-of-flight) without deflection or energy modification as they approach the target. For these reasons, in organic FAB[17], where the sample molecules

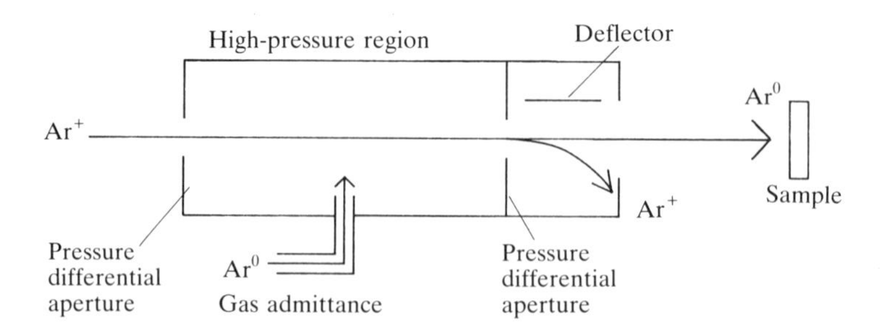

FIG 4.8. Schematic representation of a charge exchange cell incorporating residual ion removal.

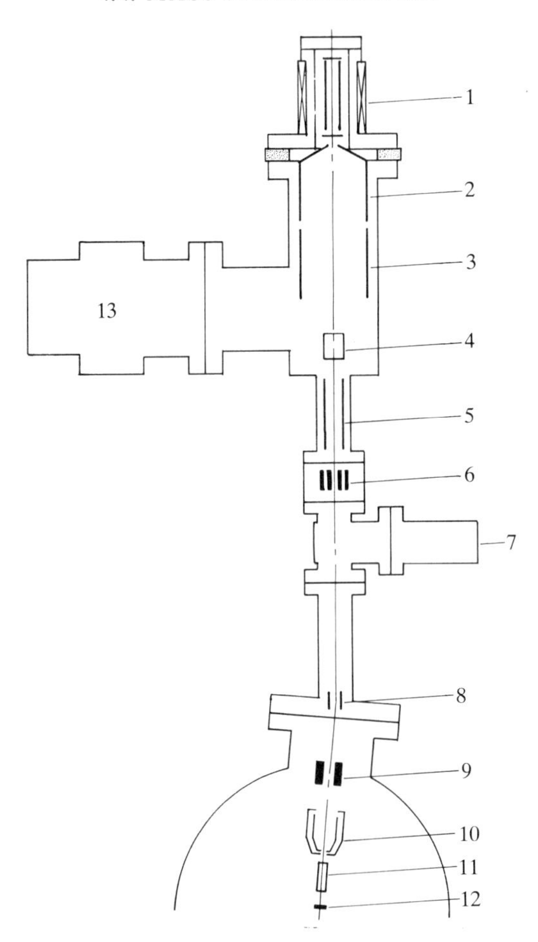

FIG 4.9. Schematic of an ion/atom gun for SIMS imaging. (1) Source; (2) extractor; (3) accelerator; (4) beam alignment plates; (5) Wien filter; (6) stigmator; (7) valve; (8) neutral removal plates; (9) scan rods; (10) final lens; (11) charge exchange cell; (12) target; (13) turbomolecular pump. (After Eccles, A. J. *et al.* (1986). *J. Vac. Sci. Technol.*, **A4**, 1888.).

are suspended in a liquid matrix, e.g. glycerol, neutral beams are found almost exclusively.

4.4. Mass spectrometers

The other essential part of a SIMS system is a mass spectrometer to measure the mass of the secondary ions sputtered from the sample. The different types of SIMS analysis have different requirements for the characteristics of the spectrometer.

4.4.1. General principles

4.4.1.1. Static SIMS Here the requirement is to minimize the total primary ion flux density incident on the sample. Large area irradiation is therefore desirable, which in turn requires a large area acceptance of the secondary ion collection optics that focus the ions into the analyser. Also, the transmission of these optics and of the mass spectrometer should be high to maximize the sensitivity of the system. Applications to the analysis of biomedical materials may require a high mass range ($m/z > 10^3$) to allow access to the large fragment clusters that contain so much valuable information about complex molecular structures; resolution and transmission must remain acceptable even at these high masses.

4.4.1.2. Dynamic SIMS The majority of depth profiling experiments are directed at finding low mass elemental impurities in electronic materials such as silicon, gallium arsenide, etc. High transmission is thus still important to give the lowest possible detection levels, but a high mass range is frequently not necessary. However, high mass resolution can be needed to separate molecular and atomic ions at the same nominal mass, e.g. SiH and P. The transfer optics may also incorporate some form of optical gating to suppress background signals and crater edge effects.

4.4.1.3. SIMS imaging In an ion microprobe system, the small analysis areas inherent in high-resolution imaging mean that acceptable current densities and spot sizes can only be achieved by low primary beam currents. A high transmission secondary ion system is thus very important. A high mass range is usually not required, since mapping is usually, but not exclusively, done using the most intense ions, and these tend to be either elemental or small cluster species. Mass resolution requirements tend to be very low because the mass of interest is frequently an isolated peak with no interfering neighbours. Peak shape is also unimportant, so the spectrometer resolution may be detuned to give the maximum possible transmission. In an imaging microscope the requirement of a focusing secondary ion system becomes the overriding factor in the choice of mass spectrometer.

4.4.1.4. MS–MS SIMS Finally, mention should be made of tandem mass spectrometry techniques which in essence utilize two mass spectrometers placed in series. This experimental configuration allows elucidation of the fragmentation pathways of unstable molecular ions. Analyses can benefit by identification of the original ion when the mass alone is inconclusive. Alternatively, pure data on ion lifetimes can be generated. The most obvious method is to tune the first spectrometer to the mass of interest, and scan the second to produce a spectrum of the daughter ions. A second method is to scan the first spectrometer while the second is tuned to a particular daughter, thus identifying all the precursor parent ions. A further method involves scanning both spectrometers with a fixed mass offset between them, which identifies all reaction pathways that produce a given fragment ion with a mass equal to the offset. A collision chamber of high gas pressure is included between the two analysers so that the majority of fragmentation occurs at this point.

4.4.2. Mass spectrometer components

The demands of high sensitivity and high mass resolution have meant that the mass spectrometer of choice for dynamic SIMS has been the magnetic sector instrument. The quadrupole analyser has been mainly chosen for its compact size and low price and relative insensitivity to extraction geometry. Until recently it was the analyser primarily used in SSIMS. This reflects the emergence of SSIMS as an additional technique to existing surface analysis systems and then in multi-technique systems. However, as SSIMS has matured, the emphasis has shifted to dedicated systems and the quadrupole's limitations have begun to limit the power of the technique. Attention is therefore turning to the alternatives, notably magnetic sector instruments (featuring high mass resolution) and time-of-flight systems (featuring high transmission). It is difficult to generalize about the component parts of the mass analyser itself, since the three basic types are totally dissimilar. However, the analyser will always be preceded by some form of collection optics, and followed by an ion detector. The performance of these parts has an important role to play, the transmission efficiency of the former being particularly critical.

The collection optics can be classified, according to the strength of the extraction field, into high- or low-voltage types. The advantages of a high extraction field ($> 10\ \mathrm{kV\ cm^{-1}}$) are two-fold; first, the strong field means that ions are drawn from the sample into the spectrometer with high efficiency. Second, as the ions are accelerated to high energy, the inherent energy spread of sputtered ions becomes relatively smaller, so that good resolution can be obtained in the spectrometer. In a low extraction field ($< 100\ \mathrm{V\ cm^{-1}}$) system, the energy spread must be reduced by energy filtering of the secondary ions, which means a loss in signal intensity. The ion collection efficiency from the sample will also be low. The choice of high or low extraction field depends largely on whether the mass spectrometer requires high- (keV) or low-energy (eV) ions for analysis, so that low-field versions are only found on quadrupole systems. Even here, the trend is now towards high-field systems followed by a deceleration step at the quadrupole entrance. A practical point is that low extraction fields are generated by biasing the secondary ion optics, but this becomes impractical for high extraction–acceleration fields and the sample is then usually biased instead.

Ion detection is generally achieved by pulse counting using a single-channel electron multiplier. This is placed off-axis in quadrupole systems and the ions are deflected into it. If this is not done, neutral particles and high-energy ions are also detected, resulting in high background signals. Interference from electrons is also suppressed by incorporating a weak magnetic deflection field in this region; the field strengths used do not affect the heavier ions. Ion arrival rates can exceed the count rates of these devices (10^6 cps) so care must be exercised, especially in dynamic SIMS. However, microscope systems require a detector with spatial resolution. This is simply achieved by a channel plate (to convert ions to electron showers), followed by a phosphor (to convert the electrons to light), followed by a TV camera or similar to digitize the image for storage. A better but more expensive system is a detector array; a signal event can be recorded in terms of its x and y position and hence digitized directly.

Channel electron multiplier detection efficiency decreases as the incident ion velocity falls (i.e. low energy or high mass). Many systems therefore include an acceleration step (up to 20 keV) just prior to the detector to minimize this effect.

4.4.3. Types of mass spectrometer

4.4.3.1. Quadrupole mass analysers An ideal quadrupole (see Fig. 4.10), consists of four hyperbolic rods, connected together as two opposite pairs. In practice, the difficulty of manufacturing elliptical rods means that circular rods are used: this gives an acceptable on-axis approximation to the ideal case if the rod diameter: quadrupole diameter $\simeq$ 1:1.145 (rod diameter $\sim$1 cm).

A potential consisting of a constant (d.c.) component + an oscillating (r.f.) component is applied to one pair of rods. An equal but opposite voltage is applied to the other pair. The rapid periodic switching of the field sends most ions into unstable oscillations of increasing amplitude until they strike the rods and hence are not transmitted. However, ions with a certain mass: charge ratio follow a periodic but stable trajectory of limited amplitude and therefore are transmitted. By increasing the d.c. and r.f. fields whilst maintaining a constant ratio between them, this resonant condition is satisfied for ions of each ascending mass in turn, allowing the collection of a complete mass spectrum.

The mass resolution and transmission of a quadrupole are interrelated by a complicated series of equations. Increasing the d.c.:r.f. ratio improves the absolute mass resolution over the entire mass range (i.e. Δm is constant).

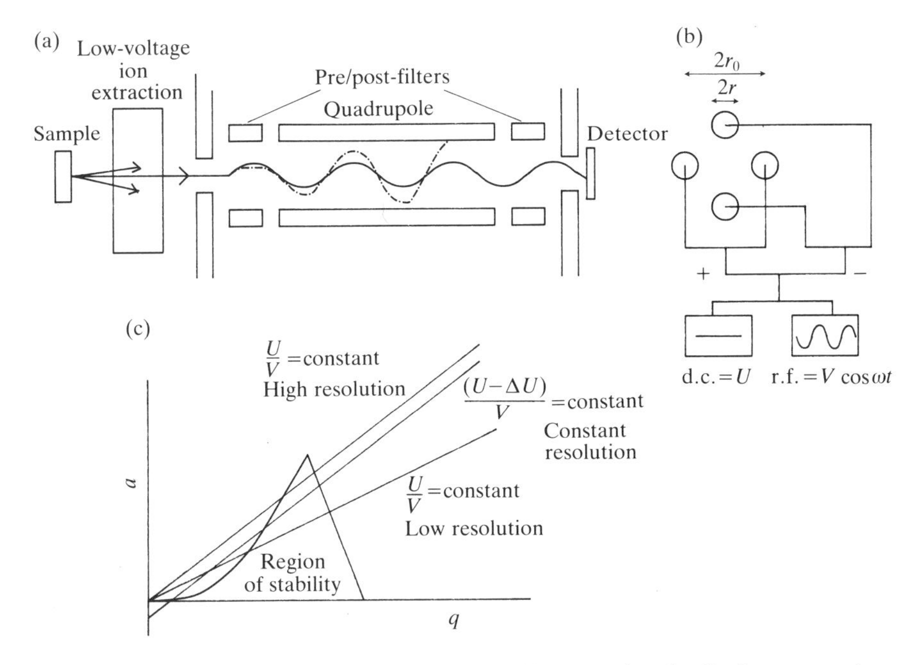

FIG 4.10. Operation of a quadrupole mass filter: (a) longitudinal cross-section showing stable and unstable trajectories; (b) radial cross-section, showing applied voltages; (c) Ion trajectory stability diagram—ion trajectories are a function of the two dimensionless parameters $a = (8U/r_0^2\omega^2)(z/m)$; $q = (4V/r_0^2\omega^2)(z/m)$

Applying a small offset d.c. voltage improves mass resolution at lower masses. However, the two adjustments also reduce the transmission of higher mass and lower mass ions respectively. The quadrupole is usually set up to give constant $(m/\Delta m)$ with acceptable (but decreasing) transmission over the entire mass range. Other ways of increasing mass resolution and transmission are the use of larger quadrupole (and rod) diameters and higher-frequency r.f. field components. The price paid is in the higher specification then needed for the electronics, which generally translates to a restricted mass range ($m/z \simeq 250$). Aberrations are also introduced by the fringe fields at the entrance and exit of the device and these can be improved by adding short pre- and post-filter rods carrying a percentage of the main rod r.f. voltages.

Quadrupoles are therefore generally characterized by a low transmission (1 per cent), medium mass range ($m/z < 1000$). Additionally, the ions for analysis must be of low energy ($\simeq 10$ eV) to allow an adequate length of time to be spent inside the filter ($\simeq 15$ cm long) for effective separation. This is the correct energy regime for secondary ions, allowing operation of the system with the target at or near earth potential, which is an advantage in multi-technique systems. However, simple low-field collection optics have limited performance, giving an overall transmission <0.1 per cent. More complex optical systems are therefore coming into use, incorporating higher extraction fields (up to 1000 V cm^{-1}), lenses, apertures, and focusing electrostatic sectors (see next section) to improve transmission. The extra complexity usually makes possible optical gating or energy analysis of the ions, and overall transmission is increased by an order of magnitude to 1 per cent.[18]

4.4.3.2. Magnetic sector mass spectrometers When charged particles move through a magnetic field (see Fig. 4.11), they experience a force orthogonal to both the direction of motion and the magnetic flux lines, resulting in a circular trajectory. The force acting on the particle, and hence the radius of its path, depends on the velocity of the ion. If ions of all masses are accelerated to a given potential, then the resultant velocity is dependent on the mass:charge ratio of each ion. Thus, subsequently passing the ions through a region of constant magnetic field will separate ions of different masses.[19] In theory it is possible to detect all masses simultaneously, but in practice the more convenient solution of a point detector and scanning of the field strength by an electromagnet is adopted; laminated magnets allow rapid switching between masses. The dispersion of adjacent masses, i.e. resolution, decreases with increasing ion mass, and is proportional to the radius of curvature of the ion. Large magnets (radius 10–100 cm) are therefore required for good mass resolution at higher mass. The instrument has some degree of focusing, but is nonetheless sensitive to variations in entrance angle (and energy) of the ions, which is seen as a broadening on the axis of dispersion. This degradation in resolution can be alleviated by the use of shaped magnetic sector pole faces, but this is better achieved by the use of double focusing instruments. Sectors of 180° eliminate exit boundary effects, but require large amounts of magnetically homogeneous material, which is expensive and has the disadvantages of excessive size and weight, sectors of 60 and 90° are therefore also found in practice.

A radial electrostatic field, such as is generated by a pair of cylindrical sector electrodes produces no spatial dispersion for different masses, but does disperse different energies. A suitably sized slit at the exit will therefore allow energy filtering, and this monoenergetic beam could be injected into a magnetic sector. The resulting spectrum would then not suffer from degraded resolution due to the initial energy spread of the secondary ions. In practice, the focusing action of the electrostatic sector is also exploited by placing the sample at its focal point, and carefully matching the magnetic and electrostatic sector angles (usually 60 and 90° respectively) and the intermediate distances (see Fig. 4.11 (c)).[20] Ions of a given mass are then brought to a focus irrespective of their initial energy and angle of emission (to first-

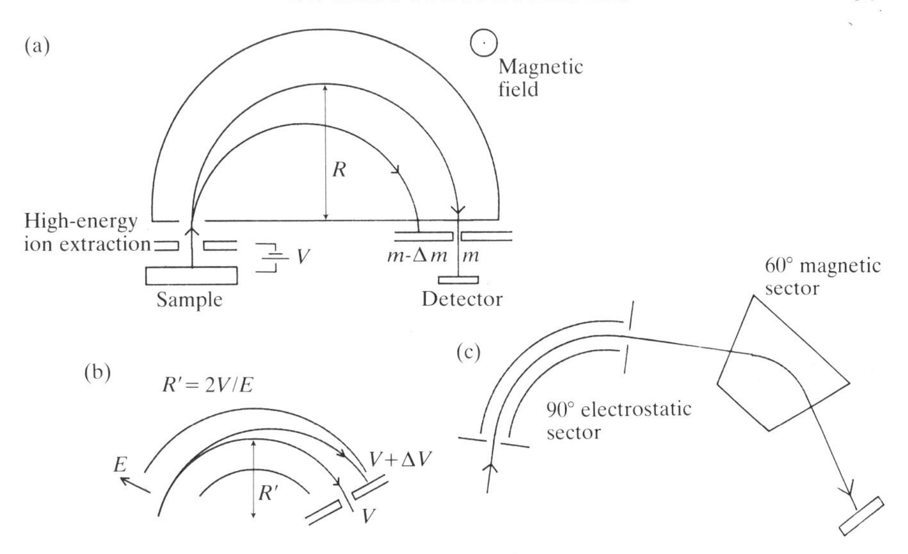

FIG 4.11. Operation of a magnetic sector mass analyser; (a) longitudinal cross-section of magnetic sector, showing mass dispersion; (b) longitudinal cross-section of electrostatic sector, showing energy dispersion; (c) double focusing mass spectrometer with spatial and energy focusing.

$$\begin{aligned} \tfrac{1}{2}\, mv^2 &= zV \\ BzV &= mv^2/R \end{aligned} \quad \Rightarrow R = \frac{1}{B}\left(\frac{2mV}{z}\right)^{1/2}.$$

order approximation), giving excellent mass resolution. Variable apertures allow a trade-off between resolution and transmission in the analyser.

Magnetic sectors thus have medium transmission (10–50 per cent), large mass range ($m/z >$ 10 000) and unequalled mass resolution ($m/\Delta m > 10^4$). They are also the most expensive and largest type of mass spectrometer. High-energy ion extraction is used, giving an overall transmission of 1–10 per cent. Their characteristics make them the best choice in biomedical (high-mass) applications, and they are thus the industry standard in the closely related field of FAB using a liquid matrix. The high mass resolution also means that they are very attractive for SIMS analysis in semiconductor research. One other important feature of magnetic sector instruments is the capability for focusing of the secondary ion beam discussed above, which preserves the lateral resolution of ions emitted from the surface and hence allows the construction of ion microscopes as described earlier.[21]

4.4.3.3. Time-of-flight (ToF) mass spectrometers When ions are accelerated to a given potential so that they have the same kinetic energy then ions of different mass: charge ratio will have different velocities. If these ions then pass through a region of field-free drift space, they will spread out in time, with the higher mass ions arriving later (see Fig. 4.12). A time-sensitive detection system is then all that is needed to produce a mass spectrum. However, the secondary ions must be produced at a definite point in time, which is best achieved by pulsing the primary ion into short bursts of <10 ns—the time-scale of secondary ion emission after impact is negligible ($<10^{-12}$ s). The beam is most usually pulsed by rapid deflection across a small axial aperture, but one system uses off-axis deflection, followed by a curved magnetic field to compress the pulse in space. The emitted secondary ions are then accelerated to about 5 kV so that the flight time over a distance of ~ 2 m is reasonable ($\simeq 100$ μs). This time, plus the time needed for data-processing (again $\simeq 100$ μs) must elapse before the next pulse of primary ions is sent. The arrival time of the ions at the detector has to be measured to ≤ 1 ns accuracy.

The transmission of a linear flight tube system should approach 100 per cent. Mass range is theoretically unlimited, but at the expense of long flight times, which means slow pulsing frequencies and hence long analysis times. Slow-moving ions are also difficult to detect, so that an auxiliary acceleration to, for example, 20 keV, may be required at the end of the flight tube. Mass resolution depends on many factors; the length of the primary ion pulse or

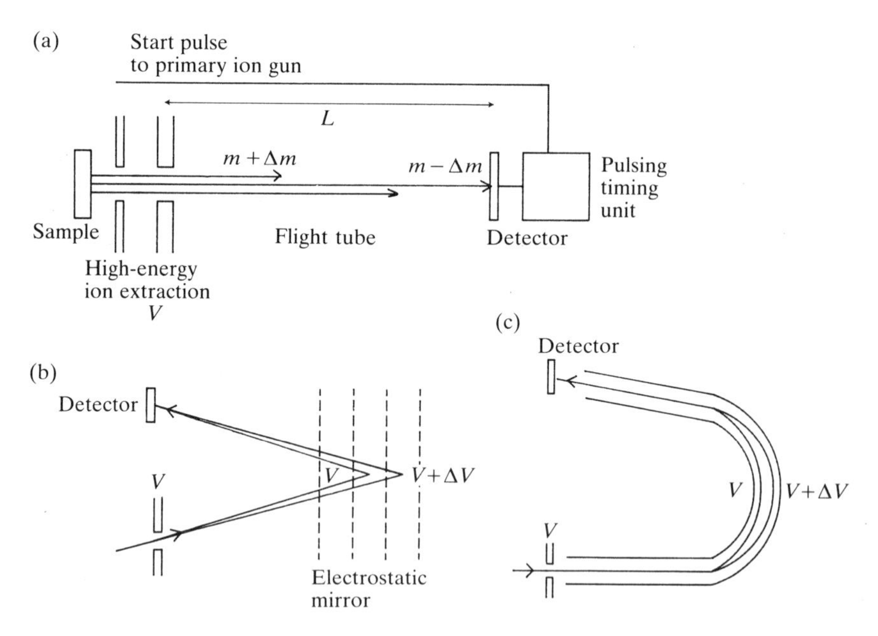

FIG 4.12. Operation of a mass analyser: (a) longitudinal cross-section of flight tube, showing time (mass) dispersion; (b) energy-compensating mirror design; (c) energy-compensating electrostatic sector design. $\frac{1}{2}mv^2 = zV$; $t = L(m/2zV)^{1/2}$.

the time resolution of the detector system may be limiting. Restricting the entrance angle of the spectrometer improves mass resolution at the expense of transmission. However, the major problem is generally the initial energy spread of the secondary ions. This effect is reduced by high accelerating fields at the sample (i.e. high extraction voltages and small (mm) extraction gaps). More complex time-of-flight systems further compensate for this energy spread by using non-linear flight tubes. One design has a curved electrostatic path so that the more energetic ions are forced round the outer part of the bend.[22] Figure 4.13 is a detailed schematic diagram of a SIMS instrument

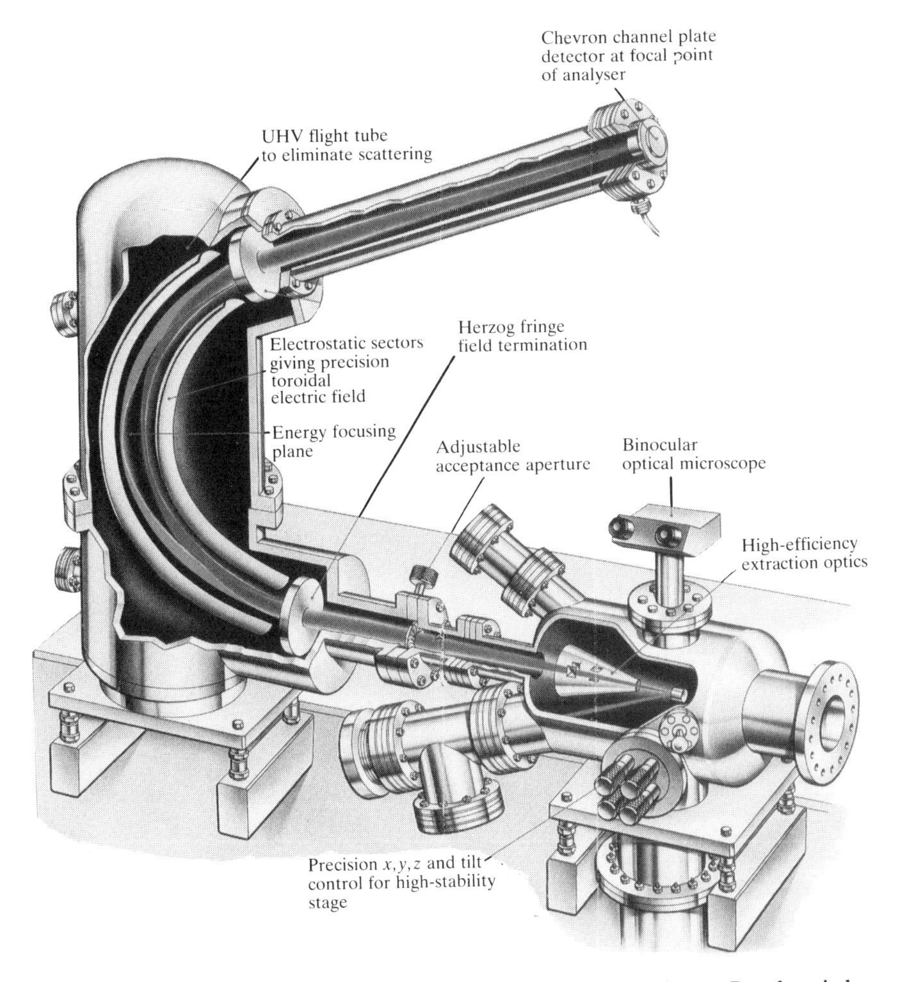

FIG 4.13. Schematic diagram of a ToF-SIMS instrument using a Poschenrieder energy-compensating analyser.

using this type of analyser. Another design incorporates an electrostatic 'mirror' in which the more energetic ions penetrate more deeply before reflection.[23,24] In both designs, the faster ions have a longer flight path to offset their increased velocity, and all ions of the same mass arrive at the detector simultaneously.

An interesting feature of linear ToF systems is that once the ions have completed the initial extraction–acceleration stage (< 1 μs), the ions are in a field-free drift region. If the ion is metastable and dissociates in the flight tube (usually giving one charged and one neutral daughter to conserve charge), the principle of conservation of momentum means that the centre of mass of the pair has the same velocity as the parent ion. In the absence of any external fields, the daughters will therefore strike the detector at the same time as would the original ion had no dissociation occurred, and thus appear as a peak at the original mass. The observed spectrum is then that of the secondary ion distribution $\simeq 1$ μs after emission. In comparison, ions must survive intact until after mass analysis (> 100 μs) is completed to be detected in a magnetic sector or quadrupole system, and so metastable ions are effectively lost from the spectrum.

The preceding argument must be slightly modified because a real fragmentation will involve minor changes in energy. This is seen as a broadening of peaks and a general increase in unresolved background noise. The fragmentation will also partition the kinetic energy unequally between the products. Deliberately introducing an electrostatic field and additional 90°-offset detector at the end of the flight-tube deflects the charged daughters out of the beam and disperses them in time by virtue of their unequal energies. This allows their mass to be determined. The neutral fragment is unaffected and strikes the original detector as above. By monitoring the coincidence of the two signals, the daughter ions produced by a particular parent ion fragmentation can be deduced; by altering the length of the flight tube the ion lifetimes can also be found.

Time-of-flight systems thus have a transmission of 50–100 per cent, which, coupled with the non-scanning parallel mass detection, gives a massive increase in effective sensitivity of the analysis (e.g. 10^5 over a quadrupole). Mass range is limited to between $m/z = 5000$ and 10 000 by practical considerations and mass resolution $m/\Delta m$ is $\simeq 10^3$ for a linear system, 2×10^3 for a Poschenrieder type, and 5×10^3 for a mirror design. The high extraction voltage means that overall transmission > 10 per cent is possible. The main disadvantages are the restrictive extraction geometry near the target and the cost of the computer system for rapid synchronized timing, data storage and processing: this latter factor is becoming less important as electronics design advances. The system is thus well suited to analyses where minimization of the total primary beam dose is the prime consideration; for example, polymer analysis, trace concentration analysis, and small (μm) area analysis and

TABLE 4.2

Mass spectrometer comparison

	Resolution	Mass range	Transmission	Mass detection	Relative sensitivity
Quadrupole	10^2–10^3	$\leq 10^3$	0.01–0.1	Sequential	1
Magnetic sector	10^4	$> 10^4$	0.1–0.5	Sequential	10
Time-of-flight	$> 10^3$	10^3–10^4	0.5–1.0	Parallel	10^4

imaging. The low duty cycle of the pulsed primary beam does mean that depth profiling would be time consuming unless continuous ion beam operation was used for etching with periodic pulsed beam operation for analysis.

The performance characteristics of the three types of mass spectrometer are compared in Table 4.2.

4.5. Conclusions

In Fig. 4.14 a typical SSIMS instrument is depicted. It should be clear that there is a wide range of SIMS equipment available to allow each system to be configured for individual applications, within budgetary limits. The recent advances in electronics have led to the widespread introduction of mini-computers, initially for data storage but increasingly for post-analysis data manipulation. This is especially true in imaging applications where vast amounts of data are generated. The trend is also towards full computer control of the instrument during the experiment, allowing control of the major instrument parameters from the computer keyboard.

At present, SIMS instrumentation is still developing rapidly. The technique has some way to go before it reaches the instrumental maturity of the older electron spectroscopies. However, the limits of the SIMS process itself are already closely approached in terms of lateral and depth resolution (though good spatial resolution at lower primary beam energies would be desirable), and the transmission efficiency of mass spectrometers could soon be close to 100 per cent. The techniques of post-ionization seem to offer the ultimate goal of quantification and full use of sputtered material, but further research in this area is needed.

However advanced the instrumentation, the role of the machine–analysis interface, normally termed the operator, is still crucial. In such a relatively young field, experience in operation and interpretation is crucial. As systems

FIG 4.14. A typical static SIMS instrument.

become more sophisticated and 'user friendly', and techniques such as neutralization of insulating samples are better understood and hence become more automated, the operator quality becomes less important. As this happens, and the advantages of mass production and competition reduce the price of equipment, then hopefully a SIMS instrument will become as familiar a sight in the surface analytical laboratory as a scanning electron microscope.

References

1. Wilson, R.G. and Brewer, G.R. (1979). *Ion Beams*. John Wiley, New York.
2. Banford, A.P. (1966). *The Transport of Charged Particle Beams*. E & F.N. Spon, London.
3. *Secondary Ion Mass Spectrometry, SIMS III* (1982). Springer Series in Chemical Physics, Vol 19 (ed. A. Benninghoven, J. Giber, J. Laszio, M. Riedel, and H.W. Werner). Springer Verlag, Berlin.
4. *Secondary Ion Mass Spectrometry, SIMS IV* (1984). Springer Series in Chemical Physics, Vol 36 (ed. A. Benninghoven, J. Okano, R. Shimizu, and H.W. Werner). Springer Verlag, Berlin.
5. *Secondary Ion Mass Spectrometry, SIMS V* (1986). Springer Series in Chemical Physics, Vol. 44, (ed. A. Benninghoven, R.J. Colton, D.S. Simmons, and H.W. Werner). Springer Verlag, Berlin.
6. Hanlon, J.F. (1980). *A User's Guide to Vacuum Technology*. John Wiley, New York.
7. Liebl, H. (1977). *Adv. Mass Spectrom.*, **7**. 751.
8. Magee, C.W. (1983). *Int. J. Mass Spectrom. Ion Phys.*, **49**, 211.
9. Levi Setti, R., Crow, G., and Wang, Y.L. (1986). In *Secondary Ion Mass Spectrometry, SIMS V*, Springer Series in Chemical Physics, Vol 44 (ed. A. Benninghoven, R.J. Colton, D.S. Simmons, and H.W. Werner), p. 132. Springer Verlag, Berlin.
10. Prewett, P.D. and Jefferies, D.K. (1980). *J. Phys. D: Appl. Phys.*, **13**, 1747. Mair, G.L.R. and Mulvey, T. (1984). *Scanning Electron Microsc.*, **IV**, 1531; (1985) *ibid* **III**, 959.
11. Hunt, C.P., Stoddart, C.T.H., and Seah, M.P. (1981). *Surf. Interface Anal.*, **3**, 159.
12. Wittmack, K. (1979). *J. Appl. Phys.*, **50**, 493.
13. Slodzian, G., Chaintreau, M., and Dennebouy, R. (1986). *Secondary Ion Mass Spectrometry, SIMS V*, Springer Series in Chemical Physics, Vol. 44 (ed. A. Benninghoven, R.J. Colton, D.S. Simmons, and H.W. Werner), p. 158. Springer Verlag, Berlin.
14. Lub, J., van Velzen, P.N.T., van Leyen, D., Hagenhoff, B., and Benninghoven, A. (1988). *Surf. Interface Anal.* **2**, 53.
15. Brown, A. and Vickerman, J.C. (1984). *Surf. Interface Anal.*, **6**, 1.
16. Lai, S.Y., Briggs, D., and Vickerman, J.C. (1989). *Surf. Interface Anal.*, in press.

17. Barber, M., Bordoli, R.S., Sedgwick, R.D., and Tyler, A.N. (1981). *J. Chem. Soc. Chem. Commun.*, 325.
18. Wittmaack, K. (1982). *Vacuum*, **32**, 65.
19. Dempster, A.J. (1918). *Phys. Rev.*, **11**, 316.
20. Nier, A.D. and Roberts, T.R. (1951). *Phys. Rev.*, **81**, 507.
21. Castaing, R. and Slodzian, G. (1962) *J. Microsc.*, **1**, 395.
22. Poschenrieder, W.P. (1972). *Int. J. Mass Spectrom. Ion Phys.*, **9**, 357.
23. Mamyrin, B.A., Karatajev, V.J. Schmikk, D.V., and Zagulin, V.A. (1973). *Sov. Phys. JETP (Engl. Transl.)* **37**, 45.
24. Tang, X., Beavis, R., Ens, W., Lafortune, F., Schueler, B. and Standing, K.G. (1989). *Int. J. Mass Spectrom. Ion Phys., in press.*

5

SIMS DEPTH PROFILING OF SEMICONDUCTORS

5.1. Introduction

SIMS depth profiling is a surface analysis technique capable of determining the elemental concentrations of dopant and impurity atoms within a material as a function of depth. The technique is normally used to determine concentrations in the range of 10^{13}–10^{20} atoms cm^{-3} lying at depths of up to 10 μm. To generate a SIMS depth profile a mono-energetic beam of mass filtered ions, the **primary ion beam**, is scanned over the sample surface. Each primary ion that penetrates the solid generates an intense but short-lived collision cascade which involves the breaking of bonds and the displacement of many target atoms from their lattice sites. Some of the displaced atoms are ejected from the sample as low-energy secondary species and the sample surface is gradually eroded. Whilst most of the sputtered atoms are (usually) neutral, a small fraction are ejected as positive or negative ions. It is possible to collect some of these ions and transfer them to a mass spectrometer where, after mass filtering, the ions of the desired mass may be counted. Since the material is being continuously eroded it is possible to build up chemical information as a function of depth (Fig. 5.1).

Accurate chemical analysis of semiconducting samples has been the main impetus for the development of the technique but there are a host of other material science problems to which it can be applied, including the analysis of geological samples, fossils, teeth, ceramics, superconductors, polymers, and metallic alloys.[1]

The strengths and weaknesses of the technique are listed in Table 5.1. It is the very high sensitivity of the technique to dopant and impurity atoms within the matrix, at parts per billion in favourable cases, together with the good depth resolution, usually between a few nanometres and a few tens of nanometres, that make it so attractive to the material scientist. The depth profile parameter is defined in Fig. 5.2. It should be noted, however, that the conditions that will favour high sensitivity, namely high primary beam current and energy and large area analysis, are exactly the conditions to be avoided if good depth resolution is required. High sensitivity and good depth resolution are, therefore, mutually exclusive. This problem is due to the destructive nature of the analytical technique; a finite volume of material has

TABLE 5.1

The strengths and weaknesses of SIMS depth profiling

	Strengths
1 Very good sensitivity to dopants–impurities	Can be p.p.m. or even p.p.b. in favourable cases
2 Good depth resolution	Can be a few nanometres
3 Good dynamic range	Can follow changes in concentration of six or more decades in favourable cases
4 Mass spectrometric	Can detect all elements and isotopes
5 Rapid analysis	Typically between a few minutes and hours for a depth profile
6 Can be quantitative	Concentration–depth plots accurate to 5 per cent on both axes
	Weaknesses
1 Destructive technique	Can never repeat the experiment on the same part of the sample
2 Mass interferences	Can make the achievable detection limit far worse than the sensitivity
3 High sensitivity and good depth resolution mutually exclusive	High sensitivity requires a large analytical volume per data point, good depth resolution the exact opposite
4 Large variations in ionization probabilities necessitate the use of ion-implanted standards	A separate standard is required for every dopant in every matrix present in the depth profile
5 Beam-induced mixing	Leads to a redistribution of target atoms prior to their emission as ions, a major limitation to the depth resolution
6 Chemical effects	Segregation and radiation-enhanced diffusion can effect the depth resolution and invalidate the quantification
7 Microtopography and macrotopography	Can lead to a loss of depth resolution with sputter depth
8 Charging effects	Can change the instrumental transmission and thus give rise to changes in signal levels

to be removed per data point. Another related problem is the speed of analysis, which is usually between 0.1 and 10 μm h^{-1}. Rapid analysis (high sputter rate) minimizes instrumental drift but may give an unacceptably large depth increment between the data points.

In addition to good sensitivity, the technique can follow changes in concentration of many orders of magnitude in a single depth profile, due to the high

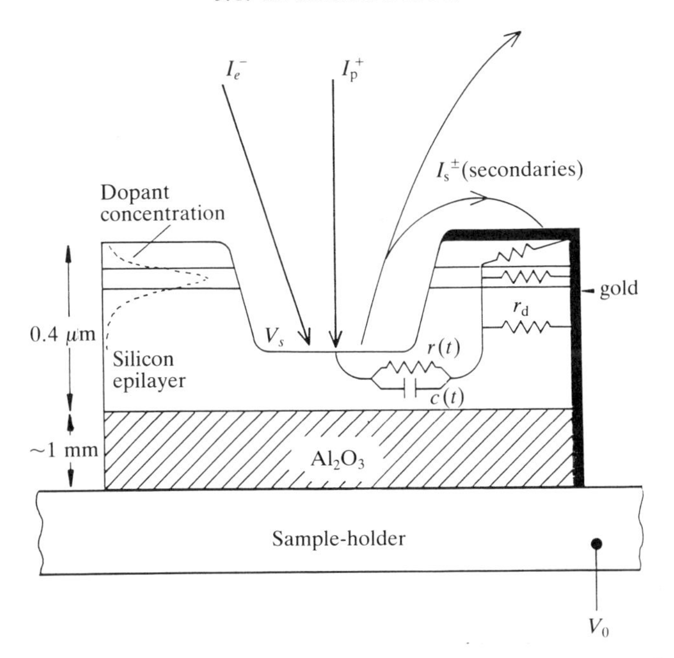

FIG 5.1. Scheme of a SIMS depth profile in a silicon-on-sapphire epilayer. An energetic primary ion beam (I_p) is used to generate low-energy secondary ions (I_s) as the material is sputtered away, thus producing chemical information as a function of depth. Accumulation of surface charge must be avoided as it will modify the surface potential and lead to a change in instrumental sensitivity. Two possible methods of charge compensation, electron beam flooding (I_e) and gold coating are shown.

dynamic range, an important attribute when analysing ion implants, for example. SIMS can detect *all* elements of the Periodic Table and since it is a mass spectrometric technique it can distinguish between different isotopes.

5.1.1. *Quantification*

SIMS depth profiling is *quantitative*; converting from raw data (count-rates as a function of time) to quantified data (concentrations as a function of depth) involves the use of ion-implanted standards and crater depth measurements. Quantification can often be made accurate to better than 5 per cent for both the concentration and depth scale, provided the dosimetry of the ion-implanted standard is accurate and the rate of sputtering throughout the profile is known.

One of the main parameters in SIMS is the fraction of atoms that are sputtered as ions (positive or negative). The ionization probability or ion fraction α is very sensitive to a number of factors. Leta and Morrison[2] determined the number of ions detected per atom sputtered (the useful ion

TABLE 5.2

Positive ion yields for lithium and chlorine in a variety of matrices under oxygen primary beam bombardment (from Ref. 2)

Ion	Substrate	Ion yield (ions per atom)
Li^+	Silicon	8×10^{-3}
	Gallium arsenide	7×10^{-3}
	Gallium phosphide	4×10^{-3}
	Germanium	3×10^{-3}
	Indium phosphide	3×10^{-3}
Cl^+	Silicon	1×10^{-7}
	Gallium arsenide	2×10^{-6}
	Gallium phosphide	1×10^{-6}
	Germanium	3×10^{-6}
	Indium phosphide	1×10^{-6}

yield) for several different dopants in several different matrices by analysis of ion implanted standards (Table 5.2). The variations they found reflected the different ionization probabilities of the different elements. In the case of oxygen primary beam bombardment, the useful ion yields for lithium monitored as a positive ion (Li^+) were between 3×10^{-3} and 8×10^{-3}, whereas for chlorine (Cl^+) they were between 1×10^{-6} and 1×10^{-7}. Thus there is a variation of four orders of magnitude across the Periodic Table in the ionization probability for positive ions. Similarly, the ionization probabilities for negative ions are far higher for elements on the right-hand side of the Periodic Table than on the left. For these reasons, elements on the left-hand side of the Periodic Table are usually monitored as positive ions with an oxygen primary ion beam, and those on the right-hand side of the Periodic Table as negative ions, using a caesium primary ion beam. It is important to note, furthermore, that the ion yield for an element varies from one matrix to another. For Li^+ the value of 3×10^{-3} reported above was for lithium in germanium and the value of 8×10^{-3} was for lithium in silicon. Another related problem is that the ionization is very strongly influenced by the surface chemistry. This can be used to good advantage; for example, a jet of oxygen gas is often directed at the sample surface to increase the positive secondary ion yield of electropositive species. By the same token, however, a trapped oxygen-rich mono-layer at an interface will play havoc with the secondary ion yields across that interface; this is a chemical effect. Finally, and most importantly, there is as yet no comprehensive theory of secondary ion emission that can explain these variations or predict the ionization probability for a new dopant–matrix combination. By far the most accurate way to achieve quantification is to have the element of interest ion-implanted into the host matrix at a suitable fluence and energy.

Another problem that becomes apparent in multilayer analysis is that different materials sputter at different rates, i.e. have different partial sputter yields (see Chapter 2). For example, Meuris *et al.*[3] measured the sputter yields of $Al_x Ga_{1-x} As$ as a function of composition (x). They found that the sputter yield decreased linearly with aluminium concentration x and that the sputter yield of GaAs was about twice that of AlAs.

5.1.2. *Depth resolution*

Another important parameter in a depth profile is the ability to resolve abrupt interfaces and multilayer structures. This capability is called the **depth resolution**, and is a 'figure of merit' for a depth profile. It is the precision with which a particular dopant distribution can be resolved, for example, an abrupt doping interface. The broadening introduced by the analysis technique arises from a number of different physical processes, instrumental problems and sample-related problems. Thus the depth resolution can often be improved by changing some of the experimental

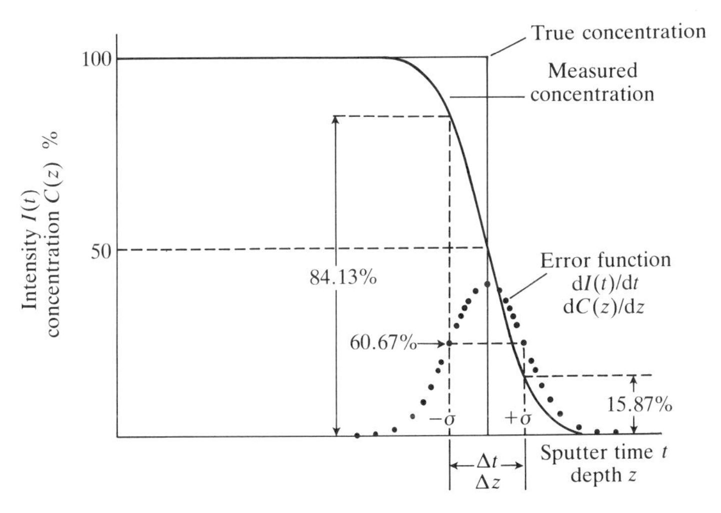

FIG 5.2. The usual definition for depth resolution. Absolute depth resolution — $\Delta t(\Delta z)$; relative depth resolution — $\Delta t/t(\Delta z/z)$. This is not always appropriate; see later in the text.

parameters, for example, the beam energy and type and the angle of incidence of the primary ion beam.

5.1.3. Parameters influencing quantification and depth resolution

There are various problems associated with quantification and depth resolution (see Section 5.3). It should be noted that there are many sources of profile distortions in SIMS depth profiling, including beam-induced broadening effects, chemical effects (e.g. segregation), the development of microtopography and macrotopography, surface charging effects, and instrumental drift. Thus skilled interpretation of the data is an important part of the experimental technique. Two basic limitations to the depth resolution arise from the statistical nature of the sputtering process and the depth of origin of the secondary ions. The base of the crater can never be perfectly flat since removal of some atoms from the top monolayer exposes the underlying layer to the primary ion beam. This and the fact that the secondary ions are believed to originate from the top few monolayers would limit the depth resolution of the technique to ~nm. Unfortunately, there are other processes at work which make the situation far worse. The process of sputtering involves a mixing or randomization of the target atoms over a depth similar to the range of the primary ions (7 nm in the case of 2 keV O^+ ions), involving two distinct mixing processes, random atomic mixing and recoil implantation. If we could freeze the depth profile at any moment in time and identify all the atoms at the bottom of the crater we would discover atoms that were originally several nanometres above that level and had been mixed downwards, together with atoms originally several nanometres below that level that had been mixed upwards (and of course atoms incorporated from the primary ion beam). The progressive mixing of a buried dopant 'marker' layer prior to sputtering is illustrated in Fig. 5.3. The result of the redistribution of dopant atoms prior to sputtering in a depth profile is that we begin to detect dopant ions before the crater reaches the position of the marker layer and continue to detect them after it has passed the position of the marker layer. Other 'fundamental' mixing effects that can spoil the depth resolution include segregation and radiation-enhanced diffusion. Finally, there are a variety of 'instrumental' effects, including uneven etching across the sample plane, which degrades the depth resolution.

5.2. The experimental arrangement

5.2.1. The primary ion beam

To generate a SIMS depth profile a primary ion beam is scanned over an area of the sample surface causing sputter erosion of the sample (see Fig. 5.1).

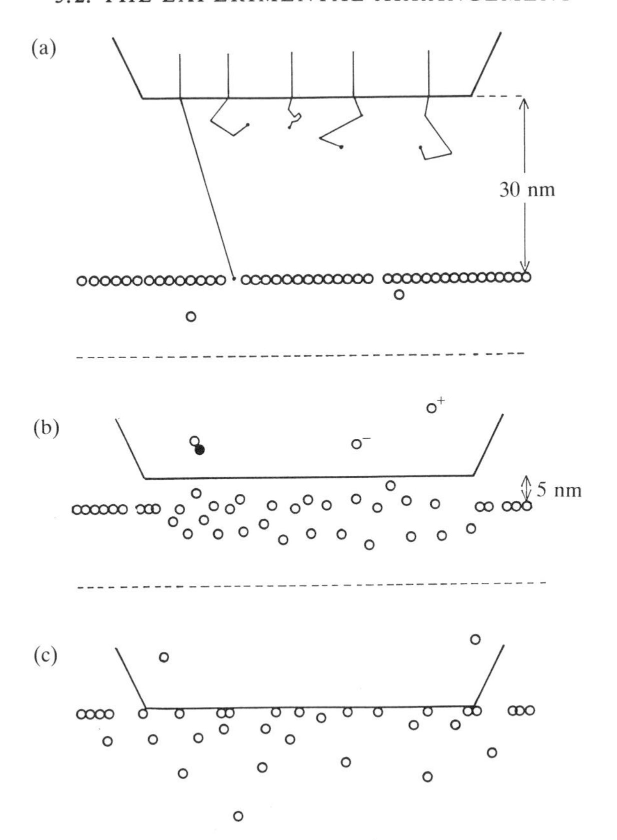

FIG 5.3. Three stages in the sputtering of a monolayer containing some impurity atoms. (a) With the bottom of the crater 30 nm above the marker layer. The primary ions have insufficient energy to reach the dopant atoms which are 'unaware' of the cascade mixing above them. (b) 5 nm above the marker layer. The primaries have sufficient energy to reach the dopant atoms. Some are knocked upward, many downward, and a few have already been sputtered. (c) With the crater bottom at the depth of the original marker layer. There are very few of the original dopant atoms left. Some have been sputtered; most have been mixed downward. It is this redistribution prior to sputtering that defines the limit of the depth resolution in SIMS depth profiling.

Ideally, the primary beam will be mass filtered, to avoid implantation of impurity atoms into the sample, and mono-energetic in order to avoid chromatic aberrations in the ion optical components. The primary beam line contains lenses and deflection plates for focusing and steering the beam and quadrupole raster scan plates for scanning it. The ions are bent through a small angle near the end of the beam line to eliminate any neutral component (see Chapter 4). Typically, the primary beam current is between 0.1 and 10 μA of O_2^+, Ar^+ or Cs^+ at an energy of between 0.5 and 20 keV. The choice of beam type is important. Species that exhibit a low positive ion ionization probability under oxygen ion bombardment (α^+) often show a high negative ion ionization probability under caesium bombardment (α^-) and vice versa. Dopants on the right-hand side of the Periodic Table are usually studied as negative ions using a caesium beam, those on the right as positive ions with an oxygen beam. The crater is usually a square of side 100–2000 μm.

5.2.2. Secondary ion collection

The secondary ions are emitted from the sample surface in all directions and with a range of energies. The energy distribution for atomic ions has a peak at around 10 eV and usually shows a high-energy tail extending to a few hundred eV (see Fig. 5.4). For these reasons, it is not usually possible to collect all the sputtered ions and the range of energies passed to the mass

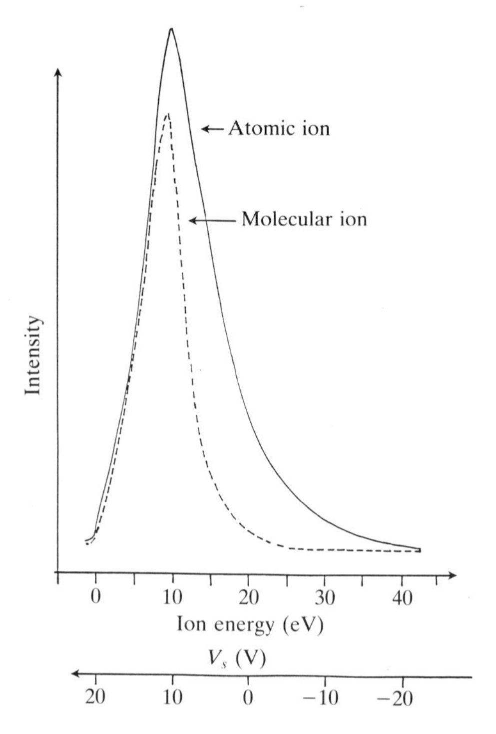

FIG 5.4. The secondary ion current as a function of sample bias (V_S) for an atomic ion and a molecular ion. The atomic ion exhibits a 'high-energy tail'.

spectrometer is termed the **band pass**. For quadrupole-based instruments, such as the Atomika, that employ a low extraction potential (100–200 V) the band pass energy is about 10–20 eV, whereas for a magnetic sector instrument, e.g. the Cameca, which employs a high extraction potential (4500 V) the band pass can be several hundred electron-volts. Collection of secondary ions will be maximized if the band pass embraces the peak of the secondary ion energy distribution, and tuning of the secondary ion optics is often required to achieve this. The transmission of secondary ions to the mass spectrometer will change if the sample potential at the surface V_s(x,y) changes. Indeed, the sample potential (V_0) is often made more negative to reduce the intensity of a very intense secondary ion (and thus protect the detector). V_s(x,y) can also change unintentionally, however, due to the accumulation of surface charge. Such charging effects will give the illusion of a build-up or depletion of dopant and must be avoided. If the sample is insulating or semi-insulating, some method of charge compensation will be required. Figure 5.1 shows two possible methods of achieving this; electron beam flooding and gold coating of the sample surface (see Section 5.4).

Only those ions emitted from the central area of the crater are collected for analysis. This is called gating. Use of a small gate will ensure that the perturbing effects of any crater macrotopography are minimized but at the expense of a reduction in the secondary ion signal. Two methods of gating are commonly used. With electronic gating, where a rastering primary beam is employed (see Fig. 5.5), the counting system is only enabled when the primary ion beam is in the centre of the crater, whereas with optical gating the

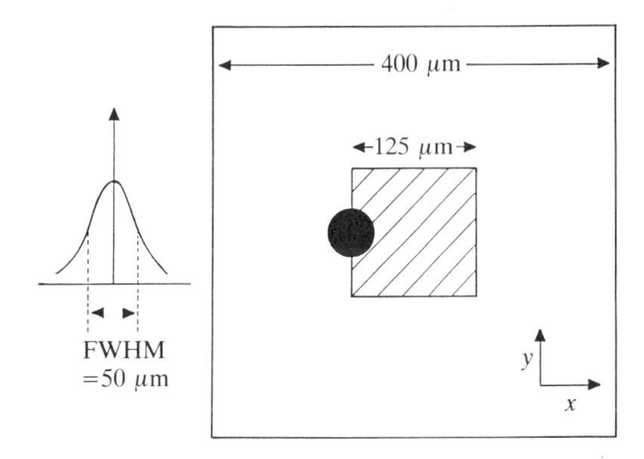

FIG 5.5. With electronic gating the counting system is only enabled when the centre of deflection of the primary beam is within a defined area, usually a square at the centre of the crater. Note, however, that when the beam is at the edges of the gated area one-half of the secondary ions are generated from points outside it and thus that beam shape is of critical importance in such a system.

ion optics of the secondary ion column determine the field of view of the crater base. After transfer to the mass spectrometer the ions of the appropriate mass are passed to a detector, for example, a channeltron or Faraday cup, the signal amplified and then counted.

Gating improves the dynamic range over which the ions of interest can be monitored accurately. Figure 5.6 demonstrates how the combination of lenses, beam rastering, and electronic gating give an improved dynamic range as compared with solely a rastered beam.

5.2.3. *Data logging and instrumental control*

Generally a microcomputer is interfaced to the instrument and will perform the operations of control and data logging automatically. This facilitates the inclusion of several mass channels in the depth profile, the computer cycling

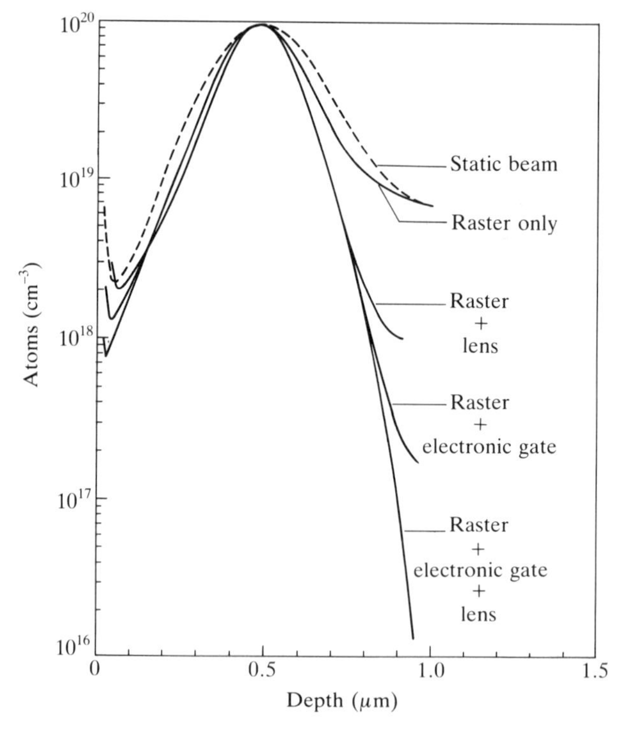

FIG 5.6. Boron (150 keV) implanted into silicon, profiled with increasingly sophisticated primary ion beam techniques and secondary ion optics demonstrating improved dynamic range. Primary ion beam 5 keV Ar^+. Sputtering rate, 6.2Å s^{-1}; time needed to sputter 1μm, 25 min; oxygen jet on.

the mass spectrometer between them in turn. The data collection periods and sample biases for each mass channel may also be under computer control. The data collection period is called the frame time, *FT* (the time spent accumulating counts of ions of a particular mass); it is usually between 1 and 100 s. The sum of the frame times for all the mass channels together with the 'dead time' taken for the mass spectrometer to switch between the mass channels is called the cycle time, *CT*. The raw data (counts per frame $N(F)$ vs. frame number F) are usually saved on disc or tape for processing. Typically counts are in the range 0–10^8 ions per frame.

5.3. Quantification of the data

Quantification is the process of converting the raw data ($N(F)$ vs. F) for each mass channel into concentrations as a function of depth.

5.3.1. Quantification of the depth scale

5.3.1.1 Depth calibration and its pitfalls Calibration of the depth scale is the first step and usually involves measurement of the depth of the final crater using a Talystep or Dektak. It is then often assumed that the sample has sputtered at a uniform rate ($\dot{z}$). This assumption can be erroneous for several reasons. First, in the early stages of a profile the system is in a poorly understood pre-equilibrium period. During this period, components of the matrix with high sputter yields are preferentially sputtered until the surface layers are depleted of them. Furthermore, if a reactive primary ion beam such as oxygen or caesium is used, the surface chemical composition may change. For example, when analysing silicon with a 4 keV oxygen primary beam at normal incidence, the top 20 nm is rapidly oxidized to form an altered layer.[4] Since the sputter rates are changing during this pre-equilibrium period this leads to a systematic error in the depth calibration. The pre-equilibrium period and thus the magnitude of this error are a function of primary beam energy. If a sample is repeatedly analysed at different primary beam energies with respect to the surface and the raw data are then analysed assuming a constant sputter rate throughout the profile, the apparent depth of a feature (e.g. the depth of the peak of an implant) will shift. This effect is known as the differential shift;[5] it arises because the width of the pre-equilibrium period is different at different energies. Thus the size of the error involved in assuming a constant sputter rate is different at different primary beam energies.

The sputter rate may also have varied because the profile has passed through several different matrices. Clearly, it is important to know or to determine experimentally the sputter yields of all the layers traversed and to stretch and compress the depth scale as appropriate to correct for them.

Finally, the sputter rate may have altered due to variations in the primary beam current. One method of correcting for this is to reference all the mass channels to a matrix channel, but it is better if possible, to complete the analysis in a reasonable period of time (<1 h) to avoid instrumental problems of this type.

5.3.1.2. The relationship between sputter rate and sputter yield The rate of recession of the sample surface is called the sputter rate, $\dot{z}$, often expressed in $\mu m\ h^{-1}$, $nm\ s^{-1}$, or $Å\ s^{-1}$. Different materials sputter at different rates, as mentioned above, i.e. have different partial sputter yields, y_x. The partial sputter yield, y_x, is defined as the number of target atoms sputtered per incident primary atom and is a function of the primary beam type, energy and angle of incidence.[6] The partial sputter yield of silicon under oxygen bombardment is high at large primary beam energies and at grazing angles of incidence.[7] In a round robin analysis of a boron implant in silicon[8] it varied between 0.18 (4 keV O_2^+ at normal incidence) and 3.4 (5.6 keV O_2^+ at 64° to the normal). Note that at (dynamic) equilibrium the rate of incorporation of primaries into the sample is equal to the rate at which they are sputtered but that the sputtered component is not included in the definition of y_x. If the primary beam current is I_p of ions type A_n^+ ($n = 2$ for $O_2{}^+$) then the number of primary atoms striking the surface,

$$N_p\,(A) = I_p \cdot n/e \text{ atom s}^{-1} \tag{5.3.1}$$

where e is the electronic charge. It then follows from the definition of y_x that the number of target atoms sputtered per second,

$$\dot{N}_s = y_x \cdot I_p \cdot n/e \text{ atoms s}^{-1} \tag{5.3.2}$$

Writing the concentration of atoms in the target as ρ atoms cm^{-3}, then the volume of sample removed per second,

$$\dot{V} = (y_x \cdot I_p \cdot n/\rho \cdot e)\ \text{cm}^3\,\text{s}^{-1}. \tag{5.3.3}$$

The sputter rate $\dot{z} = \dot{V}/A$, where A is the area of the crater (in cm^2) can now be related to the partial sputter yield y_x from eqn (5.3.3):

$$\dot{z} = (y_x \cdot I_p \cdot n/\rho \cdot e \cdot A)\ \text{cm s}^{-1}. \tag{5.3.4}$$

The sputter rate can be increased by reducing the crater area but the sputter volume $\dot{V}$ will remain constant. In practice the sputter yield is calculated [from eqn (5.3.4)] using the experimentally determined values for the sputter rate $\dot{z}$, the analytical area A and the beam current I_p. For example, in the analysis of boron in silicon, under conditions of a crater area A of 600 μm × 600 μm, a primary beam current of 0.3 μA of $^{16}O_2^+$, and a sputter rate (from the crater depth and experimental time) of 0.13 $\mu m\ h^{-1}$, the atomic concentration of silicon $\rho_{Si} = 5 \times 10^{22}$ atoms cm^{-3}, $e = 1.6 \times 10^{-19}$ C and $n = 2$, the resulting value for the partial sputter yield is $y_{Si} = 0.17$, indicating that most of the sputtered material was oxygen incorporated by the primary ion beam.

5.3.2. The concentration scale: the relationship between the secondary ion count-rate and the dopant–impurity concentration

To quantify the concentration scale it is necessary to know the relationship between the secondary ion count-rate and the dopant concentration, i.e. the sensitivity of the analysis. This is usually achieved by depth profiling an ion-implanted standard and the samples of interest using the same analysis conditions. Since the ionization probabilities vary so much and are difficult to predict theoretically it is necessary to depth profile ion-implanted standards for each dopant–matrix combination. It should be noted, furthermore, that any sensitivity constant thus derived will not be applicable to systems where the impurity–dopant concentrations are very high (i.e. $>10^{20}$ atoms cm^{-3}), for then the ionization probability of the species is no longer constant (dilute regime) but becomes a function of the concentration of the species.

The dopant concentration can be related to the secondary ion count-rate, in the dilute concentration regime, as follows. Suppose the target contains a dopant type X, present in the solid at some atom fraction n(X). Then, since N_s (total) target atoms are sputtered per second,

$$N_s(\mathrm{X}) = N_s\,(\mathrm{total}) \cdot n\,(\mathrm{X}) \text{ atoms (X) s}^{-1} \tag{5.3.5}$$

atoms of type X are sputtered per second. This equation will not be valid if, for example, there is segregation of the impurity away from the analysis region. Let us assume that we decide to monitor singly charged positive ions of type X^+. If we tune the mass spectrometer to this mass and count X^+ ions we will detect a flux $N_s\,(X^+)$ dopant ions per second. This will be lower than N_s (X) for a number of reasons. First, only a fraction $\alpha\,(X^+)$ of the dopant atoms X are sputtered as the ions of interest X^+. Second, the use of gating, either electronic or optical, reduces the secondary ion flux in the ratio a/A, where a is the gated area and A the crater area. Third, only a fraction (η) of the gated ions can be successfully collected, transmitted to the detector and then counted. $\alpha(X^+)$ is called the ionization probability and $\eta(X^+)$ the instrumental transmission. Their product:

$$\tau_u\,(\mathrm{X}^+) = \alpha(\mathrm{X}^+) \cdot \eta\,(\mathrm{X}^+) \text{ secondary ions per atom} \tag{5.3.6}$$

is termed the useful ion yield. τ_u is the ratio of the ions detected to atoms sputtered (within the gate) and its importance is that it can be determined directly, using ion-implanted standards, whereas α and T cannot. (There exists potential for confusion in the literature as Werner[9] and others define the ion yield as the number of ions detected per incident primary atomic ion.)

Thus the number of ions of type X^+ detected per second is from (5.3.5) and (5.3.6):

$$N_s(X^+) = N_s\,(\text{total}) \cdot n\,(X) \cdot \tau_u\,(X^+) \cdot (a/A) \text{ ions s}^{-1}. \quad (5.3.7)$$

The atom fraction n(X) can be expressed as the ratio of the concentration of the impurity atoms in the solid, ρ(X), to the total concentration ρ of atoms of the solid, i.e. $n(X) = \rho\,(X)/\rho$. Thus [from eqn (5.3.7)]:

$$\rho\,(X) = [\rho \cdot / (N_s\,(\text{total}) \cdot \tau_u\,(X^+) \cdot (a/A)\,)]\, N_s\,(X^+) \text{ atoms cm}^{-3}. \quad (5.3.8)$$

Substituting for N_s (total) from (5.3.2) yields the relationship between the concentration of X in the solid and the number of secondary ions X^+ detected per second:

$$\rho(X) = [\rho \cdot e/y_x \cdot I_p \cdot n \cdot \tau_u\,(X^+) \cdot (a/A)]\, N_s\,(X^+) \text{ atoms cm}^{-3}, \quad (5.3.9)$$

or, from (5.3.4):

$$\rho(X) = [1/\dot{z} \cdot a \cdot \tau_u\,(X^+)] \cdot N_s\,(X^+) \text{ atoms cm}^{-3}. \quad (5.3.10)$$

The sputter rate ($\dot{z}$) is determined by crater depth measurements on the sample of interest, and the useful ion yield $\tau_u(X^+)$ by depth profiling an ion-implanted standard under identical conditions.

The sensitivity may now be calculated. If the frame time is FT seconds then the number of ions of type X^+ detected per frame is $N_s(X^+)$. FT or (by rearranging (5.3.10):

$$N_s(X^+) \cdot FT = \rho\,(X) \cdot \dot{z} \cdot a \cdot \tau_u\,(X^+) \cdot FT \text{ ions per frame}. \quad (5.3.11)$$

The sensitivity (in atoms cm^{-3}) is the concentration of X, $\rho'(X)$ that corresponds to a nominal number of counts of X^+ ions per frame. Setting $N_s(X^+)$. $FT = 100$, corresponding to statistical fluctuations of 10 counts per frame, then the sensitivity

$$\rho''\,(X) = 100/\dot{z} \cdot a \cdot \tau_u(X^+) \cdot FT \text{ atoms cm}^{-3}. \quad (5.3.12)$$

To achieve a good sensitivity (low ρ'' (X)) it is necessary to sputter a large volume of material per second and thus to use a high sputter rate $\dot{z}$ (high primary beam energy and primary beam current, shallow angle of incidence) and a large analytical area a. The data collection period FT should be as long as possible, and the useful yield as close as possible to unity. The best reported useful yield for boron in silicon, $\tau_u = 2 \times 10^{-2}$, is that of Williams.[10] His measurement was made on a Cameca IMS3F using an oxygen primary ion beam together with oxygen flooding. Williams suggested that this figure consisted of an ionization probability of 0.2 and a transmission of 0.1. Clegg *et al.*[8] reported a value, for boron in silicon, of 3×10^{-4} measured on an Atomika DIDA. This would appear to suggest that the transmission of the Cameca is two orders of magnitude better than the Atomika. Some caution must be exercised, however, as the ionization probability is itself a function of primary beam energy and angle of incidence, and the experiments were not conducted under the same conditions. A typical value for the sensitivity can be derived from the values quoted in Ref. 8:

Instrument	$\dot{z}/(\mu\ \mathrm{mh}^{-1})$	$a\ (\mu\mathrm{m}^2)$	τ_u	ρ'' (atoms cm^{-3})
Quadrupole	0.13	200×200	2×10^{-5}	$3.5 \times 10^{18}/FT$
Magnetic sector	2.8	8 μm diameter	2×10^{-3}	$1.27 \times 10^{18}/FT$

The large useful ion yield of the Cameca (magnetic sector) together with the high sputter rate in the analysis more than compensate for the small gated area used in the analysis. If the Cameca analyst had used a 50 μm gate the sensitivity would have improved to $\rho''(\mathrm{X}) = 3.25 \times 10^{16}/FT$ atoms cm^{-3}.

The relationship between analytical volume sputtered per frame $\dot{V}(FT) = \dot{z}.A.FT$ and sensitivity can be written, from (5.3.12), as

$$\log \rho''(\mathrm{X}) = 2 - \log \dot{V}(FT) - \log \tau_u. \tag{5.3.13}$$

It is plotted in Fig. 5.7 where two examples are given. Suppose the useful yield is 10^{-3}, if a large analytical area, 1000 μm $\times$ 1000 μm and large depth increment per second, 10 nm s^{-1}, are used then a sensitivity of 10^{13} atoms cm^{-3} is realized. However, if the analytical area is reduced to 10 μm $\times$ 10 μm and sputter rate to 1 nm s^{-1}, i.e. for microvolume analysis, then the sensitivity is 10^{18} atoms cm^{-3}. This fundamental limit to microvolume analysis arises from the limited number of dopant atoms available for sputtering from a given microvolume.

5.4. Description of a typical depth profile

5.4.1. Choice of experimental conditions

As an example, the analysis of a silicon epilayer grown by molecular beam epitaxy (MBE) is described, including the choice of experimental conditions, the steps in quantification, and interpretation of the data, taking into account any effects due to sputtering and process mixing.

The 0.8 μm thick epilayer was grown in a VG Semicon V80 silicon MBE growth chamber and 11 boron-rich layers were incorporated by co-evaporation of boron. The intended structure is shown in Fig. 5.8. The growers were primarily interested in the concentration of boron in the layers, the inter-peak concentration, and the sharpness of the doping transitions.

The sample was analysed on an ion-microprobe; an electronically gated low extraction field system fitted with a quadrupole mass spectrometer (EVA 2000). The primary beam chosen was $^{16}O_2^+$, as oxygen is known to enhance the ionization probabilities of electropositive species such as boron.

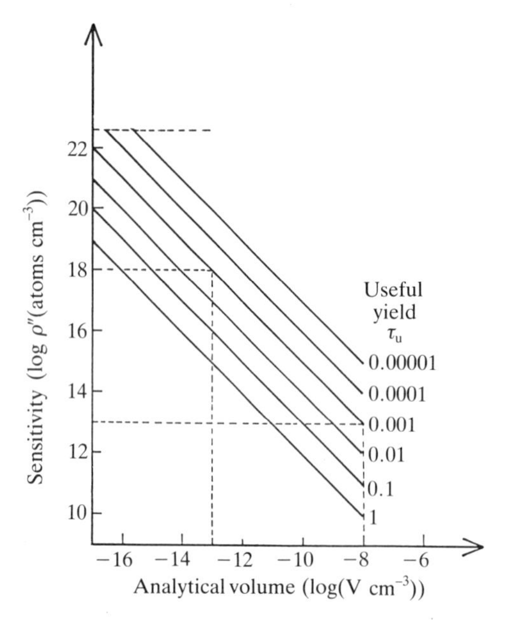

FIG 5.7. Relationship between the amount of material removed per data point (the analytical volume) and the instrumental sensitivity ρ'', Two examples are shown, see dashed lines both for a useful ion yield of 0.001. If a large volume is sampled s^{-1} (1000 μm × 1000 μm at a sputter rate of 10 nm s^{-1}) then a concentration of 10^{13} atoms cm^{-3} will generate 100 secondary ions s^{-1}. If a small volume is sampled, however (10 μm × 10 μm at a sputter rate of only 1 nm s^{-1}) then a concentration of 10^{18} atoms cm^{-3} will be required to generate the same 100 secondary ions s^{-1}. This represents a fundamental limitation to microvolume analysis.

These ions were extracted from an oxygen cold cathode discharge type source. The choice of primary beam energy, current, crater size and data acquisition period (frame time) is always a compromise between completing the analysis in a reasonable period of time (thus minimizing instrumental drift) and achieving adequate data density, depth resolution, dynamic range, and sensitivity. Generally, one tries to operate the primary beam column at the lowest energy at which adequate beam current can be realized, in order to minimize the beam-induced mixing processes which degrade depth resolution. In this instance a well-focused 200 nA spot was attained at 4 keV per molecular ion. This energy is often quoted as 2 keV per atomic ion. The primary beam had a Guassian intensity profile of full-width-at-half-maximum (FWHM) of 50 μm. This was established by imaging a copper grid and confirmed by running a Talystep scan over a hole produced by the static beam (see Fig. 5.9). For the analyses the beam was scanned over a square of side 400 μm to produce a uniform primary beam current density

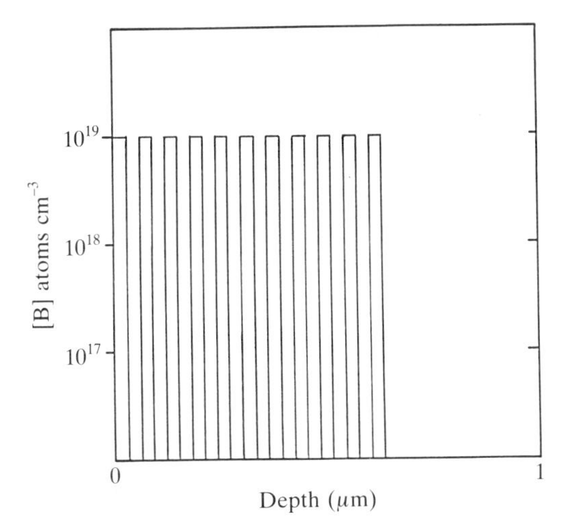

FIG 5.8. Intended doping distribution for the boron-in-silicon test structure grown by silicon MBE.

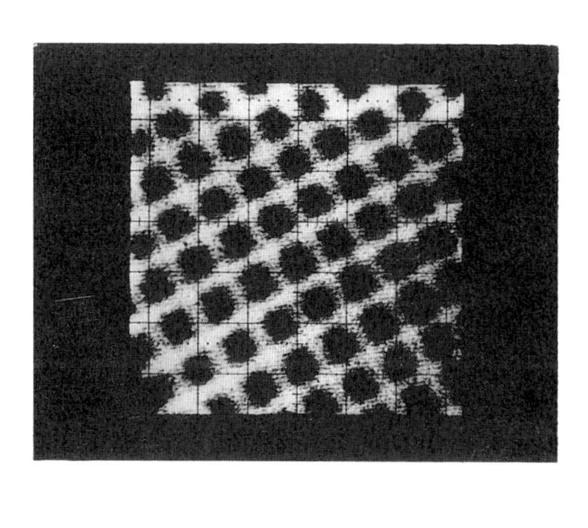

FIG 5.9. Secondary ion images of a copper mesh used to check the primary beam focus prior to the depth profile. The mesh in these images consisted of 20 μm bars spaced 50 μm apart and this image suggested a beam width of 50 μm or less. This was subsequently confirmed by stylus measurements on a pit in the sample produced by the static beam.

and thus a flat-bottomed crater. The real flux at the surface was $\phi = 7.8 \times 10^{14}$ molecular ions $cm^{-2}\,s^{-1}$.

Electronic gating was used to restrict the area of the crater from which the secondary ions were being collected. This was necessary because ions were being emitted from all points on the crater including its walls, with the contribution from the edge of the crater greatest when the primary beam was near the end of its traverse in x or y. For accurate concentration–depth information this crater edge signal had to be eliminated. In this instance, the counting system was only enabled when the centre of deflection of the beam was in a central area 125 μm × 125 μm. Note, however, that when the beam was at the edge of the gate one-half of the current fell outside it and ions were being collected from these points. For this reason, beam shape is of critical importance in an electronically gated system. If the beam is extended or has wings it is possible to collect secondary ions from the crater walls or even the original sample surface when the gate is open, thus reducing the depth resolution and dynamic range. Beam shape is less critical in optically gated instruments, where it is impossible for ions outside the gate to describe the trajectories necessary to enter the secondary ion column.

Having achieved conditions of primary beam current, energy, focus, and crater size suitable for the analysis, one is left with the choice of the mass channels and the frame times to be associated with them. Boron has two isotopes at masses 10 (20 per cent) and 11 (80 per cent) and since the dopant was introduced by thermal evaporation the isotopes should be present in the sample in this ratio. This is not the case when ion-implantation is used. Since there is no mass interference at mass 11 (e.g. from the silicon matrix and/or residual gas species such as hydrogen, carbon, nitrogen, or oxygen) the more abundant isotope was selected as mass channel 1. The secondary ion column energy filter[11] was then tuned to ensure maximum transmission for this ion, i.e. such that the band pass fell over the peak of the secondary ion energy distribution (see Chapter 4). Tuning was achieved by setting the target bias to 10 V, sputtering into a piece of the standard and adjusting the energy filter for maximum transmission.

It is important during the depth profile to monitor any changes in instrumental sensitivity. These may occur for a variety of reasons (variations in beam current, surface charging, electronic drift) and can be detected by monitoring a matrix species such as $^{30}Si^{+}$ throughout the profile. The lowest abundance silicon isotope is selected to avoid saturating the detector with silicon ions and the intensity is reduced still further by applying a sample bias offset to this mass channel (e.g. − 100 V), thus sampling the high-energy tail of the energy distribution. One disadvantage of this approach is that the matrix line becomes very insensitive to charging effects, i.e. the fractional change in signal intensity per volt of sample charging is low. One alternative, used here, was to run two Si_2O^{+} mass channels. The molecular ion has a narrow symmetric energy spectrum (see Fig. 5.4), and by having one mass channel tuned to the high-energy side of this distribution and the other tuned to the low-energy side one can distinguish between a drift in beam current and surface charging.[12] If the current varies, the matrix channels will rise or fall concurrently, whereas if charging occurs one channel will rise and the other fall. Since antimony had been evaporated as an n-type dopant in previous growth runs, it was also of interest to include an antimony isotope in the profile to investigate memory effects. A frame time of 10 s was selected for all four mass channels. Conditions used are given in Table 5.3 and schematically in Fig. 5.10.

The ion-implanted standard ($^{11}BF_2 \rightarrow Si$ implanted at a dose of 10^{15} cm^{-2} and an energy of 40 keV into silicon) and the MBE sample were depth profiled under the same conditions. Variations in the primary beam current were <5 per cent throughout. The raw data, counts per frame $N(F)$ vs. frame number F, were saved on disc for subsequent analysis. The original data are shown in Fig. 5.11. The matrix channels are stable and the antimony is count-rate limited. Thus the concentration of antimony is at or below the instrumental detection limit (10^{16} atoms cm^{-3}).

5.4.2. *Quantification of the data*

The data from the ion-implanted standard can be used to find the useful ion yield and thus the instrumental sensitivity for boron-in-silicon in the MBE sample. The standard had been implanted with a fluence ϕ of 10^{15} ^{11}B ions cm^{-2} and then depth profiled with a gate area $a = 1.56 \times 10^{-4}$ (cm^2). Thus, provided no segregation occurred, $\phi \cdot a = 1.56 \times 10^{11}$ ^{11}B atoms were sputtered during the depth profile, whilst the counting system was enabled. However, the mass spectrometer was only tuned to the mass 11 for

TABLE 5.3

Instrumental set-up conditions

Primary ions
0.2 μA of 4 keV $^{16}O_2^+$ ions incident normal to sample surface. Crater size 400 μm $\times$ 400 μm. Electronic gate 125 μm $\times$ 125 μm. Beam shape Gaussian FWHM 50 μm.

Secondary ions (+)

Channel	Mass (m/z)	Species	Target potential (V)	Frame time (s)
1	11	B	10	10
2	121	Sb	10	10
3	72	Si_2O	5	10
4	72	Si_2O	17	10

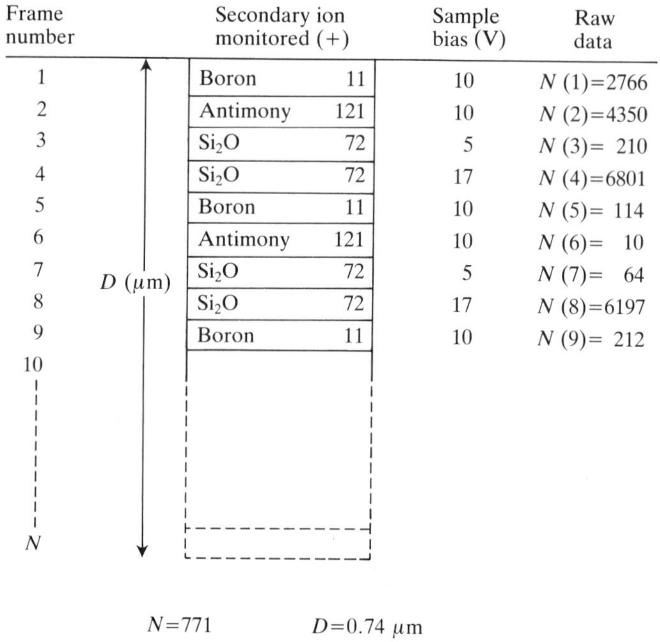

Frame number	Secondary ion monitored (+)		Sample bias (V)	Raw data
1	Boron	11	10	N (1)=2766
2	Antimony	121	10	N (2)=4350
3	Si_2O	72	5	N (3)= 210
4	Si_2O	72	17	N (4)=6801
5	Boron	11	10	N (5)= 114
6	Antimony	121	10	N (6)= 10
7	Si_2O	72	5	N (7)= 64
8	Si_2O	72	17	N (8)=6197
9	Boron	11	10	N (9)= 212
10				
N				

D (μm)

N=771 D=0.74 μm

FIG 5.10. Schematic representation of the depth profile showing how successive mass channels and sample biases are selected as the depth profile proceeds.

one-quarter of the analysis (FT/CT) and only 3.9 × 10^{10} boron atoms were sputtered during this period. The total number of boron ions detected in mass channel 1 during this period, $IN(^{11}B^+)$, was 498 598 ions, giving a useful ion yield for boron in silicon, for these experimental conditions, of 1.27 × 10^{-5}, or more generally,

$$\tau_u(^{11}B^+) = IN(^{11}B^+)/(\phi \cdot a \cdot FT/CT). \tag{5.4.1}$$

We now assume that this ion yield, derived from the standard, is applicable to the MBE sample. Rewriting eqn (5.3.11), substituting $Ns(^{11}B^+)$. $FT=1$, and writing $\rho=\rho^*$ gives

$$\rho^*(^{11}B) = 1/\dot{z} \cdot a \cdot \tau_u(^{11}B^+) \cdot FT. \tag{5.4.2}$$

ρ^* is the concentration of boron in the MBE test sample equivalent to 1 count of a ($^{11}B^+$) ion per frame, $\dot{z}$ is the sputter rate of the MBE sample, and a the gated area. Substituting the experimental values for the sputter rate $\dot{z}$, the gated area a, and the frame time FT into eqn (5.4.2), together with the value for the useful ion yield τ_u measured above, gives a ρ^* value of 5.23 × 10^{15} atoms cm^{-3} per boron ion per frame (crater depth 0.74 μm, number of frames 771, frame time 10 s => $\dot{z}$ = 9.59 × 10^{-9} cm s^{-1}: gated area 125 μm × 125 μm → gated area a = 1.56 × 10^{-4} cm^2). The boron ion counts per frame for the MBE sample can now be converted into concentrations of ^{11}B

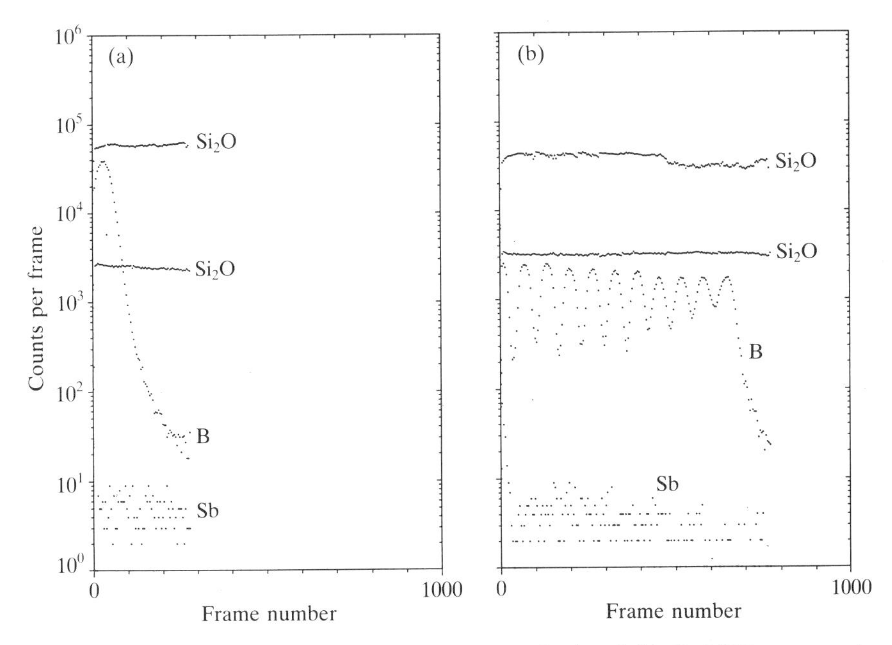

FIG 5.11. The original data for (a) the boron standard and (b) the MBE test sample (counts per frame vs. frame number). (a) 10^{15}cm^{-2}, 40keV^{11}BF$_2$→Si; primary ions: 4 keV $^{16}O_2^+$, 0.2 μA; total frames: 279: (b) Primary ions: 4 keV $^{16}O_2$, 0.2 μA; total frames: 771.

simply by multiplying by ρ^* (^{11}B) and the total boron concentration then found by correcting for the isotopic abundance, i.e. by multiplying the concentrations by 1.25. The quantified data are plotted in Fig. 5.12. The dynamic range is poorer than might have been expected and appears to decrease with depth. The typical causes of this are discussed in Section 5.

The accuracy of the quantification (ρ^*) can be seen from eqns (5.4.1) and (5.4.2) to depend upon the accuracy of the ion-implantation dosimetry ϕ, the crater depth measurement, and the value of the integral secondary ion count-rate IN. Dosimetry is often suspect, especially if the available current in the implanter is low and thus difficult to measure, or if there is a mass interference. Meuris *et al.*,[13] for example, reported dosimetry errors when implanting silicon isotopes into gallium arsenide of up to 40 per cent, due to the co-implantation of $^{14}N_2^+$ with $^{28}Si^+$ and $^{14}N^{15}N^+$, $^{10}B^{19}F^+$, and $^{11}B^{18}O^+$ with $^{29}Si^+$. Clearly it is useful to run two or more standards prepared on different implanters and to check the dosimetry by chemical analyses. Crater depths are usually determined by surface profiling (e.g. Dektak or Talystep) or by interferometry. Generally, both techniques can be considered accurate to 5 per cent. Interferometry cannot be used for very shallow craters (<500 nm) whereas the Talystep can resolve craters only a few nanometers deep. The base of a SIMS crater is rarely flat and an unevenness of 1 per cent is

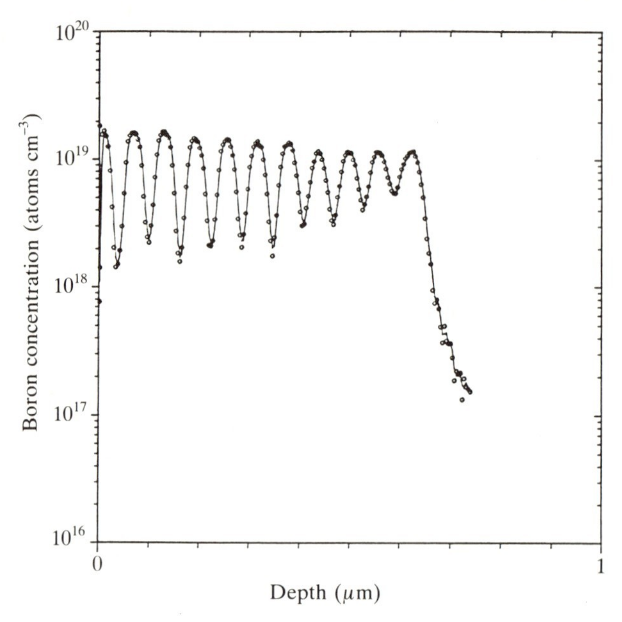

FIG 5.12. The quantified data for the MBE test sample, showing the total concentration of boron (all isotopes) as a function of depth.

considered good. Finally, care must be exercised when integrating the secondary ion count-rate for the standard and then applying the calibration constant to the sample. The steps described assume that each ion detected at mass 11 is a $^{11}B^{+}$ originally present in the solid as a ^{11}B atom. However, the observed signal may also include contributions from mass-interferences, crater edge effects, ions produced by electron stimulated desorption, and electronic noise. In the case of boron in silicon, for example, a very high signal level is observed in the first few frames due, it is believed, to surface contamination. Fortunately, the contamination signal contains both boron isotopes in the appropriate isotopic ratio (80 per cent: 20 per cent) and can be removed from the data by monitoring a ^{10}B channel, i.e.

$$^{11}\text{B (true)} = {}^{11}\text{B (observed)} - 4 \times {}^{10}\text{B (observed)}. \qquad (5.4.3)$$

5.5. Further examples of SIMS depth profiles

5.5.1. *Stationary beam analysis—a useful diagnostic technique*

SIMS depth profiling is an expensive technique and it is important, therefore, not to waste analysis time. Often the sample may be different from the specification; for example, it may have a masking layer on the surface that was meant to have been removed prior to SIMS analysis, or very occasionally, the sample may have been loaded upside down. One useful diagnostic technique, therefore, is to drill a hole into the material with a stationary beam (i.e. with the raster scanner disabled) and to obtain images for the dopant and matrix channels of interest. The gallium and aluminium images in Fig. 5.13 were taken on an ion microscope, drilling with a stationary caesium beam. The images show the various layers in the multi-layer sample, as expected, and also indicate that there are particulates between the layers. This technique also generates a record of the primary beam shape since the shape of the hole is determined by the primary beam current density.

5.5.2. *Mass interferences in SIMS depth profiling*

Generally the secondary ion mass spectrometer (whether it is a quadrupole or a magnetic sector) will be run in a low mass resolution mode in order to maximize the secondary ion transmission, and under these conditions all ions with the same nominal mass will be transmitted. Since we are only interested in those ions that were originally present within the solid as dopant–impurity atoms, the remainder of the signal represents mass interference. Mass interferences may arise due to:

1. *Contaminants*, either ions or neutrals, transported down the primary beam line, generated from apertures in the sample region, or produced in the secondary ion column. The contaminants may be generated in the primary ion source or by sputtering, either from the apertures and lenses in the primary beam line (by the primary ions), or from the secondary ion extraction plate (by energetic secondary ions) and other components of the secondary ion column. The interferences will be characteristic of the materials used to fabricate the ion source, the beam line, and the secondary ion extraction optics (e.g. Al, Fe, Ta).
2. *Species in the residual vacuum*. Since a vacuum of 10^{-6} Torr will lead to an arrival rate at the crater surface of approximately one monolayer per second, ultra-high vacuum is mandatory when analysing for C, H, N, and O. High sputter rates and small area analysis are also helpful.
3. *Memory effects from previous analyses*. For example, the analysis for silicon in gallium arsenide the day after analysing silicon matrices may reveal high backgrounds unless components such as the extraction plate are cleaned or replaced. Careful planning of the experimental programme may be required.
4. *Matrix interferences*, e.g. $^{56}Fe^{+}$ is difficult to analyse in silicon due to the interference $^{28}Si_2^{+}$.
5. *Mixed interferences*. For example, hydrogen in the residual vacuum with a silicon matrix can form the secondary ion $^{30}SiH^{+}$ which, makes phosphorus analysis difficult.

Mass interferences lead to a high background count-rate in the mass channel and thus to poor detection limits and a loss of dynamic range. They may often be overcome by eliminating the mechanism producing them (e.g. by improving the residual vacuum), or by separating the interference from the

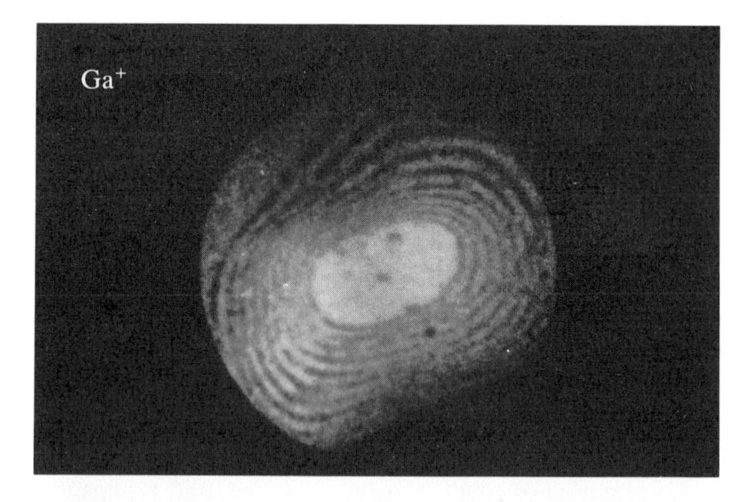

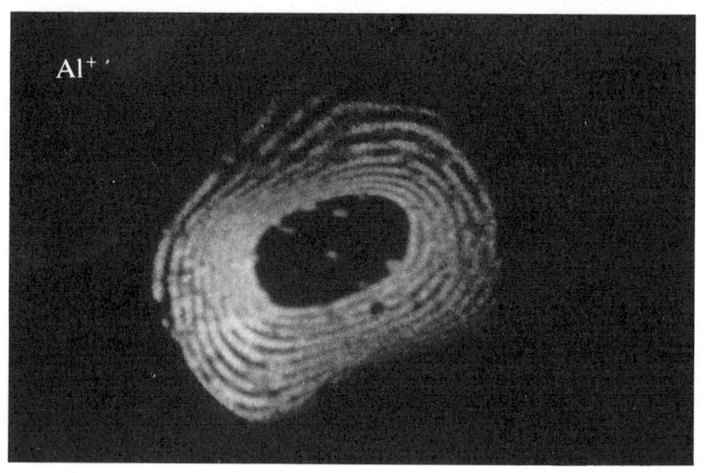

FIG 5.13. Secondary ion images of a GaAs–AlAs multi-layer sample obtained on an ion-microscope by switching the raster scanner off and drilling a hole into the material. The entire process required 1 min. This is a useful method of checking the expected structure and, indeed, of checking that the sample has been loaded the correct way round, and as such can save hours of wasted analysis time (which is expensive).

ion of interest, either on the basis of a slight difference in mass or on the basis of a difference in secondary ion energy spectrum. Leta and Morrison have compiled tables of sensitivities and detection limits for a large number of dopant–matrix combinations.[2]

Separation on the basis of differences in energy spectra is illustrated in Fig. 5.14, where we see a depth profile of an arsenic implant in silicon, analysed with an oxygen primary ion beam. The most intense mass 75 channel, i.e. the one tuned for optimum transmission (sample bias +10 V),

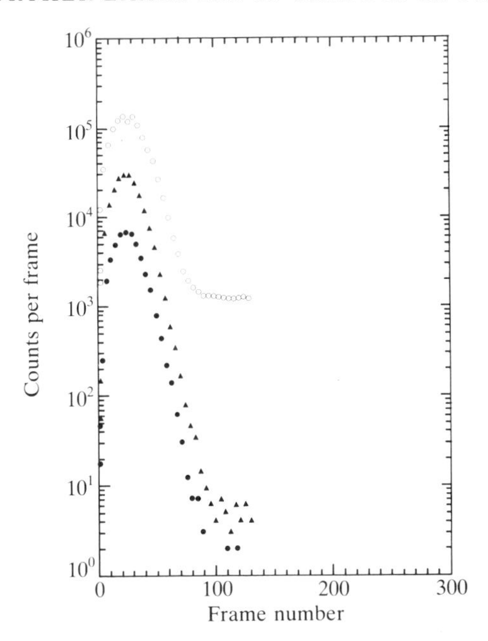

FIG 5.14. Raw data for a multichannel depth profile of an arsenic implant into silicon (10^{16}/40), showing how the dynamic range in the mass 75 (arsenic) channels improves with sample bias offset due to the exclusion of the mass interference $^{29}Si^{30}Si^{16}O^+$. Primary ions: 4 keV $^{16}O_2^+$, 0.3 μA; total frames: 131; run time 1320 s. V_0: ○, + 10 V; ▲, − 20 V; •, − 40 V.

peaks at 10^5 counts per frame, corresponding to a concentration of 10^{21} atoms cm^{-3} and a ρ^* value of 10^{16} atoms cm^{-3}. Unfortunately, the signal levels out at 2000 counts per frame, corresponding to an arsenic concentration of 2×10^{19} atoms cm^{-3}. This background signal is in fact the molecular ion $^{29}Si^{30}Si^{16}O^+$ formed by association of the matrix and the primary ion beam. Some method of eliminating this mass interference is required and those analysts with access to a high mass resolution instrument could resolve the two on the basis of their slight difference in mass, at the expense of some loss in signal intensity (a mass resolution of 3250 is required).[9] An alternative, however, is to separate the ions on the basis of their different energy spectra, for molecular ions tend to have narrow symmetric energy distributions whereas monatomic ions exhibit a high-energy tail (see Fig. 5.4). The two other mass 75 channels in the profile correspond to sample biases of − 20 and − 40 V respectively. As the sample bias is increased, the relative contribution of the Si_2O^+ ion is reduced. The − 20 V offset leads to a dynamic range of four decades corresponding to a detection limit of 10^{17} atoms cm^{-3}, at the expense of some loss of secondary ion signal. Further sample bias offset (− 40 V) reduces the count-rate at the peak of the dopant distribution to $< 10^4$. Thus the optimum sample bias for the analysis, i.e. the one that produces the maximum dynamic range, is − 20 V. It should be noted, however, that the technique fails to separate the mass interference ^{30}SiH from the analytical ion ^{31}P as their energy spectra are similar. This problem of separation can be overcome by using a magnetic sector instrument with a resolution capability of $M/\Delta M > 4000$, or by using MS–MS, as discussed in Chapters 4 and 7.

5.5.3. The depth resolution in SIMS depth profiling

The depth resolution is the precision with which we can state the original depth beneath the surface of a sputtered atom or ion. It can be measured in a number of ways, for example, as the width of a broadened marker layer or the sharpness of an interface, as shown in Fig. 5.2.

Recently,[14] assessment of the optimum experimental conditions for high depth resolution SIMS depth profiling was performed by studying a boron-in-silicon modulation doping structure grown by silicon MBE. The test structure was a 1.6 μm thick silicon epilayer and contained 31 boron-rich layers spaced 50 nm apart. The concentration of boron in the layers was intended to be 10^{20} atoms cm^{-3}. The material was depth profiled using five different instruments (two Cameca IMS3Fs, one Cameca 4F, one Atomika, and EVA 2000). The analysts were asked to use a low primary beam energy to minimize beam-induced mixing effects and to choose conditions such that the loss of depth resolution with depth was minimized.

The results of seven depth profiles are shown in Fig. 5.15. Also plotted are the interface widths for the up-slope and down-slope of each peak (leading edge LE = 16 – 84 per cent of peak height, trailing edge TE = 84 – 16 per cent of peak height) and half-widths of the peaks HW (FWHM) as a function of depth. The near-surface peaks appeared asymmetric (TE>LE) in all the profiles, due, it is believed, to atomic mixing by the primary ion beam. In two experiments (A1 and B) these interface widths remained constant throughout the profile, i.e. there was no loss of depth resolution with depth, proving that all 31 peaks were of similar width. However, peak broadening was observed in the remaining five experiments as the analyses proceeded, sometimes affecting the LE and sometimes the TE the most. The degradation in depth resolution, measured as the peak half-width (FWHM = dz), was proportional to the sputtered depth (z) and thus the depth resolution could be written:

$$\mathrm{d}z = M + Uz. \tag{5.5.1}$$

The U term reflected an instrumental problem and chemical images taken during experiment D1 revealed an uneven breakthrough pattern to a boron rich layer due to unevenness in the bottom of the crater (see Fig. 5.16(a)). Talystep measurements confirmed the suspected topography of the crater base (see Fig. 5.16(b)). A significant unevenness (>0.1 per cent) was also

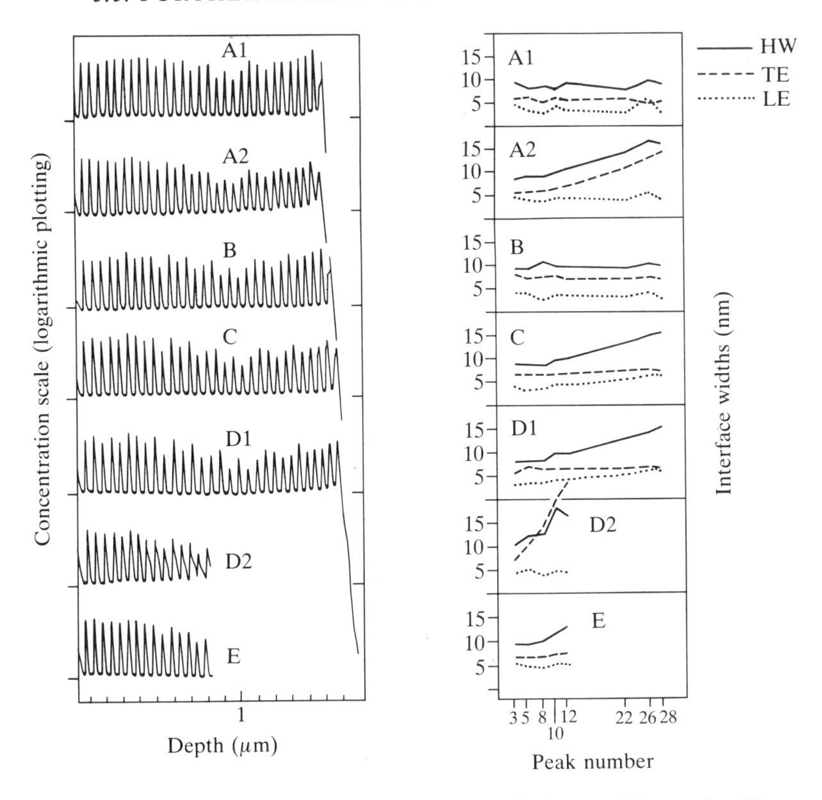

FIG 5.15. The quantified results of seven analyses of a 31 – peak boron-in-silicon test structure together with the measured peak interface widths as a function of depth. A1 and A2 from a Cameca IMS 3F using two different optically gated areas (8 and 60 μm) respectively; B from a Cameca IMS 4F using a 62 μm optical gate; C from a Cameca IMS 3F using a 35 μm optical gate; D1 and D2 from EVA 2000 using a 125 μm electronic gate and E from an Atomika DIDA ion microprobe using a 160 μm gate. Reproduced with permission from Ref. 14

found in the craters from experiments (A2, C, D2 and E) whereas the bases of craters A1 and B were flat (within the gated area) to better that 0.1 per cent. In all cases, the unevenness within the gated area of the crater was similar to the loss of depth resolution with depth. The unevenness is believed to arise from variations in primary beam current across the gated area of the crater due to faults in the primary beam column such as non-linear output from the raster scan generator and misaligned scan plates.

It was of interest to model the effects of uneven etching (crater macro-topography) on a dopant distribution to explain how uneven etching could in some cases broaden the LE and in other cases the TE of a dopant peak. The numerical model developed[15] simulates the effect of passing an uneven crater

5.16(a)

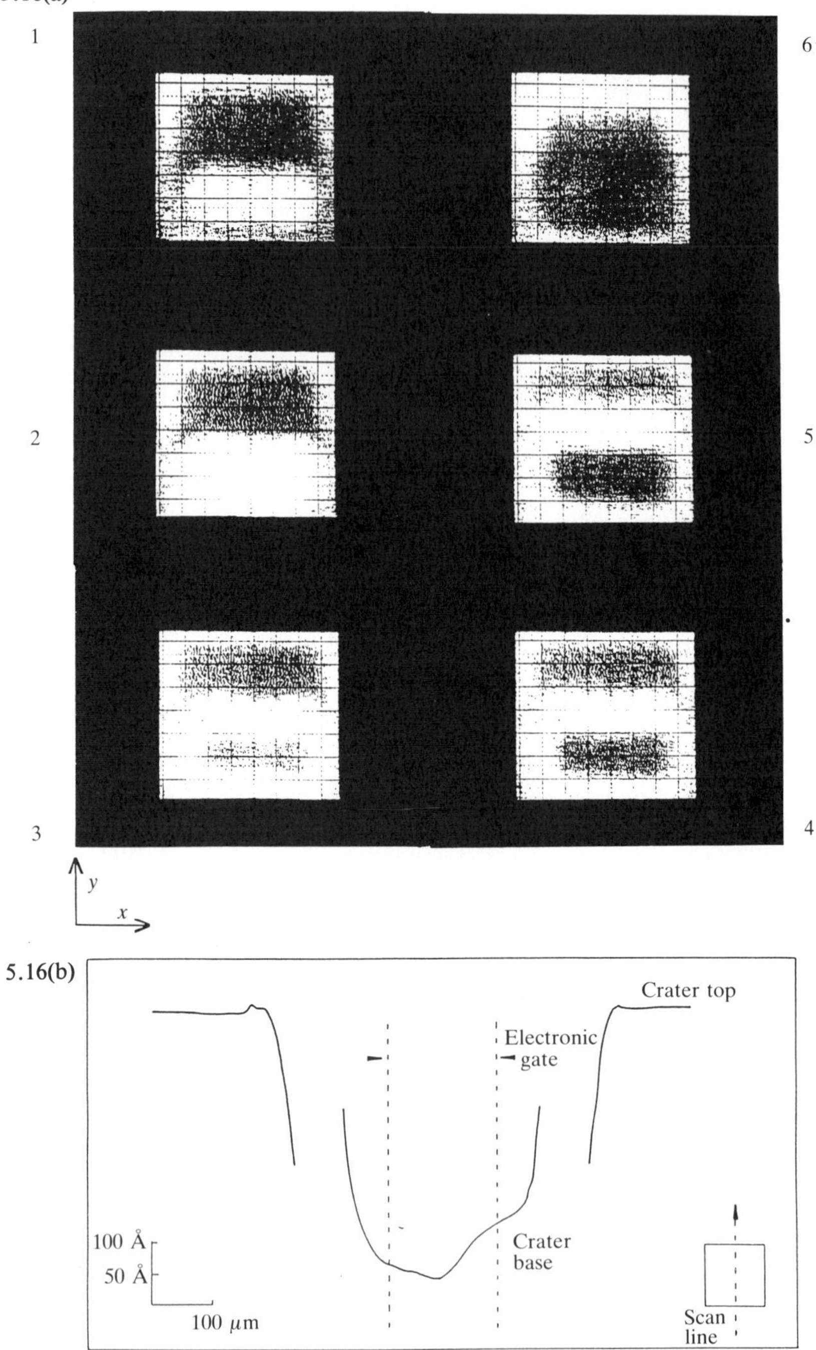
1
6
2
5
3
4
y
x
5.16(b)
Crater top
Electronic gate
100 Å
50 Å
100 μm
Crater base
Scan line

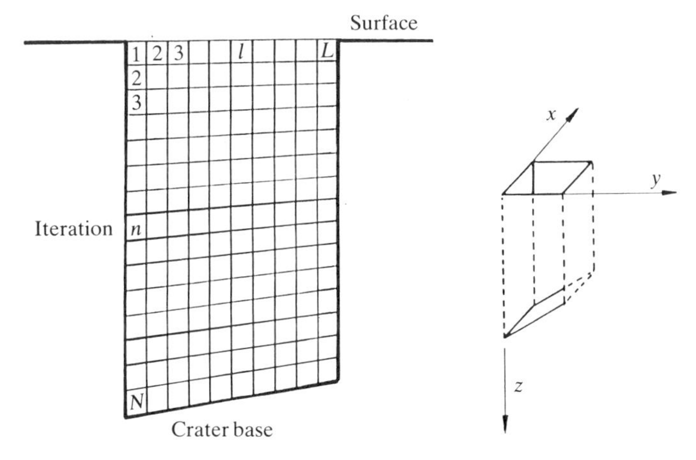

FIG 5.17. Numerical model developed to simulate the effects of uneven etching. The special case of unevenness in one direction only (y) is shown. In general, the surface (x,y) is divided up into a grid and the sputter rate of each column is determined by an unevenness function $f(x,y)$ deduced from crater depth measurements. The simulation profile proceeds in a series of equal-depth iterations (1. . .n. . . .N) with the SIMS signal in the nth iteration as the sum of the contributions from the nth element in every column. Profile broadening arises becauses the elements n are at different depths in different columns and because the volume elements are of slightly different size. Reproduced with permission from Ref. 14

through a known laterally homogeneous dopant distribution $p(z)$ (see Fig. 5.17). The crater unevenness function $f(x,y)$ is found by Talystep or Dektak measurements with the simulation producing the number of secondary ion counts $C(n)$ per frame n according to the formula:

$$C(n) = \alpha T \Sigma_x \cdot \Sigma_y (D/N) f(x,y)\, p(z)\, \mathrm{d}x\mathrm{d}y \qquad (5.5.2)$$

where D is the total crater depth and N the total number of iterations. The modelling of experiment D2, for example, correctly predicted that in that instance most of the broadening would be in the TE of successive peaks (see Fig. 5.18). This is an important result as depth resolutions of SIMS instruments are often determined by measuring the width of a single interface. If the resolution of the instrument in experiment D2 had been measured on the LE of a feature it would have been concluded that the instrument was

---

FIG 5.16. (a) Six successive boron secondary ion images showing the non-uniform breakthrough pattern to a boron-rich layer. This suggested that the crater base was uneven and (b) a crater contour map built up from Talystep scans confirmed the suspected topography.

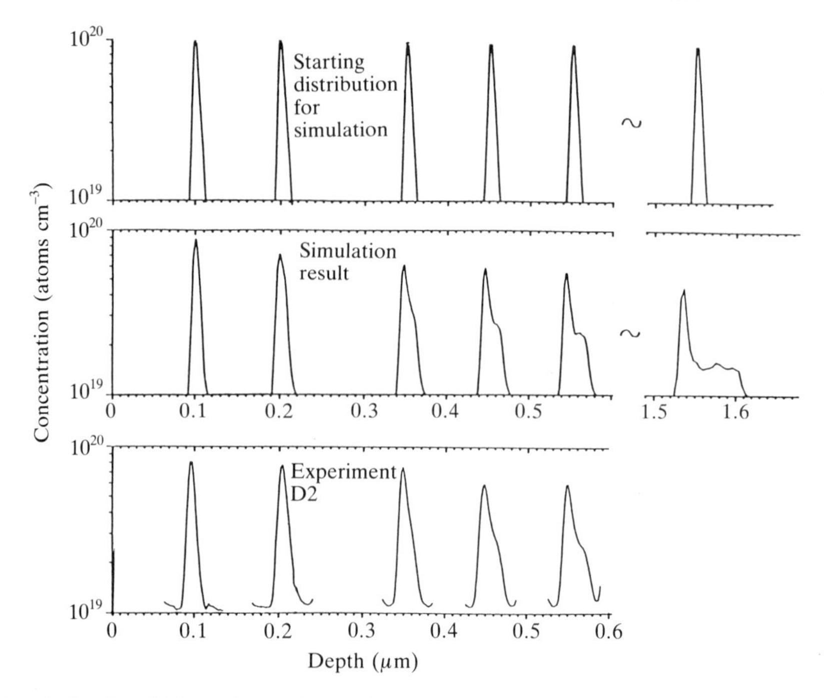

FIG 5.18. Modelling of experiment D2, using an unevenness function derived from the crater shape, successfully predicts that most of the broadening is in the trailing edge. Reproduced with permission from Ref. 14

working well, whereas measurement on the TE of a feature would reveal the gross problem in fact present.

The model was also tested on two of the types of doping distribution often encountered in depth profiling, periodic doping distributions, and a Gaussian distribution. The results of 1 and 10 per cent unevenness on a sinusoidal distribution are shown in Fig. 5.19. In both cases, the distributions are bounded by symmetric amplitude envelopes (they appear distorted due to the logarithmic plotting). In the case of 1 per cent unevenness the envelopes meet after 100 peaks (not shown) and in the case of 10 per cent unevenness they meet after 10 peaks, and then exhibit a series of beats. These results are independent of the absolute spacing of the layers but for thin layers the effects of atomic mixing must also be considered.

The effect of 10 per cent unevenness on a Gaussian distribution, an approximation to the distribution produced by ion-implantation, is very small (see Fig. 5.20). Five decades down on the TE the distribution is only broadened by 3 per cent. Thus ion-implants can be accurately profiled using

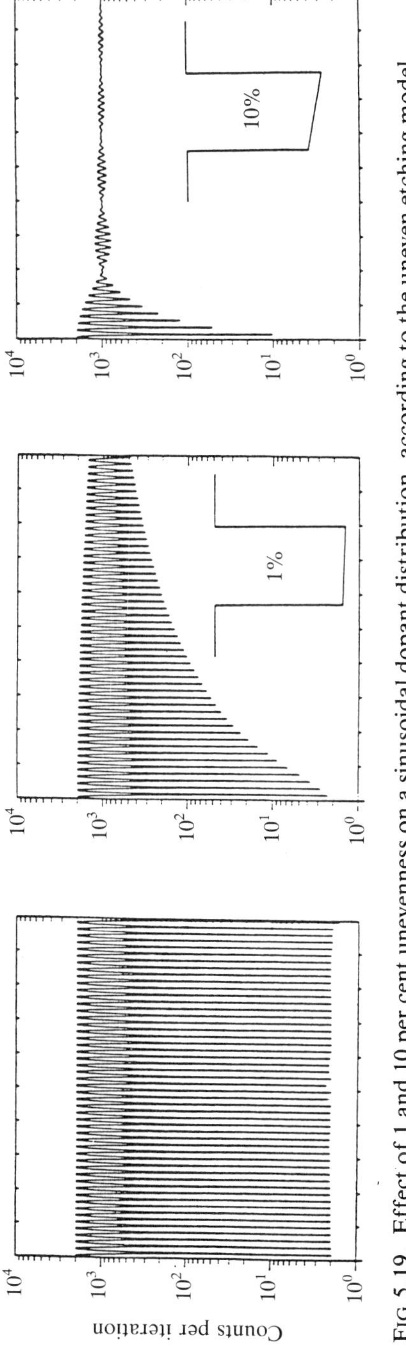

FIG 5.19. Effect of 1 and 10 per cent unevenness on a sinusoidal dopant distribution, according to the uneven etching model.

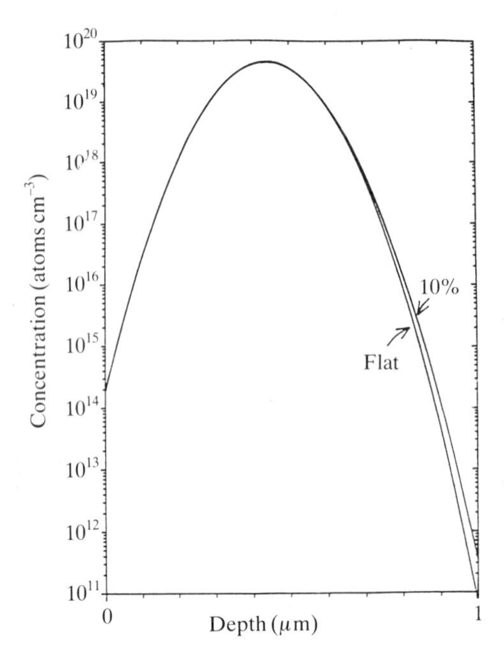

FIG 5.20. Effect of 10 per cent unevenness on a Gaussian distribution, according to the uneven etching model.

an instrument that produces rather poor craters. The explanation is that for every point above the average crater depth, where the concentration is higher than average, there is one below where the concentration is lower than average. Ion implants are not, therefore, suitable dopant distributions for assessing the depth resolution of a SIMS instrument.

The M term in eqn (5.5.1) above represents the true shape of the peaks together with any broadening due to SIMS processes such as atomic mixing that are independent of the depth of the analysis. Comparison of the peak shapes in experiments A1 and B above revealed that the average peak widths in experiment B were slightly larger than in A1; i.e. for the TE interface widths they were 7.1 ± (0.3) nm for B and 6.3 ± (0.4) nm for A1 respectively. This reflects the slightly different primary beam energy used in experiment A1 (4 keV as opposed to 5.5 keV) and suggests that by measuring the peak widths at a series of energies one may be able to extrapolate the true width at zero primary beam energy. The results of such experiments, conducted on the Cameca IMS3F used for experiments A1 and A2, are shown in Fig. 5.21, where the interface widths and decay lengths are plotted as a function of primary beam energy. As the energy is reduced, the peaks become sharper

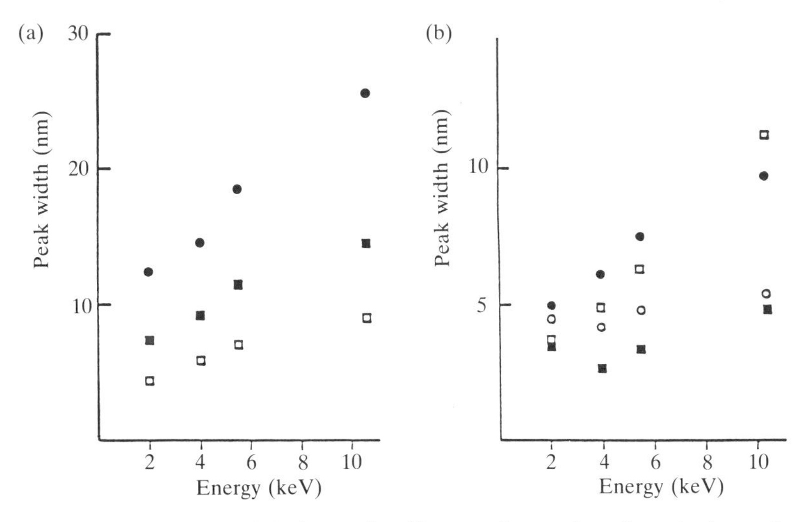

FIG 5.21. Decay lengths of the boron-in-silicon peaks as a function of primary beam energy, measured on a Cameca IMS 3F. Note that as the beam energy is decreased the angle of incidence increases (with respect to the sample normal) in this instrument. (a) Full peak width at three-quarters height, ●; at half-height, ■; at quarter height, □. (b) ●, LE; ○, TE; □, *LD*; ■, *LV* (see over for definition). Reproduced with permission from Ref. 14.

and appear to become symmetric. The up-slope and down-slope are approximately linear when plotted logarithmically, with true decay lengths of <4 nm per decade.

Thus the resolution of the peaks in these analyses were limited by beam-induced mixing effects, which produced an asymmetric broadening (M) independent of the depth of the peak, and uneven etching (U) which became worse as the profile proceeded. Mixing effects can be minimized by sputtering at very low primary beam energies and uneven etching by using a small gated area, i.e. by use of optical gating. Unfortunately, the Cameca, which incorporates optical gating together with high secondary ion transmission, cannot be operated at primary beam energies much below 1 keV per atomic ion. This is because the high extraction field forces the incoming low-energy primary ions to describe parabolic trajectories, thus moving their point of impact on the sample plane out of the field of view of the secondary ion column. One possible solution would be to use a scanned neutral primary beam in the Cameca[16] as experiments could then be conducted at low primary beam energies in a system with optical gating and high secondary ion tranmission.

In the seven analyses illustrated in Fig. 5.15 the peak dynamic range (the ratio of peak height p_{max} to inter-peak level p_{min}) was one decade or less. It is important to determine in a situation such as this, whether this was a growth-related problem or a problem of the analysis. In general, and in the absence of any uneven etching to complicate the issue, the dynamic range in a SIMS depth profile $D = (p_{max}/p_{min})$ between two doped layers can be estimated as shown in Fig. 5.22. The mixing introduced by the SIMS analysis is assumed to broaden atomically abrupt dopant interfaces into distributions that vary exponentially with depth, i.e. such that the up-slope and down-slope are linear on a logarithmic concentration plot with slopes (decay lengths) of LU and LD (nm per decade) respectively. Then, as a first approximation:

$$D = 0.5 \times 10^{(S'/LU+LD)} \tag{5.5.3}$$

where $S' = (S - w')$ is the distance between the TE of one feature and the LE of the next. Often LU and LD can be determined by profiling through an interfacial contamination 'spike' or a delta doping layer. In experiment A1, LU was 5.64 nm per decade and LD was 10.9 nm per decade; the layers were 50 nm apart and had a width of 3 nm, giving $S' = 47$ nm and a predicted dynamic range of 350. It would seem that the poor dynamic range actually

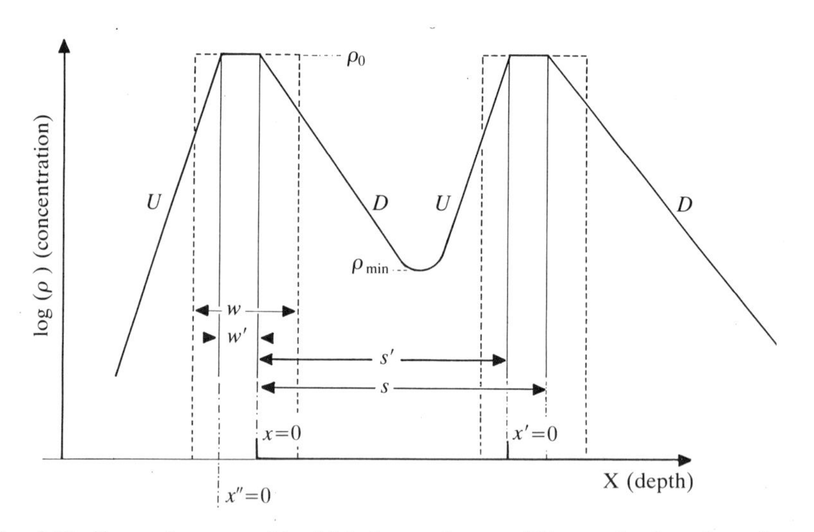

FIG 5.22. Dynamic range with which dopant layers will be resolved can be estimated by assuming that mixing produces an up-slope of the form $\rho_{\phi} = \rho_0 10(x''/LD)$ and a down-slope $\rho = \rho_0 10(-x/LD)$ where ρ_0 is the true peak concentration, and LU and LD are the decay lengths per decade. ρ_{min} is the lowest inter-peak concentration observed in the depth profile. Dashed line, true profile; solid line, SIMS.

observed (less than one decade) was a problem of growth control rather than SIMS analysis.

Atomic mixing of very thin (nm) layers will spread the dopant out over depths much larger than the feature width w. Often the apparent dopant distribution will consist of an exponentially rising LE intersecting an exponentially decaying TE (the TE is less steep than the LE). Although the integral amount of dopant under the concentration–depth curve will remain constant, the peak concentration ρ_{app} may be far less than the true layer concentration ρ_0. The magnitude of the error for boron in silicon can now be estimated (see Fig. 5.23(a)). The experiments discussed above showed that the mixing produced a broadened distribution with an exponential up-slope and down-slope characterized by decay lengths LU and LD (nm per decade). Then equating the areas under the curves:

$$\rho_{app}/\rho_0 = w \ln 10/(LU + LD) \qquad (5.5.4)$$

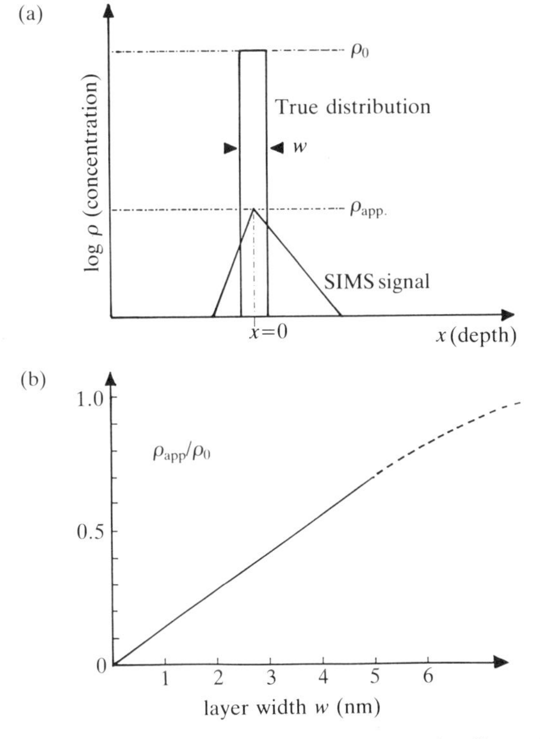

FIG 5.23. Apparent peak concentration ρ_{app} of a boron-in-silicon doped layer as a function of layer width for very thin layers. The decay lengths used in the formula eqn (5.5.4) were measured in instrument A.

Using the values from experiment A1 above ($LU = 5.64$ nm per decade and $LD = 10.9$ nm per decade) yields the curve shown in Fig. 5.23(b). The apparent peak concentration in a layer 1 nm thick would be 14 per cent of its true value, with these analysis conditions.

5.5.4. Distinguishing between uneven etching during analysis and diffusion during growth

It is often difficult to know whether an observed feature in a depth profile is 'true' or merely an artefact of the analysis technique. For example, the profile shape obtained in Fig. 5.12, a series of peaks of decreasing dynamic range bounded by an amplitude envelope, could have arisen from uneven etching during analysis or dopant diffusion during growth, with the deepest peaks having diffused the most. The uneven etching model suggested that an unevenness of at least 6 per cent would be required to produce this effect, three times that observed in the final etch pit. It seemed likely therefore that diffusion was at play, and this was tested by use of the SUPREM III model[17] of dopant diffusion and by thermal cycling experiments. For thermal cycling the samples from the wafer were re-heated to the growth temperature (780 ± 20°C) for various periods of time and then SIMS depth profiled again. The resulting profiles (see Fig. 5.24) strongly suggest that the growth temperature was too high; this was confirmed by the SUPREM III model (see Fig. 5.25) and led to a modification of the growth conditions.

5.5.5. Charging effects—how to detect and overcome them

Unintentional changes in the sample potential during the course of a depth profile will modify the secondary ion extraction field, produce a change in instrumental transmission and lead to a profile distortion. Such changes will occur if there is an accumulation of charge at the sample surface due, for example, to the absence of a conduction path to earth. Charge compensation can be achieved by use of simultaneous electron beam 'flooding' of the sample; by providing a conduction path, e.g. by gold coating; or by continuously monitoring for any change in the sample potential and applying a d.c. potential to the sample-holder to correct for it. The problem of charging is most acute when analysing insulators or multi-layer structures such as silicon-on-sapphire.[18]

The maximum change in sample potential that can be tolerated is determined by the band pass of the secondary ion column. The fractional change in secondary ion count-rate per volt of charging (dI_s/dV) will be least if a large (200 eV) band pass is placed on the high-energy tail of the secondary ion energy distribution and greatest if a small band pass (10 eV) is placed close to the peak of the energy distribution. For example, if a 10 eV band pass

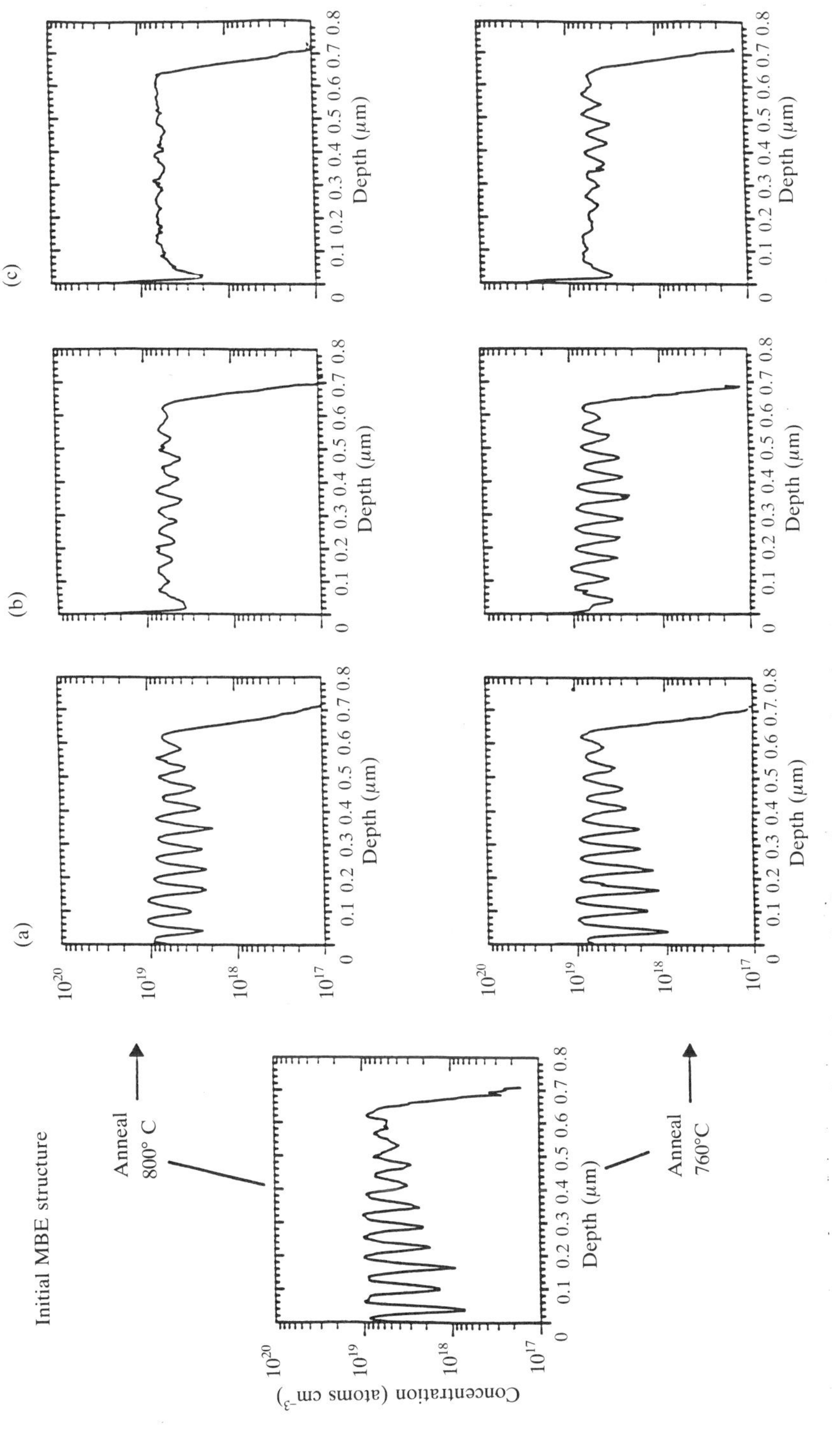

FIG 5.24. SIMS depth profiles of the samples from the MBE wafer after they had been thermally annealed under vacuum for various periods of time (5 min (a), 12 min (b), and 32 min (c)). Two anneal temperatures corresponding to the upper and lower bound on the MBE growth temperature were used.

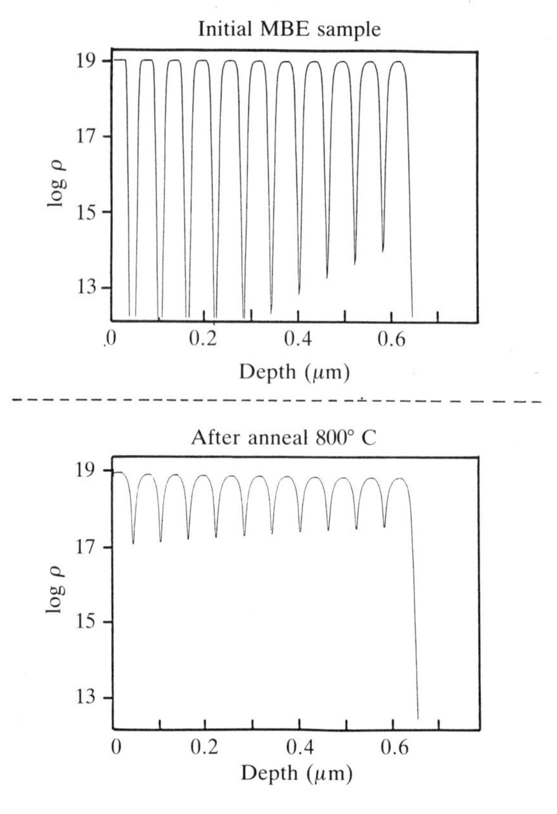

FIG 5.25. Modelling of the growth of the boron-in-silicon test structure using the SUPREM III code confirmed that significant diffusion of the dopant was likely during the MBE growth at the substrate temperature used.

is placed a few volts off the peak of the Si_2O^+ energy distribution, 1V of charging will change the signal by 30 per cent.

Charging effects can be particularly severe across interfaces. Figure 5.26 shows the results of profiling an annealed arsenic implant into a silicon-on-sapphire film with several arsenic mass channels tuned to different points on the high-energy side of the arsenic secondary ion energy spectrum. Bias offset was used to avoid the mass interference with the molecular matrix ion $^{29}Si^{30}Si^{16}O^+$ as this ion does not exhibit a high-energy tail. Low-energy electron beam flooding was used for charge compensation (see above). There was an apparent accumulation of arsenic at the interface but both the magnitude and position of that interfacial peak depended upon the sample bias. Furthermore, there was a rise in the matrix channels ($^{30}Si^+$), which were biased to reduce their intensity, as the interface was approached. This suggested that charging was taking place and images of a matrix ion (Si_2O^+)

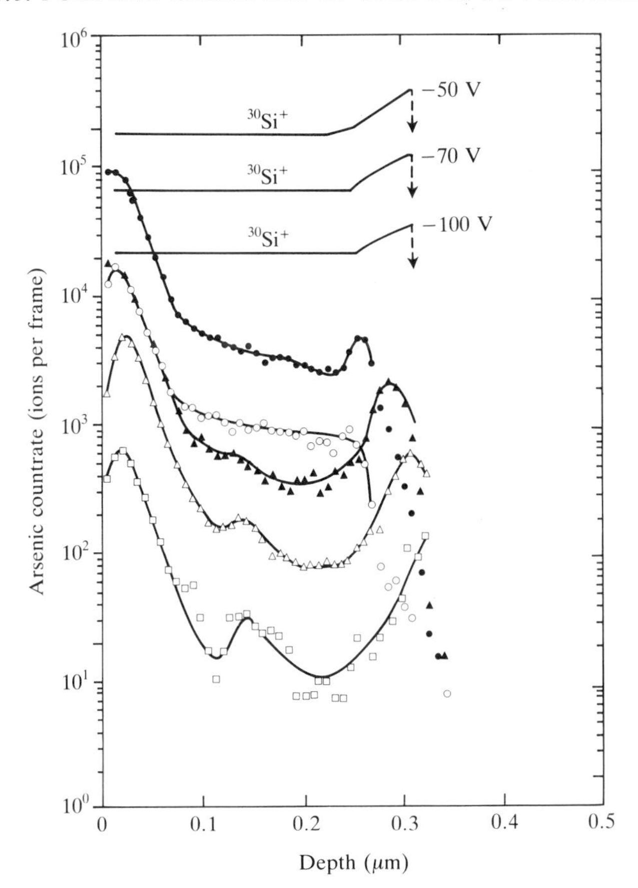

FIG 5.26. Counts per frame as a function of depth for a multi-channel depth profile of arsenic in silicon-on-sapphire. Five $^{75}As^+$ channels (○, 20 V; ●, 10 V; ▲, 0 V; △, −10 V; □, −40 V) were run tuned to different parts of the secondary ion energy distribution. The arsenic was introduced into the sample by ion-implantation followed by thermal annealing. The apparent accumulation of arsenic at the interface is in fact due to the ion $^{29}Si^{30}Si^{16}O^+$ being swept into the band pass of the instrument as charging occurs (there is no arsenic at this depth).

at six different sample biases confirmed this (see Fig. 5.27), proving that as the interface was approached the centre of the crater charged positive with respect to the sides due to the increased resistance between those points. The electron beam flooding was not adequately set up to compensate for this effect. The apparent accumulation or 'pile-up' of arsenic at the interface was in fact the result of Si_2O^+ ions at mass 75 being swept into the band pass of the instrument.

It is clearly important to provide adequate charge compensation. Another method of charge compensation is to coat the surface of the samples with a

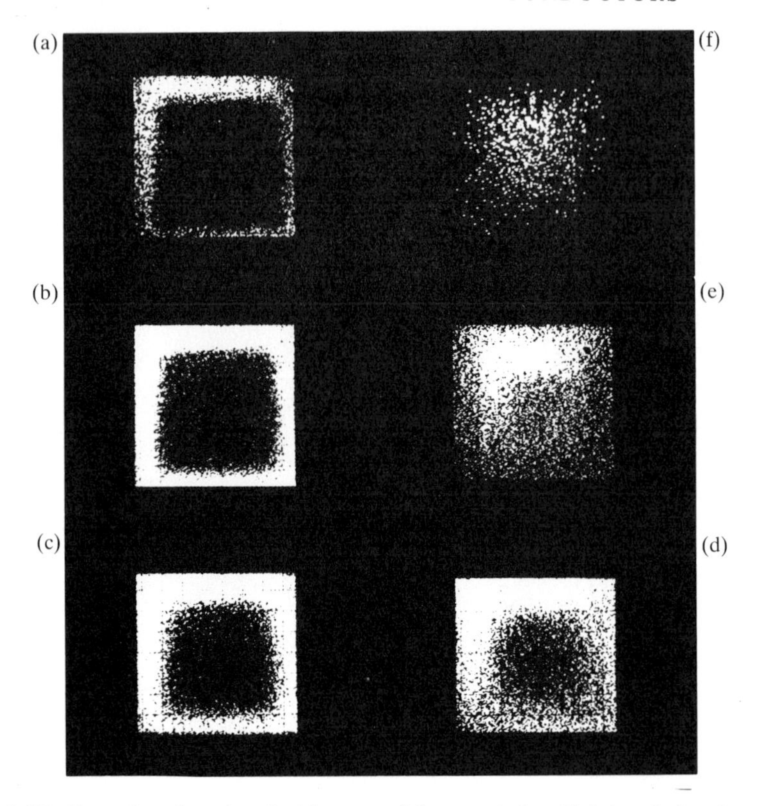

FIG 5.27. Secondary ion chemical images of the matrix ion Si_2O^+ emission from the base of the crater close to the sapphire substrate, using six different sample biases. (a) 20 V; (b) 15 V; (c) 10 V; (d) 5 V; (e) 0 V; (f) -10 V. At $+10$V there is no emission from the centre of the crater, and this suggests that the centre has charged up positively with respect to the sides, i.e. that a lateral potential gradient has been set up along the base of the crater. The image taken at a sample bias of -10 V confirms the hypothesis, for the emission is now greatest at the centre of the crater, suggesting a lateral potential gradient from the centre of the crater to the edge of 20 V.

layer of gold a few tens of nanometres thick. This method was used for the analysis of BF_2 implants in silicon and silicon-on-sapphire, where electron beam flooding was inappropriate due to the intense ESIE signal at mass 19. The accuracy of the charge compensation was monitored by running three mass channels for boron, one tuned to the peak of the energy distribution and one tuned to either side of this. Three fluorine mass channels were set up in the same way. If the sample charges positive, the mass channels tuned to the high-energy side of the energy distribution should increase and those on the low-energy side of the distribution decrease. The implant in bulk silicon was analysed first (see Fig. 5.28(a)). No charging was expected and in the

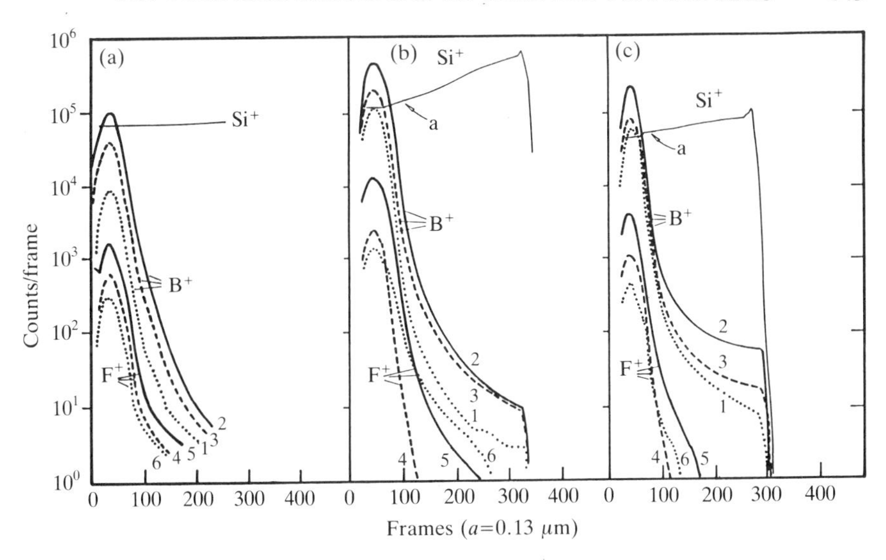

FIG 5.28. Three multi-channel depth profiles, (a) of a BF_2 implant into silicon, (b) and (c) of a BF_2 implant into silicon-on-sapphire, using gold coating for charge compensation. 1, B (+25 V); 2, B (+10 V); 3, B (−10 V); 4 F (+27 V); 5, F (+10 V); 6, F (+5 V). The three boron and three fluorine channels were concurrent in the analysis of the silicon sample, indicating that no charging had occurred. Analysis of the SOS sample using the same primary beam current and crater size (300 nA → 400 μm × 400 μm) suggested that significant charging has occurred, with the fluorine channels crossing over. The charging was minimized by reducing the primary beam current to 100 nA, the crater size to 200 μm.

resulting profile the matrix line was stable and the three boron and three fluorine channels rose and fell concurrently. Analysis of a gold-coated BF_2 implant in silicon-on-sapphire with the same primary beam current and crater size revealed significant charging (see Fig. 528(b)). The boron and fluorine channels were not concurrent and the fluorine channels crossed over on the TE, having decay lengths of between 23.3 and 41 nm per decade. Fluorine is particularly sensitive to charging as it has a narrow energy spectrum similar to Si_2O. To reduce these charging effects, believed to be associated with the resistance between crater centre and edge, the primary beam current was reduced from 300 nA to 100 nA and the crater size from 400 to 250 μm (see Fig. 5.28(c)). The profiles were now accurate.

A third method of charge compensation is to monitor the sample bias continuously by running matrix lines very sensitive to any charging (e.g. Si_2O^+). Any change in these matrix lines due to charging can be detected by the control computer. The computer can then change the sample bias (V_0 in Fig. 5.1) in the opposite sense,[12] thus stabilizing the potential at the sample surface.

5.5.6. Multi-layer analysis

Quantification of the data from multi-layer samples is a difficult task for two reasons. First, there will be variations in the sputter rate $\dot{z}$ in the different layers due to their different partial sputter yields y_x (eqn (5.3.4)). The sputter rates of all the layers must be determined in a series of separate experiments and the depth scale of the profile then scaled (stretched and compressed for the different layers) accordingly. The second problem is that the ionization probabilities of the dopants will be different in the different layers. Thus the useful ion yields τ_u will have to be determined, for example from ion-implanted standards, for every dopant in every matrix. If one is analysing for 10 different dopants in a three-layer sample, 30 ion-implanted standards will have to be analysed; however, one need only find the ion yield for one isotope of each element.

Once the sputter rates of every layer and the ion yields of all the dopants from all the layers are known then we can quantify the concentrations. The concentration of a dopant X is related to the secondary ion count-rate $N_s(X^+)$ by eqn (5.3.10):

$$\rho(X) = [1/\dot{z} \cdot a \cdot \tau_u (X)^+)] \cdot N_s(X^+). \tag{5.3.10}$$

The proportionality constant [square brackets] involves both the sputter rate $\dot{z}$ and the ion yield τ_u. Quantification of the dopant concentration in the different layers in a multi-layer sample must take account of changes in the sputter rate and/or changes in ion yield, the latter reflecting differences in ionization probabilities.

The analysis of a 2 μm thick InGaAs epilayer on an InP substrate for two matrix channels (In, P) illustrates these points. The sputter rates of the InGaAs and InP were measured in separate experiments; the sputter rate of the InGaAs by drilling a crater up to the interface; and the sputter rate of InP by drilling a crater into a piece of the substrate material supplied for this purpose. The sputter rate of the InGaAs was approximately one-half that of the InP substrate material. The depth axis was scaled accordingly and thus the data density was half as high in the substrate as in the epilayer. Figure 5.29 illustrates some hypothetical secondary ion signals across an InGaAs–InP interface taking this difference in sputter rates into account. In the first example, we see that a change in signal can occur despite the fact that the concentration of the dopant is identical in both epilayer and substrate. In the second example, we see that a constant secondary ion signal may be observed despite a change in concentration, and, in the third example, we see that a drop in secondary ion signal can be associated with an increase in dopant concentration across the interface. These matrix effects emphasize again the importance of using standards for quantification. This is one of the most serious shortcomings of the technique since the time taken to analyse

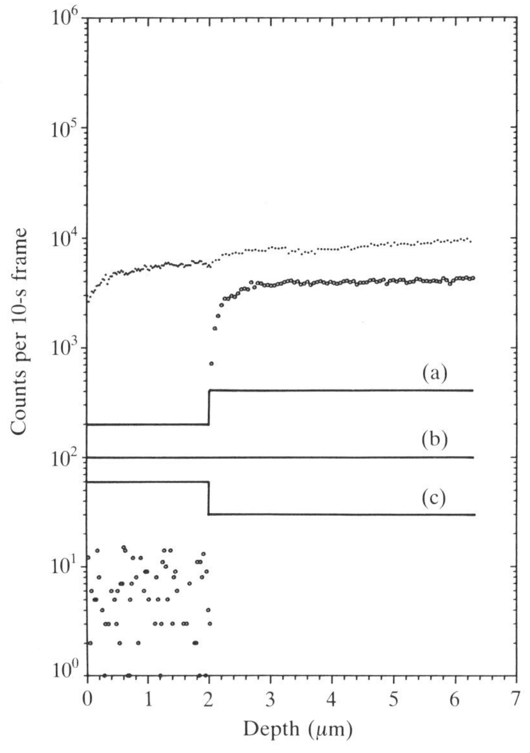

FIG 5.29. Counts per frame depth (a semi-quantified plot) for a depth profile of an InGaAs epilayer on an InP substrate, showing the matrix channel intensities as recorded. Primary ions: 4 keV $^{16}O_2^+$, 0.3 μA. Also shown are hypothetical secondary ion signals for three cases. (a) If the concentration of a dopant ρ(X) is the same on either side of the interface and there is no difference in ionization probability, i.e. α (X, InGaAs) = α (X, InP) then the secondary ion signal is seen to double, due to the change in sputter rate. (b) It follows that if the ionization probabilities are the same and if the secondary ion signal is constant, that the concentration of X in the epilayer must in fact be twice that in the substrate. (c) Finally, if the ionization probabilities are different, for example if the ionization probability of X from the epilayer is 10 times that from the substrate, then a situation can arise where the secondary ion count-rate goes down despite the fact that the concentration has gone up. In this case, the signal decreases by 50 per cent despite the fact that the concentration in the substrate is 250 per cent greater than that in the epilayer. These examples illustrate the need to determine the useful ion yields for every dopant in every matrix and to determine the sputter rates of all the layers.

standards and quantify the data can exceed the time taken to analyse the multi-layer test sample itself by an order of magnitude.

References

1. See for examples the Springer Series in Chemical Physics books on the bi-annual SIMS conferences:
   Secondary Ion Mass Spectrometry II (ed. A. Benninghoven, C.A. Evans, Jr, R.A. Powell, R. Shimizu and H.A. Storms).
   Secondary Ion Mass Spectrometry III (ed. A. Benninghoven, J. Siber, J. Lazlo, M. Riedel, and H.W. Werner).
   Secondary Ion Mass Spectrometry IV (ed. A. Benninghoven, J. Okano, R. Shimizu, and H.W. Werner).
   Secondary Ion Mass Spectrometry V (ed. A. Benninghoven, R.J. Colton, D.S. Simons, and H.W. Werner).
   All by Springer Verlag, Berlin. Also:
   Secondary Ion Mass Spectrometry VI. The Proceedings of the 1987 Conference in Versailles. John Wiley, Chichester/ New York.
2. Leta, D.P. and Morrison, G.H. (1980). *Anal. Chem.*, **52**, 514.
3. Meuris, M., Vandervorst, W. Borghs, G. and Maes, H.E. (1988). *Secondary Ion Mass Spectrometry VI* (ed. A. Benninghoven, A.M. Huber, and H.W. Werner), p 277. John Wiley, Chichester, New York.
4. Clark, E.A. Dowsett, M.G., Augustus, P.D., Spiller, G.D.T., Thomas, G.R. and Sutherland, I. (1989). *Appl. Phys.*, in press.
5. Wittmaack K. and Wach, W. (1981), *Nuclear Instrum. Methods*, **191**, 327.
6. Morgan, A.E., de Grefte, H.A.M. Warmoltz, N., Werner, H.W. and Tolle, H.J. (1981). *Appl. Surf. Sci.*, **7**, 372.
7. Wilson, I.H., Chereckdjian, S., and Webb, R.P. (1985). *Nucl. Instrum. Methods Phys. Res.*, **B7/8**, 735.
8. Clegg, J.B., Morgan, A.E., de Grefte, H.AM., Simondet, F., Huber, A. Blackmore, G., Dowsett, M.G., Sykes, D.E., Magee, C.W., and Deline, V.R. (1984). *Surf. Interface Anal.* **6**(4), 162.
9. Werner, H.W. (1974). *Vacuum*, **24**, 493.
10. Williams, P. (1985). *Scanning Electron Micros.* **II**, 553.
11. Wittmaack, K., Dowsett, M.G., and Clegg, J.B. (1982). *Int. J. Mass Spectrom. Ion Phys.* **43**, 31.
12. Dowsett, M.G., McPhail, D.S., Parker, E.H.C., and Fox, H. (1986). *Vacuum*, **11/12**, 913.
13. Meuris, M., Vandervost, W., and Maes, H.E. (1988). *Surf. Interface Anal*, **12**, 339.
14. McPhail, D.S., Clark, E.A., Clegg, J.B., Dowsett, M.G., Gold, J.P., Spiller, G.D.T. and Sykes, D. (1989). *Scanning Microsc.*, **2**, 639.
15. McPhail, D.S., Dowsett, M.G., Fox, H., Houghton, R., Leong, W.Y., Parker, E.H.C. and Patel, G.K. (1988). *Surf. Interface Anal.*, **11**, 80.
16. Degreve, F. and Lang, J.M. (1985). *Surf. Interface Anal.* **7**(4), 177.
17. SUPREM III. Unpublished work conducted by S.T.C., Harlow, Essex.
18. McPhail, D.S., Dowsett, M.G., and Parker, E.H.C. (1986). *J. Appl. Phys.*, **60**(7), 2573.

6
THE APPLICATION OF STATIC SIMS IN SURFACE SCIENCE

Over the past two of three decades basic surface science has been dominated by studies of adsorption and surface reactivity using the many techniques of surface physics. The reasons for this interest are obvious when one considers the many areas of technology which are influenced by, or dependent on, the reactivity of a solid surface. The emergence of SSIMS as a technique for surface analysis means that mass spectrometric information on the surface state is now available to add to all the other spectroscopic data. This is of course a very attractive prospect for the surface chemist. However, there has been considerable scepticism amongst the ranks of surface physicists that any useful information could be obtained. There were two reasons for this.

The first is the perceived violence of the SIMS process. It is possible to calculate that the power input to the surface from an Auger or XPS experiment is considerably more than that deposited in a SSIMS study (see Table 6.1). However, the difference is that the penetration depth of photons is considerably greater than that of ions. Most of the energy in ion bombardment is deposited in the surface layers. Thus it was thought that a primary particle depositing several kilovolts of energy in the top layers of a solid would cause so much disruption that the emitted particles would not be in any way relatable to the surface and any adsorbed species before impact. Whilst this seems a reasonable view it will be seen that, using static conditions, secondary ion emission can be directly related to the chemical state of the surface. There are a number of reasons for this:

(1) the detected secondary ions are of low kinetic energy and these ions are emitted up to 20 nm from the impact site;

TABLE 6.1

Comparison of typical primary particle flux densities and energies and the resulting power dissipated in SSIMS, LEED, and X-ray photoelectron experiments

	Primary flux (cm^{-2})	Primary energy	Power ($W\ cm^{-2}$)
SIMS	10^{10} ions s^{-1}	3 keV	3×10^{-6}
LEED	10^{15} electrons s^{-1}	50 eV	5×10^{-3}
XPS	10^{14} photons s^{-1}	1.4 keV	2×10^{-2}

(2) the final cascade collision which results in emission is relatively gentle (10–30 eV);

(3) under static conditions no spot on the surface is struck more than once; and

(4) surface annealing occurs in fempto-seconds.

Taking these points together, it can be seen that the secondary ions of interest will be emitted with high probability from undamaged surfaces. They will be lifted off the surface with relatively small amounts of energy, so we might expect that a significant yield of cluster ions should survive.

The other reason for scepticism arose from the computer modelling of the sputtering process. The various approaches have been outlined in Chapter 3. Some of the molecular dynamics models seemed to suggest that cluster particles were formed by recombination of sputtered atomic species above the surface. If this were true the relationship between surface chemistry and secondary ion emission could be rather remote. However, the results from this type of calculation do depend rather strongly on the assumptions underlying the model. There may be some situations where recombination does occur but the experimental evidence all strongly points to a direct relationship between surface chemistry and SSIMS spectra.

In this chapter we will examine the progress which has been made in using SSIMS for basic surface studies and try to define its place in the vast armoury of surface techniques.

6.1. Metal surface characterization

The cleanliness and structural integrity of a single crystal surface (sometimes polycrystalline foils are used) are essential prerequisites to any basic surface science experiment. To date, most studies have concerned metals, although increasingly investigations on semiconductor surfaces are appearing.

Usually the single crystal surface has to be subjected to cycles of oxidation and reduction to remove carbon compounds and oxides. Sometimes extensive ion etching is required to remove more stubborn contaminants such as sulphur or other metallic elements. Frequently, the bulk of the metal contains some contaminant which moves to the surface under certain conditions of temperature or ambient gas. To reduce the concentration of such elements requires cycles of heating and etching. The progress of the cleaning process is usually monitored by AES or XPS. When the contaminants concerned disappear or are at very low concentrations in the electron spectra the sample is judged to be 'clean'. The sensitivity of AES and XPS is limited to about 0.1 per cent of a monolayer; furthermore, there are situations where spectral overlap can be serious and reduce the sensitivity considerably. SIMS is considerably more sensitive and elemental spectral overlap does not occur. Its disadvantage is that if an element is detected it is more difficult to quantify the concentration.

Figures 6.1(a) and 6.2(a) show AES and SIMS spectra from an as-received Ru(0001) crystal. Carbon, oxygen and alkali metals are evident from the SIMS spectrum. The ruthenium signals are quite weak, particularly Ru_2^+, and Ru_3^+, indicating quite high levels of contamination, perhaps in excess of a monolayer. In the AES spectrum oxygen and sulphur are seen, alkali metals are difficult to detect at low levels, and the main carbon and ruthenium peaks overlap at $\sim$280 eV.

Figures 6.1(b) and 6.2(b) show the spectra after several oxidation–reduction cycles, and argon ion sputtering. AES shows a spectrum apparently clear of contaminants, although the overlap of carbon and ruthenium peaks leaves some uncertainty. With SIMS, although the surface is very much cleaner, a little carbonaceous material is still evident, along with the presence of alkali metals, to which SIMS is especially sensitive. At this spectral level, the concentration of the metal species is probably very low, in the p.p.m. region; the carbon, however, is probably about 0.01 per cent of a monolayer. It is clear that SIMS provides a very rigorous test of surface cleanliness.

After cleaning, it is necessary to check the crystallography of the surface. Sputtering will certainly have disrupted the surface. Normally the surface will be annealed and low-energy electron diffraction (LEED) is used to monitor the success of this process. However, it is well known that electron diffraction monitors the crystal structure from the surface and a few layers below; thus clear, sharp LEED spots may be obtained yet the surface layer may not be perfect. Low-energy ion scattering has shown this, and SSIMS is also able to monitor top-surface chemical structure.

In the series of SIMS spectra in Fig. 6.2(c), it can be seen that the relative intensities of the cluster ions Ru^+, Ru_2^+, and Ru_3^+ have risen significantly after annealing. Figure 6.2(d) shows the high-resolution spectra of Ru^+ and Ru_2^+. The M_x^+/M^+ ratio is a measure of surface crystallinity. In Chapter 2, Table 2.3 shows how this ratio varies with the geometry of the low index f.c.c. faces of nickel and is related to the relative probability of finding nearest neighbours in each of the surface layers. Experimentally, it is found that these ratios are very sensitive to sputter-induced damage of the surface, and are therefore extremely useful in monitoring such damage and assessing its removal by annealing.

That there should be a relationship between cluster ion yield and surface crystallography is to be expected from both the cascade theory and the molecular dynamics theories described in Chapter 3.

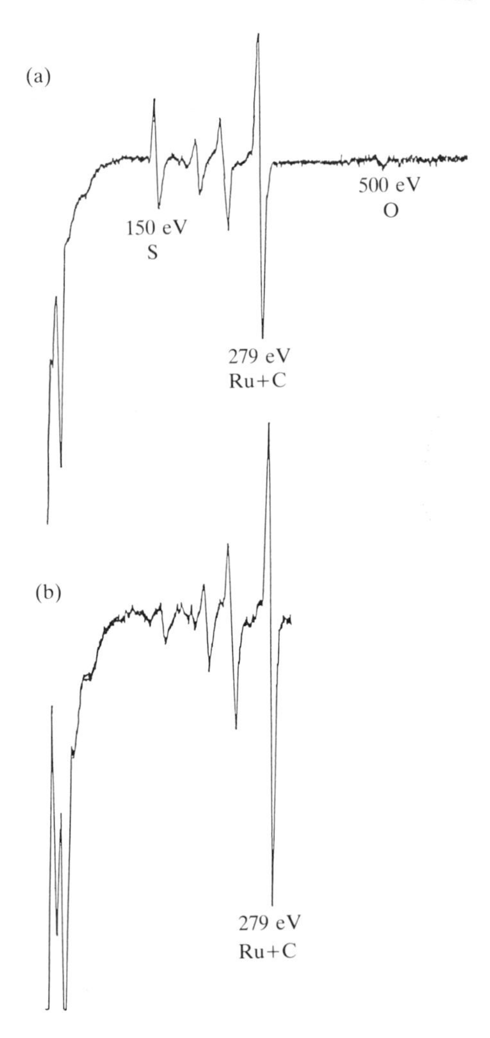

FIG 6.1. (a) Auger spectrum of a contaminated Ru(0001) single crystal surface. (b) Auger spectrum of the Ru(0001) crystal after cleaning by argon ion etching and heating in oxygen followed by hydrogen to 1073 K.

6.2. Studies of thin metal films

An understanding of the structure of metal overlayers a few monolayers thick is important in both a basic surface science and in some of the new electronic device technologies. It is necessary to know how the metal overlayer grows, how uniform the layer is, and what is the binding energy to the substrate.

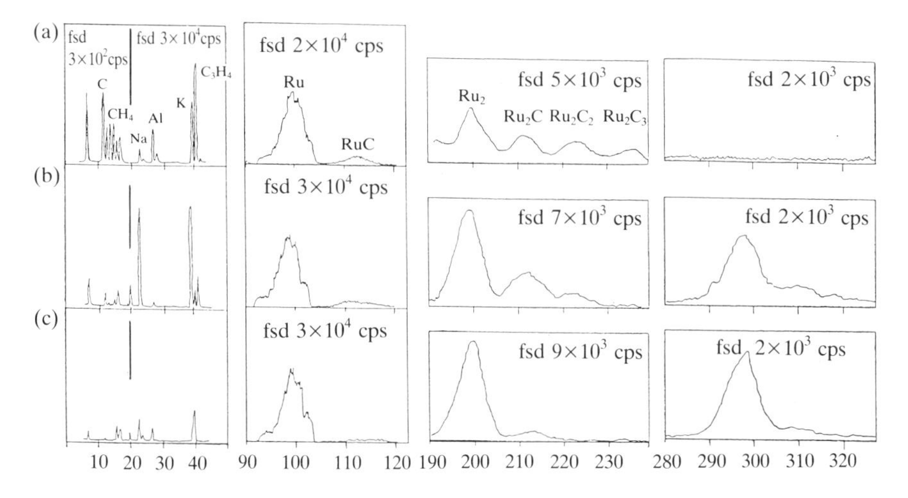

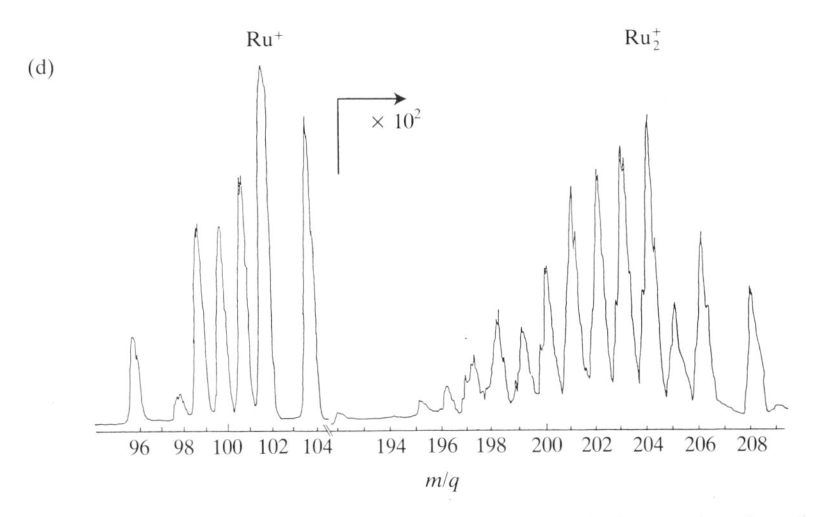

FIG 6.2. SSIMS spectra of a contaminated Ru(0001) single crystal surface showing the $m/z = 0$–40 region and the mass regions around Ru^+, Ru_2^+ and Ru_3^+. (b) SSIMS spectra of the Ru(0001) surface after ion etching and O_2/H_2 treatments. (c) SSIMS spectra of the Ru(0001) surface after annealing to 1300 K. (d) High-resolution mass spectra of the Ru^+ and Ru_2^+ isotope patterns.

Usually a combination of electron spectroscopy, thermal desorption (TD) and LEED and or reflection high-energy electron diffraction (RHEED) are used to investigate the growth process. The large literature produced by Bauer and his group illustrates the approach.[1]

Studies of the development of LEED features as overlayer coverage increases enable the researcher to decide whether growth is epitaxial, crystallographically strained or disordered. Thermal desorption experiments give a measure of overlayer coverage and the activation energies of desorption which reflect the binding energy of the overlayers to the substrate. Finally, monitoring, via XPS or AES, the variation in characteristic electron yield from substrate and overlayer as the overlayer grows can provide information on the growth mode of the overlayer. Thus, as Fig. 6.3 shows, layer-by-layer growth (known as Franck–van der Merwe growth) is characterized by a series of straight-line plots $I_{substrate}$ vs. $I_{overlayer}$ of differing gradient.[2] On the other hand, island growth (Volmer–Weber) is characterized by a smooth single-gradient line. If initially a smooth overlayer is formed followed by island growth (Stranski–Krastanov growth), the graph has a single break in it.

Clearly, a fairly complete analysis is provided by this array of information. SIMS can, however, provide further complementary and confirming data. By examining the way in which overlayer elemental and cluster ion yields vary

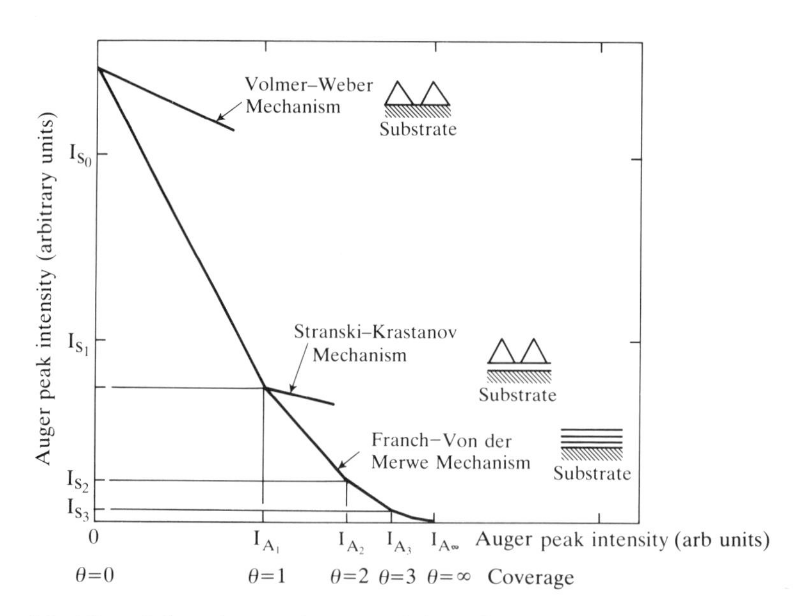

FIG 6.3. Plot of the substrate Auger peak intensity as a function of the adsorbate Auger peak intensity, illustrating how different thin film growth modes can be distinguished. Reproduced with permission from Ref. 12.

with substrate ion yields as a function of overlayer coverage, a good deal of information can be obtained. The arguments are basically statistical. The copper on ruthenium system is a useful example to illustrate the application.[3]

6.2.1. *Overlayer growth mode*

A plot of the ratio of overlayer elemental ion to substrate ion (i.e. Cu^+/Ru^+) against overlayer coverage measured by thermal desorption can help to define the growth mode. Thus in Fig. 6.4(a) the ratio rises exponentially

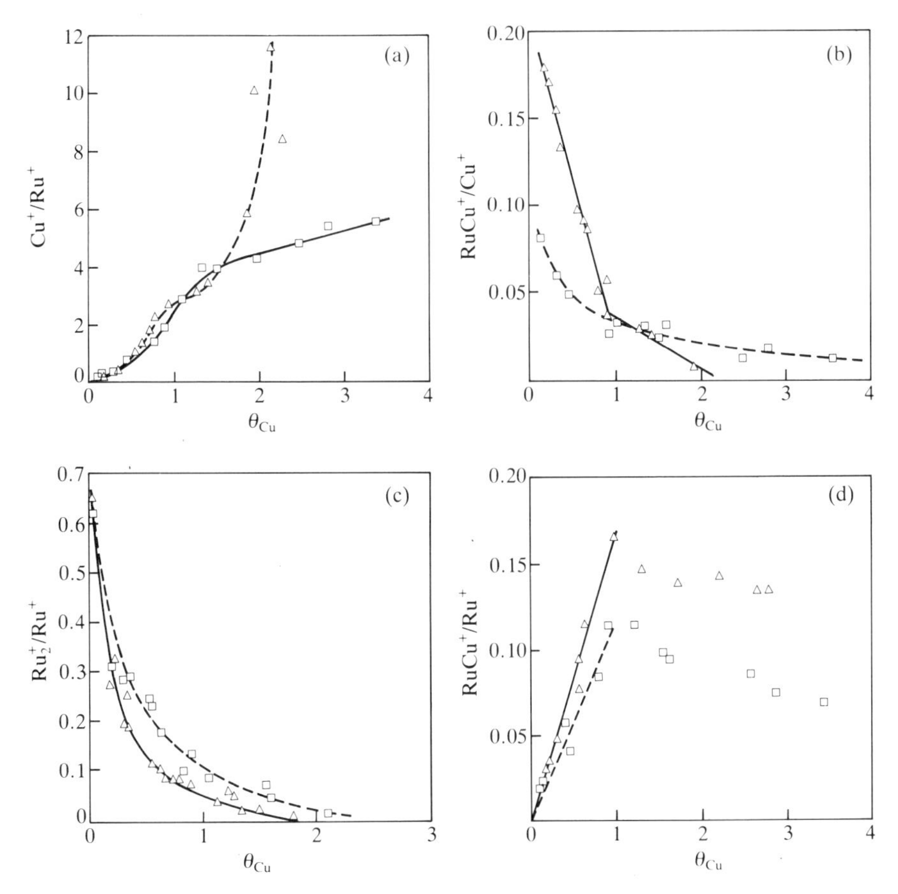

FIG 6.4. (a) Variation of SSIMS intensity ratio, Cu^+/Ru^+, as a function of Cu coverage for the 540 K (□) and 1080 K (△) series. (b) Variation of $RuCu^+/Cu^+$ for 540 K(□) and 1080 K (△) (c) Variation of Ru_2^+/Ru^+ as a function of Cu coverage for the 540 K (□) and 1080 K (△) series. (d) Plot of the secondary ion ratio $RuCu^+/Ru^+$ as a function of Cu coverage for the 540 K (□) and 1080 K (△) series. Reproduced with permission from Ref. 3.

towards two monolayers for surfaces prepared by evaporating copper onto a Ru(0001) surface held at 1080 K and then cooling rapidly to 300 K. This implies that Ru^+ emission is tending to zero and the surface is completely covered with copper, confirming the layer-by-layer growth. On the other hand, the ratio reaches a plateau at quite a low level for the surfaces prepared at 540 K, suggesting that the ruthenium surface is not completely covered and implying island growth.

Similar conclusions can be derived from an examination of the cluster ion emission. Because SIMS is very surface sensitive, a cluster ion involving a substrate atom should fall to zero within two monolayers coverage if smooth layer-by-layer growth occurs. This is seen to occur for both $RuCu^+/Cu^+$ and for Ru_2^+/Ru^+ for the 1080 K surfaces (see Figs. 6.4(b) and (c)), but not for the 540 K surfaces. Again, island growth is implied for the lower-temperature surfaces.

More detail about possible island size is available from a figure such as Fig. 6.4(b). Thus the probability of emission of a cluster involving both a substrate atom and an overlayer atom compared with the emission of an overlayer atom (in this case $RuCu^+/Cu^+$) will be related to the size of developing islands. Thus on the high temperature surface the ratio starts off high, suggesting a high dispersion of copper on the surface, whereas on the low-temperature surface the low coverage ratio is a factor of two lower, suggesting a copper cluster size of three to five atoms.

6.2.2. Overlayer coverage

An assessment of coverage can be obtained using a cluster ion involving a substrate and an overlayer atom and finding its ratio to a substrate ion. Thus if the ratio $RuCu^+/Ru^+$ is plotted against copper coverage (see Fig. 6.4(d)), in the sub-monolayer regime straight lines result which are proportional to coverage. We can assess the relative coverages attained by the two types of film. It can be seen that at a nominal single monolayer, measured by thermal desorption, the low-temperature surface has 25 per cent lower coverage than the 1080 K surface. Again, this is consistent with the idea that the overlayer is not spreading smoothly over the surface but is forming three-dimensional clusters.

It is clear that, with careful analysis, SIMS can provide extensive information about metal surface overlayers and indeed other insights into surface smoothness, etc., are accessible with further analysis.

6.3. Oxidation of metals

Much is known about the metal oxidation process from a variety of areas of study; for example, the progress of bulk oxidation has been extensively studied by investigators in the corrosion field and the initial interaction of oxygen with a metal single crystal is still an area of active investigation using LEED or ion scattering. Static SIMS made its first significant and unique contribution to a problem of surface reactivity in an investigation of the initial process whereby the first two or three metal layers participate in oxidation. Much of the work was done by Benninghoven and his group, who investigated a range of polycrystalline metal surfaces.

Two of the earliest studies were of the oxidation of chromium and copper. They illustrate the approach very well. The results were rather different, and they are early examples of the use of cluster ions to unravel the chemistry of surface processes.[4]

Surface studies of this type naturally need to be carried out under UHV conditions. Sputter-cleaned and annealed metal films were investigated and the SSIMS spectrum was monitored as a function of exposure to oxygen. The spectra were obtained using very low primary argon beam fluxes, 0.1–1 nA cm^{-2}.

6.3.1. *Initial oxidation of chromium and copper*

As the initial oxygen exposure was increased over a chromium film, the emission of Cr^+, Cr_2^+, and Cr_3^+ rose, owing to the increase in ionization probability, which is correlated with the increasing work function as oxidation proceeds. This initial rise was followed by a fall in Cr_2^+ and Cr_3^+, which indicated a decrease in the number of adjacent Cr atoms as the surface layer was broken up by oxygen atoms. However, although the O^- ion appeared from the beginning, strangely, no oxygen—containing Cr secondary ions appeared until an exposure of about 50 L oxygen. Figure 6.5 shows the variation in the intensity of a number of significant secondary ions with oxygen dose. It can be seen that the CrO_2^- ion develops first at exposures above 100 L. This passes through a maximum and the CrO^+ ion begins to develop beyond 250 L.

It appears that there are two oxide phases: phase I characterized by CrO_2^-, and phase II by CrO^+. To test this conclusion, after the SSIMS signals had saturated at an exposure of about 1200 L, the oxide layers were sputtered away progressively. Figure 6.6 shows the result. Initially CrO^+ decayed to zero very rapidly, suggesting that the thickness of the related phase is probably not greater than one monolayer. Simultaneously, CrO_2^- rose, remained constant for a long period, and then decayed away. This oxide phase must be quite thick. As CrO_2^- decays, Cr_3^+ increases, reflecting the emergence of clean metal. If the experiment is carried out after the exposure of only 100 L (see again Fig. 6.6); CrO^+ has not developed and the CrO_2^- signal begins to decay immediately. The decay is exponential, suggesting an oxide layer thickness of one monolayer. It is indeed clear that CrO_2^- and CrO^+ arise from different oxide structures.

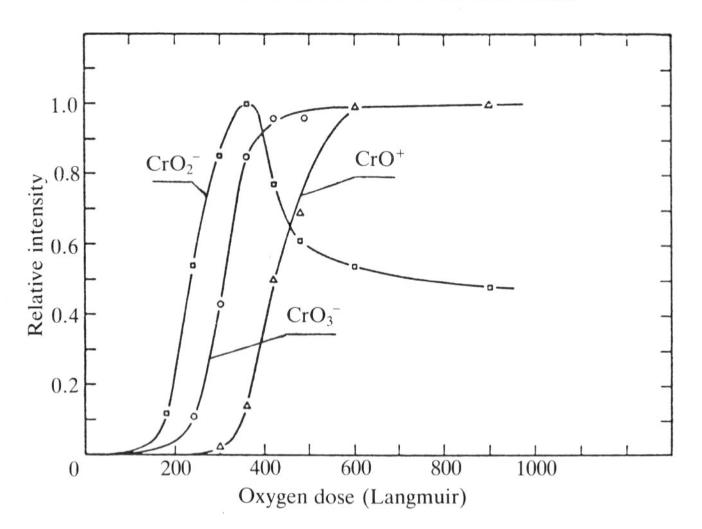

FIG 6.5. Increase of secondary ion intensities during oxygen exposure of a clean chromium surface. Reproduced with permission from Ref. 5.

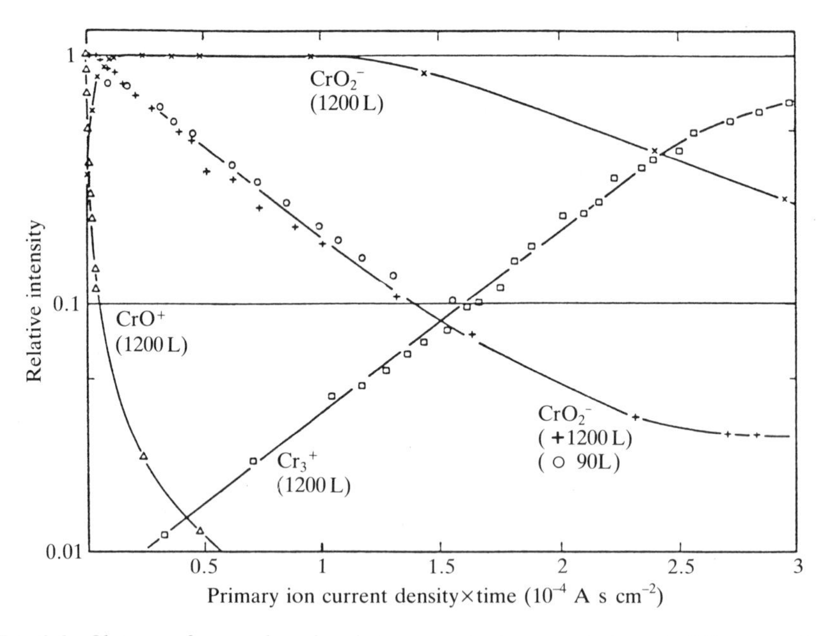

FIG 6.6. Change of secondary ion intensities during sputtering of different oxide layers on chromium. The oxide layers were formed by oxygen doses of 1200, 120, and 90 L. Reproduced with permission from Ref. 5.

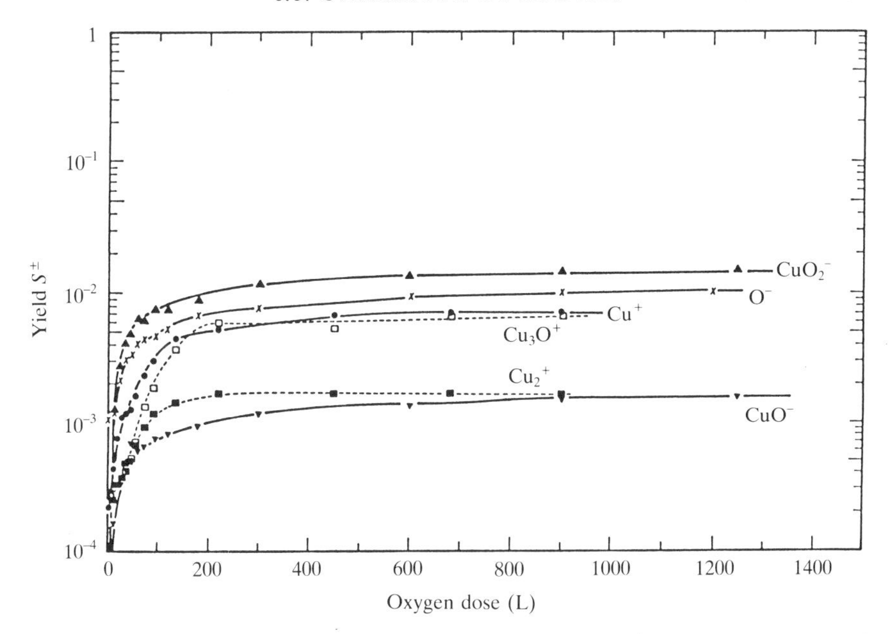

FIG 6.7. Variation of some secondary ions from a copper surface as a function of oxygen dose. Reproduced with permission from Müller, A. and Benninghoven, A. (1974). *Surf Sci.*, **41**, 493.

For the oxidation of copper, Fig. 6.7 shows that the secondary ions behave in a very different way. CuO_2^- appears from the start and rises to a plateau at an exposure of 200 L, as does Cu_2^+. None of the ions decrease at high exposure. Sputter removal demonstrates that the layer is only one monolayer thick. No further oxidation can be generated by higher exposures at 300 K at the low pressures used.

Thus on chromium the oxidation progressed beyond the simple surface adsorption to penetrate into the bulk, and characteristic fragment ions can be identified for the two phases. On copper only surface oxidation results, and the behaviour is similar for the oxidation of A1.[5]

Several other metals were studied: vanadium, tantalum, niobium, nickel, and titanium, which behaved in a similar manner to chromium, showing evidence of two phases. Two phases were also observed for the oxidation of magnesium, strontium, and barium, but very much higher oxygen exposures were required.

6.3.2. Initial oxidation of a copper–zinc alloy

Usually surface science studies require an input from more than one technique to provide a full understanding of a process. This was true for a study of the oxidation of Cu–Zn alloys (15 per cent Zn). It demonstrates that integrating data from SSIMS and XPS is necessary to obtain a fuller insight into the initial oxidation mechanism.[6]

This investigation highlights the importance of the different information depths of the two techniques. Thus the XPS spectrum of the clean annealed alloy surface suggests a mixture of Cu and Zn with some segregation of Zn to the surface. The SIMS spectra are devoid of Zn secondary ions. Taking the two pieces of information together and remembering that XPS probes several layers below the surface, it can be concluded that there is a thin surface layer of Cu followed by a Cu–Zn or Zn-rich layer.

After the exposure of the surface to oxygen at 300 K the XPS data suggest that it is Zn which is oxidized. The oxidation of Cu is difficult to determine using the 3*d* photoelectron peaks because oxidation does not result in significant chemical shifts. However, the Cu Auger peak is also little affected. These data would suggest that Zn had migrated to the surface under the influence of oxygen.

The SSIMS spectra, however, show an initial rapid rise in the emission of CuO_2^- and CuO^- which then falls off at exposures greater than 20–40 L (see Fig. 6.8). There is no evidence for Zn or ZnO_x ions which would be expected if zinc was present in significant concentrations in the top surface. Clearly this all points to the *oxidation of the Cu surface layer*, but something further happens which contrasts with the behaviour of pure Cu. Two new ions appear which do involve Zn, namely Cu_2Zn^+ and Cu_2ZnO^+. These ions must involve sub-surface Zn and possibly small quantities of surface Zn.

Thus if the XPS and SSIMS data are integrated a model develops which suggests that a surface Cu layer is maintained and initially this layer is oxidized followed by the movement of Zn to the sub-surface layer, possibly accompanied by the penetration of oxygen to co-ordinate with the Zn, which results in the modification of the structure of the oxidized Cu layer.

6.4. Studies of chemisorption

Probably the most active area of research in surface science is the study of the detailed mechanism by which molecules adsorb at solid surfaces. Metal–molecule interactions are by far the most intensively studied, and a very extensive array of surface physics techniques has been used to try to understand the crystallographic and electronic parameters which control the adsorption process.

In the last section we saw how the oxidation process could be followed from the characteristic cluster ions which were generated. In this section we will describe how the cluster ions generated as a consequence of adsorption can be used to provide valuable information which complements that obtainable from other techniques.

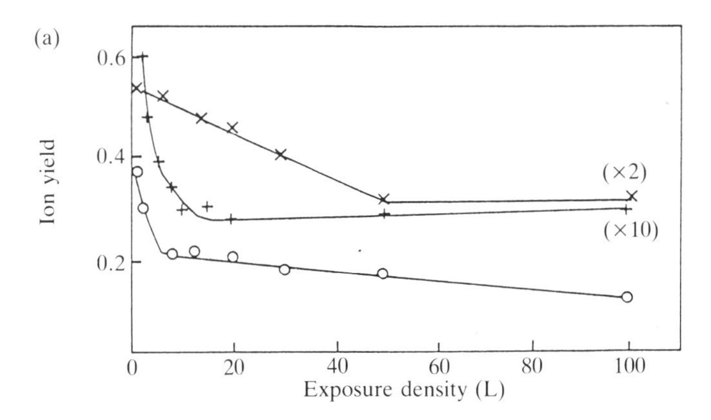

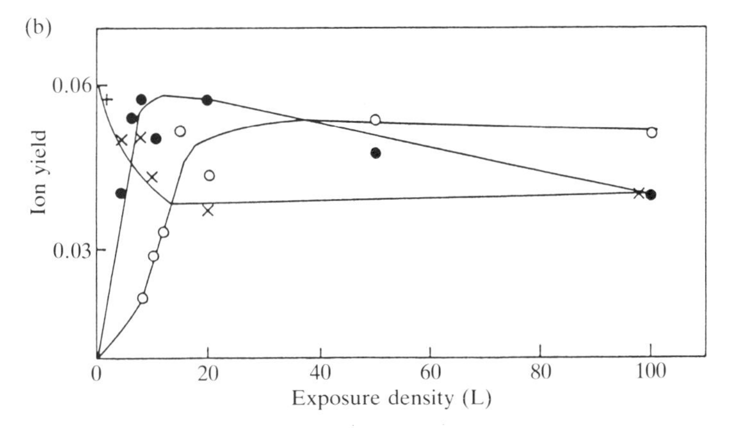

FIG 6.8. Relative intensities of some secondary ions from a Cu–Zn alloy (15 per cent Zn) as a function of oxygen exposure. (a) ×, CuO_2^-; +, CuO^-; ○, O_2^-. (b) ○, Cu_2ZnO^+; ●, Cu_2Zn^+; × Cu_2^+.

A basic surface science study of chemisorption seeks to answer the following questions:

1. Is the adsorption molecular or dissociative?
2. What coverage is obtained as a function of exposure and temperature?
3. What is the binding energy?
4. What are the electronic interactions between the surface and the adsorbing species?
5. What is the chemical structure of the adsorbate?
6. How does the adsorbate distribute itself on the surface?
7. What are the interactions between the adsorbate molecules?
8. What is the reactivity of the adsorbate?

SIMS can contribute to most of these questions except (4) and (6). The former case requires detailed electron spectroscopy studies of surface band structure and the modification of that structure by the adsorbing molecule. Synchrotron radiation has contributed greatly in this area. In the latter case, surface crystallography is required, and although SIMS is sensitive to surface structure the precision required is only provided by LEED, extended X-ray absorption fine structure (EXAFS), or low-energy ion scattering.

6.4.1. Molecular or dissociative adsorption?

It is obviously essential to know what effect the initial adsorption step has on the integrity of the molecule. This may be studied using XPS or vibrational spectroscopy. The XPS approach relies on chemical shift between the two states being significant. As we shall see below this is true for a molecule such as NO but is often not so for hydrocarbons.

Vibrational spectroscopy data usually clearly distinguish between the two states; however, high-resolution electron energy loss (HREELS) and reflection absorption IR (RAIRS) systems are not as widely used as they might be, due probably to the skill needed in their operation.

An indirect indication of dissociation may be obtainable from thermal desorption (TD) studies. If the molecule is not desorbed intact dissociation *may* be implied but care has to be used in interpreting the data.

6.4.1.1. Adsorption of CO on metal surfaces SSIMS uses the cluster ions generated as a consequence of adsorption to determine whether dissociation has taken place. One of the earliest demonstrations of this was a systematic study of CO adsorption on a series of polycrystalline transition metals.[7] It was known from other studies using electron spectroscopy that the adsorption of CO at 300 K would be molecular or dissociative depending on the adsorption energy. Where the enthalpy of adsorption is <250 kJ mol^{-1}, as it is on Cu, Pd, and Ni, the adsorption is molecular; where it is greater, as on W, it is dissociative; whilst adsorption on Fe lies between and both forms are observed.[8] Figure 6.9 shows that SSIMS can distinguish clearly between the two. Dissociative adsorption is characterized by MC^+, MO^+, M_2O^+, and M_2C^+ secondary ions; these are only observed on Fe and W surfaces at 300 K. The molecular adsorbate state on Cu, Pd, Ni, and Fe is identified by MCO^+ and M_2CO^+ ions.

6.4.1.2. Adsorption of NO on metal surfaces NO adsorption shows similar behaviour. Thus when NO adsorbs on rhodium (111) at low coverage, dissociation is known to occur and can be distinguished by XPS measurements. The N 1*s* and O 1*s* binding energies for dissociative adsorption are 397.8 and 529.7 eV respectively, whereas for molecular

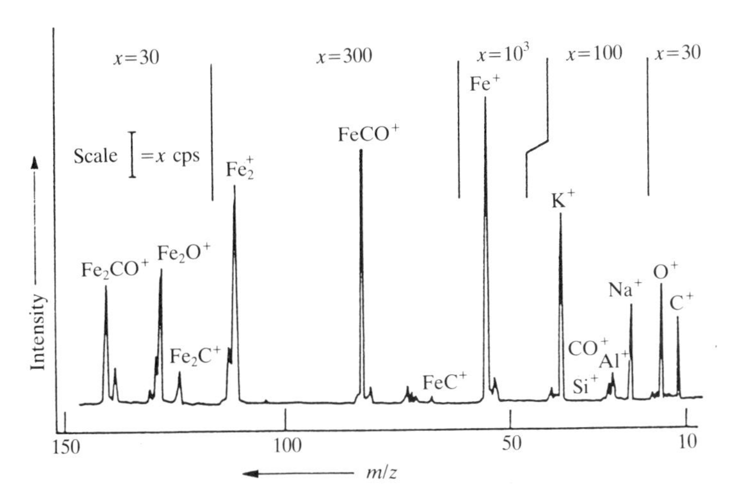

FIG 6.9. SSIMS spectrum of an iron foil exposed to 10^{-8} Torr CO. Primary ion current density 0.5 nA cm^{-2}, $E_p = 3$ keV. Reproduced with permission from Ref. 7.

adsorption they are 400.1 and 531.0 eV. Figure 6.10 shows how initial adsorption is dissociative, whereas at higher coverages molecular adsorption develops.[9] Using SSIMS, dissociation is indicated by the appearance of Rh_2N^+ and Rh_2O^+ ions, whereas, as molecular adsorption produces the NO^+, $RhNO^+$ and Rh_2NO^+ grow.

6.4.1.3. Adsorption of complex organic molecules More complex molecules can also be studied. Molecular adsorption of ethene results in simple $M_xC_2H_4^+$ ions, whereas loss of hydrogen or fragmentation of the molecule results in the loss of these ions and the generation of $M_xC_yH_z^+$ species. We will examine this further when we discuss the use of SIMS to study surface reactivity.

As molecular size increases the possibility of fragmentation occurring as the species is emitted has to be considered. As a result all the species in the observed spectrum will not necessarily directly reflect species to be found on the surface. The spectrum will be more characteristic of the fragmentation pattern found in organic mass spectrometry. Studies are then necessary to define the detail of the pattern associated with a particular adsorbate structure. Once defined the characteristic peaks can be used to monitor the adsorbate's presence and transformations at the surface. It does seem, however, that in most cases if molecular adsorption occurs a molecular ion of some description will be observed.

An interesting sequence of experiments into the adsorption of amino acids onto a number of transition metals illustrates this latter observation.[10,11] The

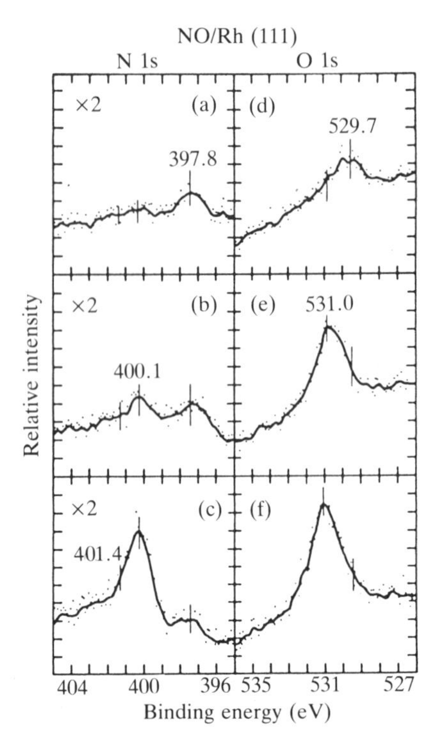

FIG 6.10. XPS N 1*s* and O 1*s* spectra for the stepwise adsorption NO on the Rh (111) surface at 300 K. Exposure to 0.4 L NO results in dissociative adsorption as shown by the single peak in both the N 1*s* region at 397.8 eV (a) and in the O 1*s* region at 529.7 eV (b). After exposure to 0.7 L NO, a second peak is detected in the N 1*s* region at 400.1 eV (c), and in the O 1*s* region at 531.0 eV (d), corresponding to molecularly adsorbed NO_{ads}. Saturation coverage is reached at 5 L exposure. Peak deconvolution of the N 1*s* line (e) suggests that 3 per cent of the overlayer is dissociated. The O 1*s* line (f) is not as well resolved. Reproduced with permission from Ref.9.

main characteristic ions for the amino acids are $(M + 1)^+$, loss of COOH $(M - 45)^+$, and $(M - 1)^-$. As the acid becomes more complex other contributing fragmentations become possible, but these three are always present (see Fig. 6.11(a)). Metals can be divided into reactive and non-reactive in terms of their interaction with the amino acids. As the coverage of amino acid increases, in the sub-monolayer region reactive metals Cu and Ni show the generation of the $(M - 1)^-$ ions *only*, with no evidence of $(M + 1)^+$ ions, whereas the 'unreactive' metals, Au and Pt, show both ions very strongly, (see Fig. 6.11(b)). Silver seems to be intermediate, in that the $(M + 1)^+$ although observed is considerably weaker than the $(M - 1)^-$ ion. If adsorption is continued at low temperature into the multi-layer region the two molecular ions are seen on all surfaces. Thermal desorption studies have demonstrated that the two ions can be associated with different adsorption states, the $(M - 1)^-$ species being significantly more strongly bound than the $(M + 1)^+$ species.

It is clear that the $(M - 1)^-$ ion is associated with a chemisorption bond in the sub-monolayer region and this type of bonding only occurs on reactive metals and possibly to a lesser extent on Ag. Weak physical adsorption is associated with the $(M + 1)^+$ ion, which always occurs with an $(M - 1)^-$. This is because the amino acids exist in a 'zwitterionic' state where the protons are mobile in the liquid phase. Physical adsorption is quasi-liquid as is multi-layer adsorption, which explains why this weak state is also observed on the reactive metals. Figure 6.11(c) summarizes the situation. Figure 6.11(d) indicates how reactivity varies across a range of surfaces.

It is evident that SIMS has considerable ability to analyse the state of the adsorbed molecule and is well able to be used to distinguish the molecular and dissociatively adsorbed state. One of its particular strengths is its ability in many cases to identify the surface atoms to which the adsorbate is attached. This is clearly of particular benefit when a mixed metal surface is being studied. We will show later that it is possible to go further and draw more detailed conclusions regarding the chemical structure of the adsorbed molecule.

6.4.2. *Surface coverage measurements*

Surface coverage of adsorbates is an important parameter. It is frequently required as a function of exposure, ambient pressure, and temperature. Its measurement is not always straightforward. Thermal desorption is frequently used and can give a measure of the total number of adsorbed molecules and an indication of the energetics of adsorption. The electron spectroscopies may be used if a characteristic element uniquely defines the species of interest. The adsorbate state can be characterized by vibrational spectroscopy data and then intensity variations used, but the intensities of such spectra are not always necessarily related directly or simply to coverage.

It is possible to use the secondary ions which characterize an adsorbate state to monitor surface coverage. In Chapter 2 the sensitivity of secondary ion yield to electronic parameters was outlined. When a molecule chemisorbs on a surface, interactions occur which change the electronic state of the system. This is often evident either via work function measurements or electronic spectroscopy studies. Thus as a molecule adsorbs on the surface the intensity of the resulting secondary ion will be a function of coverage *and* the electronic state of the surface. The latter parameter may vary in a complicated manner as coverage increases. Thus the intensity variation of the

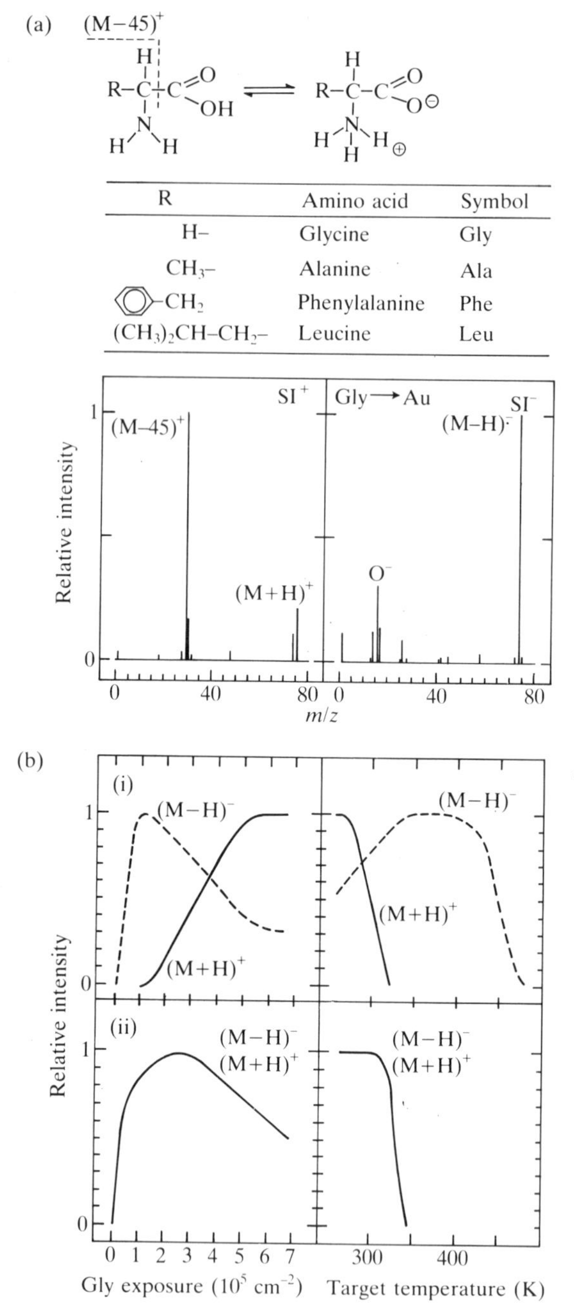

R	Amino acid	Symbol
H–	Glycine	Gly
CH_3–	Alanine	Ala
C_6H_5–CH_2	Phenylalanine	Phe
$(CH_3)_2CH$–CH_2–	Leucine	Leu

FIG 6.11. Adsorption of amino acids on metals. (a) Structure and characteristic secondary ion emission of α-amino acids. (b) $(M+H)^+$ and $(M-H)^-$ secondary ion emission during the formation of a glycine overlayer on a 'reactive', Cu (i), and a 'non-reactive', Au, metal (ii), and secondary ion emission during the thermal desorption of this layer. Two adsorption states can be observed for Cu whereas only one is evident for Au.

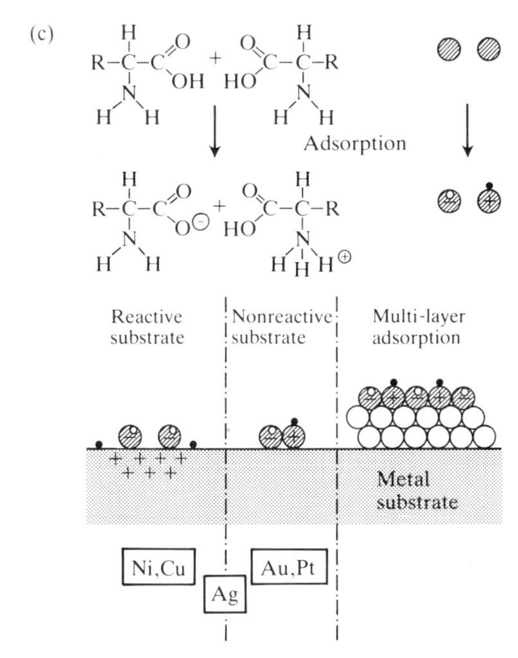

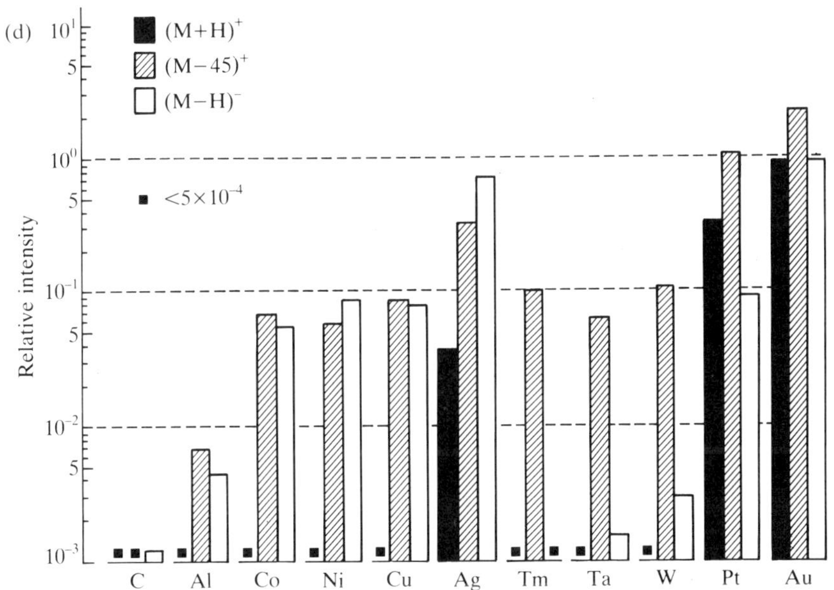

FIG. 6.11 (c) Structure of amino acid overlayers on 'reactive' (Ni, Cu) and 'non-reactive' (Au, Pt) metal surfaces. (d) Relative intensities of the secondary ions $(M+H)^+$, $(M-H)^-$, $(M-45)^-$ after deposition of a leucine layer (regions II) on several substrates $(M-H)^-_{Au} = 1$; target temperature = 300 K. $I_p = 2 \times 10^{-10}$ A (0.5 cm^2). Reproduced with permission from Ref. 10 and Benninghoven, A (1985). *J. Vac. Sci. Technol.*, **3**, 451.

secondary ion is unlikely to be linearly dependent on coverage. There is no satisfactory theoretical model which would allow the two parameters to be unravelled. An empirical approach has been found to be very successful.

The example of CO adsorption is used to outline the approach. The requirement is to relate the M_xCO^+ (usually x varies from 1 to 3) intensities to the surface coverage of CO adsorbate species. When adsorption occurs the intensity of *all* ions increase. It was argued that whilst the M_xCO^+ species were sensitive to *both* coverage and work function the $M_x{}^+$ ions also increased but only as a consequence of the work function change. It was suggested that if the ratio $M_xCO^+/M_x{}^+$ was used it would only be sensitive to coverage[12]. Since some CO is removed as MCO^+, some as M_2CO^+ and some as M_3CO^+, total coverage should be represented by

$$\Sigma_x(M_xCO^+/M_x{}^+) \propto \theta_{CO}.$$

The linear plots of this function against coverage demonstrate that SSIMS data are sensitive to adsorbate coverage (see Fig. 6.12).[13] There is a break at a relative coverage of about 0.5 in the plots for most of the surfaces. The gradient change is correlated with changes in the structure of the adlayer which usually becomes 'compressed' in this coverage region. It can be seen that the gradient of the linear relationship is surface dependent. It seems to be a function of the enthalpy of adsorption.

This approach has been used successfully for a number of simple molecules such as NO, N_2, O_2, H_2, and C_2H_4.[9,14,15] It was also shown in Section 6.2 that the coverage of metal on metal can be monitored in this way. Thus the

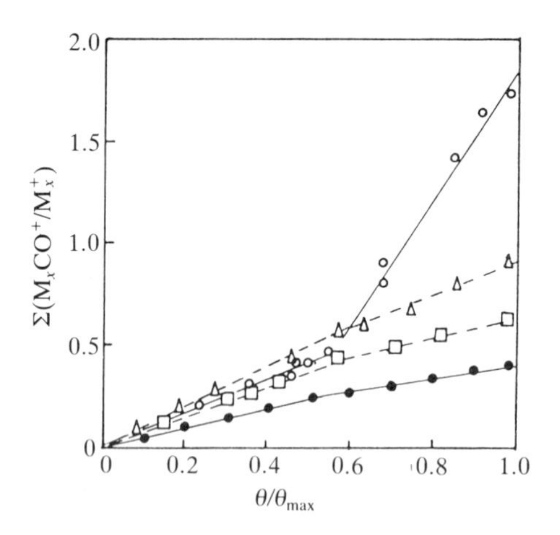

FIG 6.12. Variation of the sum of ion ratios, $\Sigma(M_xCO^+/M_x^+)$, as a function of CO coverage on ○ Pd (111); □ Ni (111); △ Ru (0001); ● Pd particles. Reproduced with permission from Ref. 13.

coverage of Cu on Ru was measured by the $RuCu^+/Ru^+$ ratio. Although the use of ion ratios yields a convenient way of defining coverage, as yet it does not give an absolute measurement. Some method such as TD or LEED has to be used to give a calibration point if absolute coverages are required.

Where adsorption is weak and little electronic interaction occurs, as for the example of the amino acids adsorbing on unreactive metals, the simple secondary ion intensity seems to be a good measure of coverage.

6.4.3. *The energetics of adsorption*

Adsorption energetics are usually measured by desorption methods. Thus temperature-programmed desorption enables the kinetics of desorption to be measured and hence, in principle, the activation energy of desorption. If the activation energy for adsorption is negligible, as it often is for metals, the enthalpy of adsorption and the activation energy of desorption should be identical.

Alternatively, if adsorption is reversible and we have a method for monitoring adsorbate coverage, isosteric methods can be used. The coverage is monitored as a function of pressure and coverage under 'equilibrium' conditions, and using the Clapeyron–Clausius equation the isosteric enthalpy of adsorption can be determined. The approach has been used for several cases where work function or electron spectroscopic measurements can be related to adsorbate coverage.

The ability to monitor automatically coverage by SSIMS enables the enthalpy of adsorption to be determined by this latter method. Thus, for the example of CO adsorption, the sum of ion ratios is monitored as a function of temperature for several ambient pressures of CO. From these plots, isosteric heats corresponding to particular coverages can be derived by plotting the log of the pressure required to attain that coverage against the reciprocal of the temperature. The enthalpy of adsorption usually varies with coverage. Figure 6.13 shows the result obtained from a SIMS study of CO adsorption on Ni(100). The result is in very good agreement with those derived by other methods.

6.5 Studies of adsorbate structure

The determination of adsorbate structure is perhaps the most significant aspect of the study of the chemisorption process. Certainly, an understanding of the chemical structure of the molecule on the surface provides considerable insights into the way it may transform, and is thus essential in providing a background for catalysis.

The detailed geometry of the adsorbate requires LEED or EXAFS

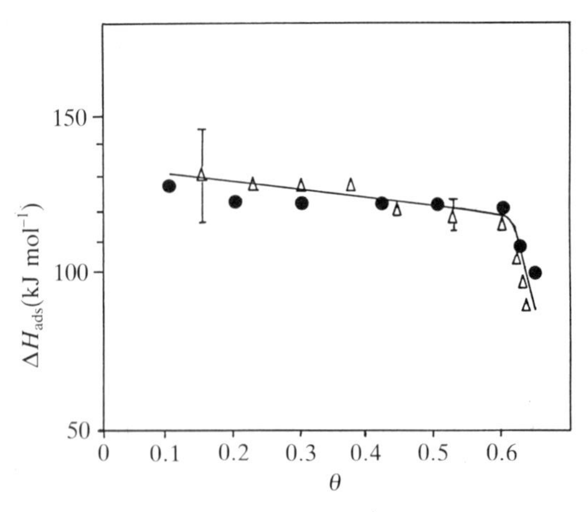

FIG 6.13. Variation of the heat adsorption of CO on Ni(100) as a function of CO coverage from SSIMS (△); and from work function measurements (●). Reproduced with permission from Bordoli, R. S., Wolstenholme, J., and Vickerman, J. C. (1979). *Surf. Sci.*, **85**, 244.

techniques, but such precise detail is often not required and it is sufficient to define the chemical structure. Vibrational spectroscopy is the technique of most obvious relevance here. For studies on single crystal metal surfaces, HREELS has been of great value. Because it is an electron reflection method, limitations on electron monochromatization have limited spectral resolution to about 40 cm^{-1}. The advent of routine Fourier transform IR spectroscopy has led to a re-evaluation of RAIRS. Using FTIR the sensitivity problems have been largely overcome and high spectral resolution is now available for surface vibrational studies.

Although these spectroscopic methods are of undoubted power in providing information about the chemical grouping present, because they do not provide an estimate of stoichiometry it is frequently difficult to know to which surface atoms the adsorbate is attached and to be sure of the overall composition of the adsorbate. Chemical structure determination also requires the input of mass spectrometry.

In Section 6.3 the sensitivity of SSIMS to the mode of adsorption has been outlined. This can be taken further. The detailed mass spectrum which arises when a molecule adsorbs can be used to identify the structure of the adsorbate. Although it does seem that standard mass spectrometry rules for molecule fragmentation can be applied in some cases, yet the SSIMS process is different from EI mass spectrometry, and considerable work has still to be done in this area to establish a systematic approach to interpreting spectra. Thus at present it is first necessary to study standard systems to determine the nature of the fingerprint spectra which are associated with different structures.

6.5.1. *CO adsorbate structure*

The method used is again illustrated by reference to the adsorption of CO. On metals CO can adsorb into linear, bridged or triply bridged surface states. By carefully selecting single crystal faces which were known from vibrational spectroscopy to adsorb CO into adsorbate states, it was possible to show by comparative studies, using vibrational spectroscopy and SSIMS, that the relative populations of the MCO^+, M_2CO^+, and M_3CO^+ ions can be used to identify the state of the CO on the surface. In other words, a SSIMS 'fragmentation pattern' of each state was defined which allows the distribution of each type of adsorbed CO on a surface to be identified and quantified (Table 6.2). Thus in Fig. 6.14 the variation in CO adsorbate geometry is monitored by SSIMS via the variation in the ratio of ion intensities $MCO^+/(\Sigma M_xCO^+)$, and by high-resolution electron energy loss spectroscopy (HREELS), by plotting the proportion of linear CO from the ratio of the intensities of vibrational bands due to linear CO (ν_1) to the total intensities of the linear plus bridged CO ($\nu_1 + \nu_2$). Since we do not know the relative extinction coefficients for the C–O stretch vibrations of linear and bridge CO, the latter ratio cannot be quantitative.[16]

Many important surface systems are composed of a number of different elements. SSIMS gives the surface scientist the possibility to identify the atoms involved in surface co-ordination. Thus, for example, in a situation where CO was adsorbed on a surface composed of Cu and Ru, whilst it was possible to show, in agreement with data from other methods, that above 300 K adsorption was predominantly only on Ru, and below 150 K extensive adsorption also occurred on Cu, it was also possible to show from the emission of $RuCuCO^+$ that between these two temperatures significant

TABLE 6.2

Secondary ion cluster emission and CO adsorbate structure

	SSIMS			CO structure
Surface	MCO^+	M_2CO^+	M_3CO^+	IR/HREELS/LEED
Cu(100)	0.9	0.1	—	Linear only
Ru(0001)	0.9	0.1	—	Linear only
Ni(100)	0.8	0.2	—	Linear + bridge
Pd(100)	0.3	0.6	0.1	Bridge only
Pd(111)	0.3	0.4	0.3	Triple bridge
Pt(100)	0.65	0.35	—	Linear + bridge

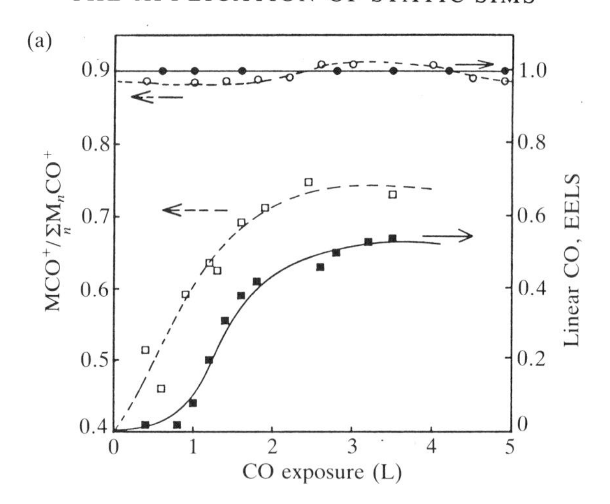

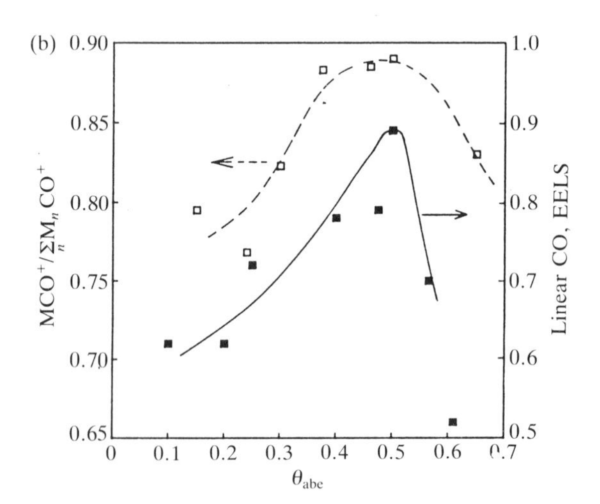

FIG 6.14. Variation of the ion fractions $MCO^+/(\Sigma M_x\ CO^+)$ with coverage or exposure for CO adsorption on (a) Ru (0001) at 300 K and Ni (111) at 195 K, and (b) Ni (100) at 300 K, compared with the proportion of linear CO calculated from EELS relative band intensities obtained from the literature. (a) o, Ru(001), SSIMS; ●, Ru (001), EELS; □ Ni (111), SSIMS; ■, Ni (111), EELS (b) Ni (100) □, SSIMS; ■, EELS. Reproduced with permission from Ref.16.

bridge-bonding adsorption occurred on mixed RuCu sites.[17] Such information on mixed sites would not be easily accessible using HREELS alone.[18]

6.5.2. *Angle-resolved SSIMS*

Many surface techniques use angle-resolved techniques to probe the orientation of adsorbates or electron bands at the surface. The information is obtained by varying the angle of the incoming radiation or measuring the angular distribution of the emitted radiation. This type of experiment is possible with SIMS although the experimental arrangement is quite complex. It is necessary to be able to move either the analyser or the primary beam with respect to each other and the sample. In principle, it allows the possibility to determine, at least qualitatively, the angular orientation of an adsorbed molecule. This information is of particular interest on surfaces which try to model real crystallites. Such single crystal surfaces are cut to expose edges and corners rather than flat planes.

An example is a study of CO adsorption on a Ni (7,9,11) surface. This face has terraces of (111) orientation five atomic rows in width with steps of (110) orientation one atomic layer high. The experiment was carried out by measuring the yields of secondary ions as a function of CO exposure as the angle of incidence of the primary beam is changed in the azimuthal direction. The type of information obtained is shown in Fig. 6.15. In this particular experiment only the $NiCO^+$ ion showed clear structure in the data. A sharp peak in the ion distribution curve is observed at $\phi = 110°$ which reaches maximum intensity and sharpness at 0.6 L. This sharp peak disappears at high CO coverage and is not so evident at higher temperature (300 K). By correlating with similar studies using HREELS, the narrow angular ion distribution is associated with CO molecules adsorbed along the step edge. All such sites will be occupied by 0.6 L exposure. As exposure is increased beyond this level the terrace sites are filled and the distinctive geometry is lost.[19]

From the foregoing, it is evident that SSIMS has considerable power for probing adsorbate chemical structure. There is some way to go before the technique is as routine in this respect as conventional mass spectrometry, but given a systematic body of work the potential is clear.

6.5.3. *Adsorbate–adsorbate interactions*

A related but important parameter which characterizes the adsorbate state is the extent to which molecules interact with one another. There are various modes of interaction: dipole–dipole and indirect chemical interactions, via

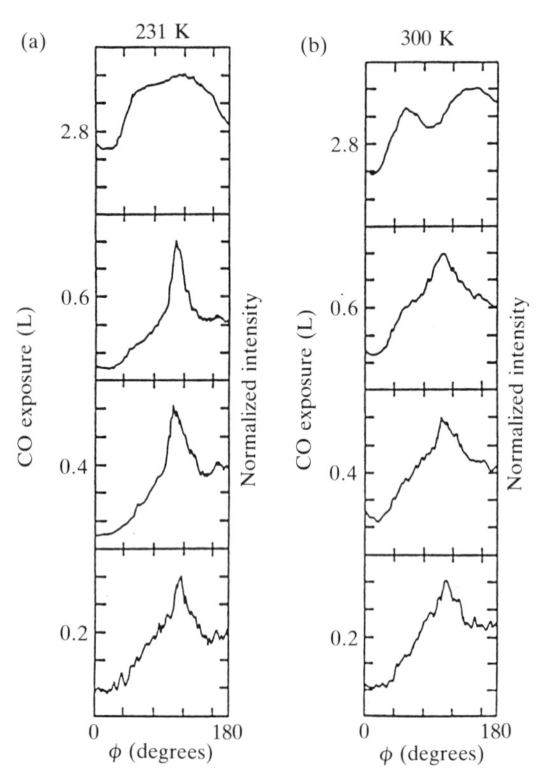

FIG 6.15. Normalized $NiCO^+$ intensity vs. azimuthal angle ϕ as a function of CO exposure. Reproduced with permission from Ref.19.

the substrate at low coverage, and direct repulsive interactions which may operate at high coverage. The usual technique to which these effects are sensitive is vibrational spectroscopy. The interactions can be monitored by frequency shifts in the molecular vibrations. In common with many of the basic surface studies, this effect has also been mainly investigated using CO adsorption and it is the C–O stretch vibration which has been studied as a function of CO coverage.

These effects may also be monitored using the SSIMS M_2^+/M^+ ratio. This can be rationalized as follows: the formation of the adsorbate–surface bond will have a statistical and chemical effect on the emission of substrate ions, M^+ and M_2^+; statistical in that if a CO molecule adsorbs onto a particular M site, the probability of its emission as MCO^+ will reduce the probability of its emission as M^+. Similarly, its chances of being emitted in an M_2^+ will be even further reduced. The chemical effects reside in the influence the formation of an admolecule–surface bond will have on the cohesion of M–M bonds in the metal surface. If the surface bonding is weakened then the probability of the emission of M_2^+ will be reduced relative to M^+. Thus a random distribution of CO on the surface could result in a steady fall of M_2^+/M^+ with θ_{CO}. If, however, the adsorbate tends to gather in islands, the fall would be much less steep or a plateau may occur until the islands coalesce, when a sharp fall may occur.

On the hexagonal surfaces there is an exact correspondence between the fall in the SSIMS M_2^+/M^+ ratio and the rise in vibrational frequency due to adsorbate–adsorbate interactions. Figure 6.16 shows the situation for CO adsorption on Ru(0001) at 300 K. For coverages below 0.33, disordered adsorption occurs. In the region 0.2–0.33, dipolar coupling begins to generate islands of a $(\sqrt{3} \times \sqrt{3})R30°$ ordered overlayer and as a consequence a plateau is observed. Beyond a coverage of 0.33 there is a loss of order in the adlayer due to repulsive interactions and this is reflected in an increase in vibrational frequency and fall in the SSIMS ratio.[16]

6.6. Surface reactivity

An understanding of the molecular processes involved in surface reactions requires that techniques be found which can follow the chemical transformations which occur in the surface layer. Three approaches are possible; we can monitor the molecules of reactant and product in the gas phase above the surface using a mass spectrometer, or a spectroscopic technique can be used to 'observe' processes on the surface, or we can combine the two!

SSIMS, vibrational spectroscopy and, in some cases, electron spectroscopy would be appropriate techniques, preferably combined, for monitoring the surface processes. The approach and the possibilities are best illustrated by two examples in which SSIMS is exploited: first, the adsorption and reaction of deuterated ethene on a Ru surface, and then the oxidation of CO over Pd.

Using techniques other than SSIMS, hydrocarbon transformations are difficult to follow. It is very difficult to distinguish the different stages of adsorbate hydrogenation and dehydrogenation. The ethene reaction demonstrates the detail accessible on hydrocarbon reactions from the mass spectra. The oxidation reaction shows that SSIMS provides the possibility of following the surface coverage of reactants and intermediates in the course of the reaction.

6.6.1. Deuterated ethene adsorption and reaction on ruthenium

Vibrational studies have suggested that ethene adsorbs on Ru at 130 K, predominantly into the di-σ bonded state although some adsorbed acetylene may be formed.[18, 20] Raising the surface temperature to 230 K results in the formation of a dehydrogenated species thought to be adsorbed ethylidyne

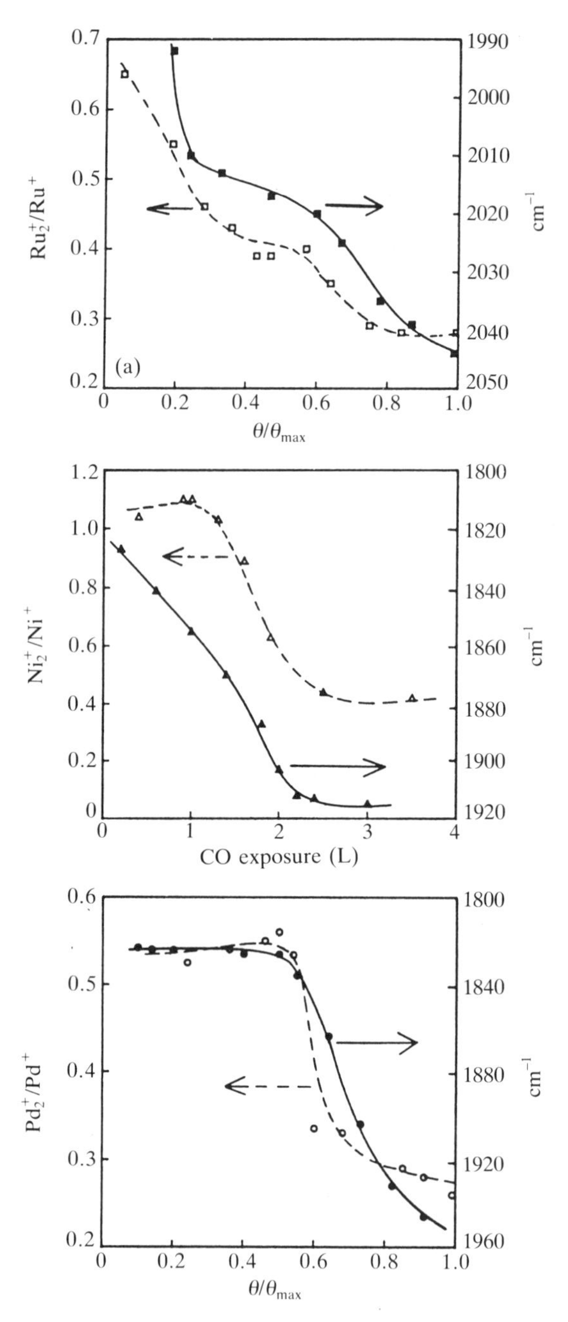

FIG 6.16. Comparison of the exposure or coverage dependence of M_2^+/M^+ from SSIMS with CO vibrational frequency from IR and EELS for CO adsorption on (a) Ru(0001) at 300 K, (b) Ni (111) at 195 K (SSIMS) and 140 K (EELS), and (c) Pd (111) at 300 K. Reproduced with permission from Ref. 16.

(C–CD_3), and acetylide (C–CD). At higher temperatures, further decomposition to methylidyne (CD) and carbon occurs. Mass spectral data were clearly needed to clarify the issue.

The adsorption of d_4-ethene at 130 K results in the CD_x^+ ($m/z = 12$–18), $RuC_2D_x^+$ ($m/z = 124$–136), and $Ru_2C_2D_x^+$ ($m/z = 220$–240) regions of mass spectral interest. We will examine the first two briefly. In the CD_x region, there is a unique pattern at $m/z = 12$, 14, and 16 related to molecularly adsorbed ethene. Decomposition of the adsorbed molecule to acetylene, ethylidyne, and beyond, will result in sequential enhancement of peaks at $m/z = 14$, 18, and 12. After adsorption at 130 K, a clear peak develops at $m/z = 18$ (CD_3); this cannot arise from molecular d_4-ethene and must indicate the formation of some ethylidyne. Figure 6.17(a) shows the variation of the relative yields of the species CD_3, CD_2, CD, and C with surface temperature. The formation of ethylidyne is at a maximum in the 180–230 K region. The formation of a CCD_2 (reflected by $m/z = 16$) species is indicated above 200 K, presumably by loss of deuterium from CCD_3. There is evidence from the $m/z = 14$ peak that acetylene may be formed at low temperature and acetylide may be formed in the 230 K region, since a significant $m/z = 14$ peak is not expected from CCD_3. Finally, carbon forms at high temperature.

If a mixture of molecular and decomposition products are present, a complex spectrum will develop in the RuC_2D_x region. Where dissociation does not occur, simple addition of $m/z = 32$ (for C_2D_4) to the isotope pattern for Ru ($m/z = 96$–104) yields the $RuC_2D_4^+$ pattern for adsorbed d_4-ethene (see Fig. 6.17(b)). This interesting feature has been observed before.[21] It might be thought that the sputtering process would cause fragmentation of the ethene molecule and clusters composed of Ru plus fragments would be observed. That this does not occur is probably due to the substrate 'buffering' the collision energy so that the $Ru_xC_2H_4$ cluster is lifted off relatively gently and with insufficient internal vibrational energy for decomposition. (See Sections 2.5.2, 2.5.3, 3.1.3.3, and 3.2.5 for further discussion of this issue.)

It is clear that $m/z = 136$ is uniquely associated with adsorbed C_2D_4; in the absence of molecular ethene $m/z = 134$ is uniquely associated with C_2D_3, ethylidene, or ethylidyne; and $m/z = 124$ will indicate the presence of adsorbed C_2D_2. Reaction to form ethylidyne between 130 and 230 K is confirmed in the $RuC_2D_4^+$ region from the variation in the ratio $I_{134}/(I_{134}+I_{136})$, which starts higher than the expected 0.63 (if molecular ethene was the only surface species) and increases to 1 at 230 K (see Fig. 6.17(c)).

The $I_{124}/(I_{126}+I_{124})$ ratio monitors the formation of C_2D_2 relative to CCD_3. A peak is evident at $m/z = 124$ even at 130 K, confirming that some adsorbed acetylene is formed in the initial adsorption process (see Fig. 6.17(d)). CCD_2 also clearly develops after CCD_3, confirming the above data. Finally, the coverage of molecular C_2D_4($I_{136}/^{104}Ru^+$) is seen to fall as a function of temperature in Fig. 6.17(c).

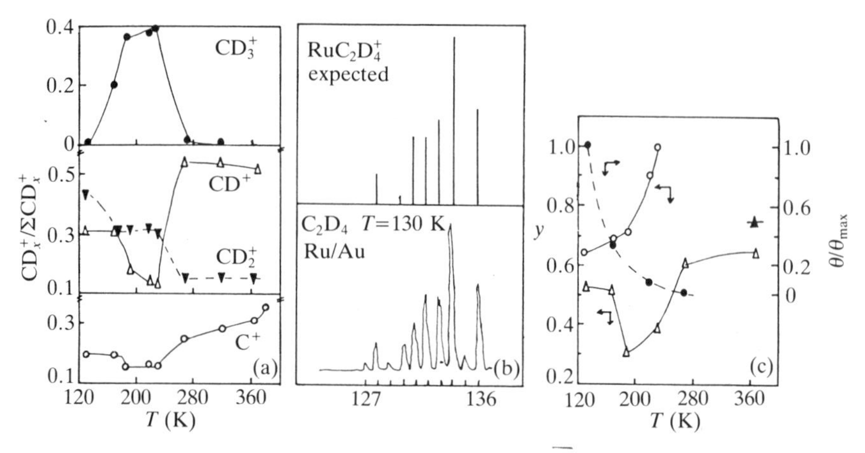

FIG 6.17. (a) Decomposition of d_4-ethene on Ru (0001) monitored by the variation of CD_x^+ secondary ions as a function of Ru surface temperature. (b) Spectrum of molecularly adsorbed d_4-ethene on a gold passivated Ru surface. (c) Reaction of d_4-ethene at a Ru (0001) surface as a function of temperature monitored via the isotope ratio (○) $y=I_{134}/(I_{134}+I_{136})$, which indicates the formation of ethylidyne; (△) $y = I_{124}/(I_{124}+I_{126})$, which monitors the formation of acetylinic species. The coverage of molecular d_4-ethene is monitored via I_{136}/I_{102} (●). Reproduced with permission from Ref.13.

6.6.2. Oxidation of CO over Pd(111)

In defining the mechanism of a surface reaction it is important to be able to measure the variation of the surface coverage of the various reactants, intermediates and, if they adsorbed significantly, the products in the course of the reaction. The SSIMS approach can be used to follow the CO coverage during the oxidation reaction.[22] As the CO oxidation reaction occurred over the Pd surface, the sum of ion ratios was used to monitor the surface coverage of CO. Thus, in Fig. 6.18 as the temperature rises (from right to left) the reaction rate rises and the CO coverage falls. It is interesting to note that maximum reaction rate occurs at a relative CO coverage of about 0.1. The surface coverage of oxygen was monitored by the ratio Pd_2O^+/Pd_2^+ (very little PdO^+ was formed) and the adsorbate density of CO–O couples could be monitored by $PdCOO^+/Pd^+$. Thus the surface coverages of each of the important participants in the surface reaction, O, CO, and CO–O couples, were studied.

The reaction was monitored under O_2-rich, CO-rich, and pure CO atmospheres, i.e. where $P(O_2)/P(CO)$ were 2, 0.5, or 0. Thus, in Fig. 6.19,

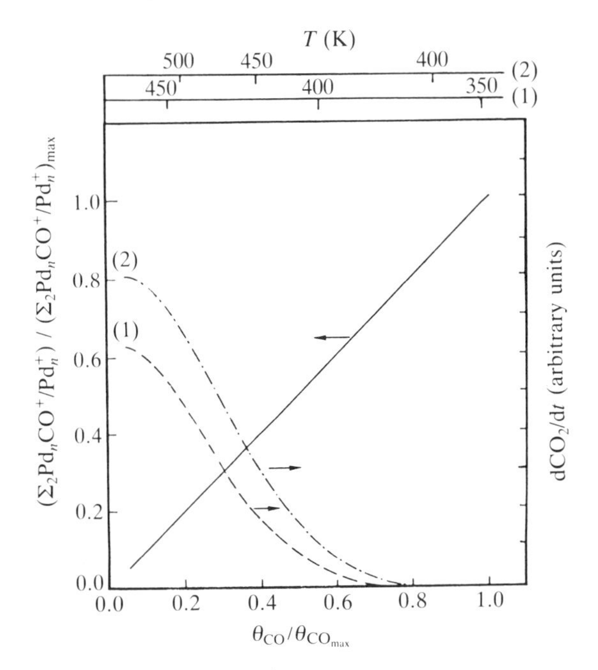

FIG 6.18. Variation of the surface coverage of CO, $\theta_{CO} = \Sigma\,(Pd_xCO^+/Pd_x^+)$ (full line), and of the catalytic turnover rate, dCO_2/dt (dashed line) as a function of temperature and or/or relative CO coverage on Pd (111) for two $CO + O_2$ mixtures: (1), $P_{O_2}/P_{CO} = 2$; (2) $P_{O_2}/P_{CO} = 0.5$. Reproduced with permission from Ref. 22.

the variation of $PdCOO^+/Pd^+$ is presented as a function of relative θ_{CO} measured by SSIMS. The coverage variations reflect the changing reaction rates and stabilities of the adsorbed species as a function of reaction temperature. In this figure the surface temperature increases from right to left but at a somewhat different rate for the different gas mixtures. Thus, at low reaction temperature (on the right), the coverage of CO–O is low for both O_2-rich and CO-rich atmospheres because CO inhibits the adsorption of O_2. As temperature rises, the coverage of CO–O increases as the reaction rate increases and the CO coverage falls, leaving sites available for oxygen adsorption. A maximum in the CO–O coverage occurs because eventually either surface O is consumed too rapidly (in the case of the CO-rich atmosphere) or surface CO is consumed too rapidly (in the case of the O_2-rich atmosphere). Both occur at the same temperature but because CO inhibits O_2 adsorption it occurs at a lower coverage of CO in the O_2-rich atmosphere.

Mechanistically it is important to know whether reaction occurs between randomly distributed CO and O or between CO and islands of adsorbed O. If the CO + O phase is homogeneous and randomly distributed,

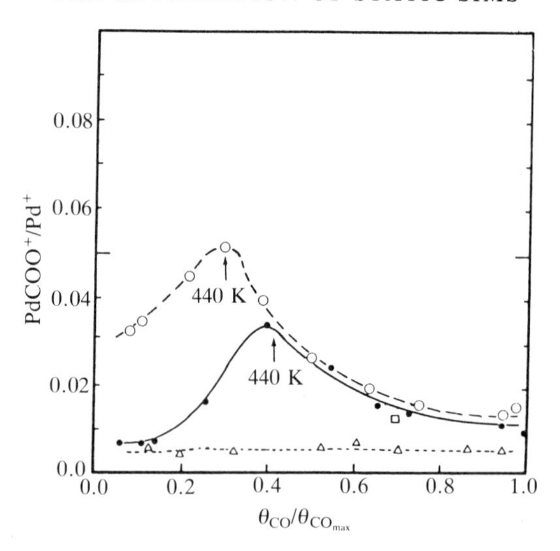

FIG 6.19. Dependence of the $PdCOO^+/Pd^+$ ratio on the relative coverage of CO for two different $CO+O_2$ mixtures: (○) $P_{O_2}/P_{CO}=2$; (●) $P_{O_2}/P_{CO}=0.5$; (△) pure CO, $P_{CO}\times 10^{-5}$ Pa. Reproduced with permission from Ref.23.

$PdCOO^+/Pd^+$ should be proportional to $\theta_{CO}\theta_O$, whereas if islands are involved it should vary as $\theta_{CO}(\theta_O)^{0.5}$. Using the SSIMS data for the coverages of CO and O it was shown that the adsorbates are randomly distributed.

6.6.3. *Other reactions*

Increasing numbers of reaction systems are being studied by SSIMS. Hydrogen isotope exchange in alkylidynes on Pd(111)[23] exploits the ability of SSIMS to mass spectrometrically detect and monitor the various surface intermediates to investigate the reactivity of the hydrogen atoms in adsorbed ethylidyne and propylidyne.

Two rather similar studies have investigated the reactions of surface carbon on Ru(0001) and Ni(111). The carbon is thought to be the intermediate in the Fischer–Tropsch reaction over several metals. The ability of SSIMS to monitor the addition of hydrogen to the surface species is again very important in these studies.[24,25]

The reduction of NO has been studied on a carbon-covered Rh(331) surface using SSIMS and XPS.[26] SSIMS was able to show the formation of adsorbed CN in a rather complex surface reaction.

Surface oxidation can occur as a consequence of the exposure of metals to other gases. This is demonstrated in an XPS, SSIMS and thermal desorption study of the interaction of d_4-methanol with clean aluminium at

273 K.[27] After an initial oxidation phase, the secondary ions OD^-, OCD_3^-, and $AlOCD_3^+$ began to appear. It was concluded that these ions reflect the formation of surface hydroxide and methoxide as methanol dissociatively adsorbed on the partially oxidized surface. The study demonstrated that a partial oxide layer is necessary before methoxide can be formed. The stability of the surface methoxide species so formed was monitored by raising the temperature of the aluminium and monitoring the secondary ion intensities. By 373 K all the methoxy species had decomposed, mostly to form methane.

The potential realized in these studies on well-characterized surfaces has been exploited on supported metal catalysts. Examples can be found in Chapter 7, p. 236.

6.7. Studies of non-metal surfaces

Although the vast majority of surface science studies have had to do with metal surfaces, under the pressure of increasing technological interest in other materials, such as ceramics, bio-polymers, semiconductors, etc., an increasing number of applied investigations using SSIMS are appearing which are directed towards understanding the surface behaviour of such materials.

The mass spectral character of SSIMS data makes it an ideal technique for the investigation of these chemically complex materials. With metal substrates, the number of elements originating from the adsorbent was one or perhaps two if an alloy was being studied. With oxides or compound semiconductors there will be at least two substrate elements and, if solid-solutions or mixed compounds are to be studied, there could be several more. When one considers using polymers as substrates the complexity rises rapidly. Clearly, the mass spectral requirements can be more stringent in terms of mass resolution. There is the possibility that the spectra become so complex that one cannot 'see the wood for the trees'. Under such circumstances there may be the need for tandem MS–MS facilities (see Chapter 7) to enable clarifying daughter ion spectra to be obtained from larger cluster ions of uncertain origin. Since most non-metal studies are technological in orientation, examples will be presented in Chapter 7 which show how the technique can contribute to our understanding of oxide and polymer surfaces.

6.8. Adsorption studies on organic surfaces

Some basic studies of adsorption on organic surfaces of the type familiar in metal studies are just beginning to appear. Obviously, with such chemically complex substrate materials it is important to be able to distinguish the contribution of the adsorbate. SSIMS and its variants can generally do this. There is an interesting example in Section 9.2 of a study using plasma desorption mass spectrometry (PDMS) to investigate the adsorption of Rhodamine on a polymer, PET.

Metal phthalocyanines are organic semiconductors whose conductivity is sensitive to the adsorption of strongly electrophilic gases. The magnitude and reversibility of the conductivity changes is dependent on the central metal ion. The reasons for these differences were not known and XPS has not been able to distinguish any clear effects. A SSIMS study of oxygen and chlorine adsorption on H_2Pc, MgPc, NiPc, CoPc, and PbPc using ToF-SIMS showed rather contrasting behaviour. Treating films of the phthalocyanines in oxygen at 470 K had no significant effect on the SSIMS spectra (see Fig. 6.20(a)). However, although treatment at 590 K left the H_2Pc, CoPc, and NiPc unchanged, in the case of MgPc and PbPc the molecular ion peaks $MPc^{\pm}$ were lost, and in PbPc, Pb_xO_y ion species were found up to high mass. Clearly, dissociation of the complex occurs in the surface layers.

Exposure of the films to 200 p.p.m. Cl_2 at 440 K had little effect on H_2Pc and NiPc, some weak (MPc + Cl) peaks appearing. However, the remaining samples all showed strong Cl^- yields. For MgPc and CoPc, ions of the general formula $MPcCl_x$ ($x = 1 - 8$) were formed (see Fig. 6.20(b)). Very significant effects were seen for PbPc in that Pb_yCl_x ions were strongly evident, with weaker peaks corresponding to H_2PcCl_x, although there was no evidence for $MPcCl_x$. Chlorine interacts with these phthalocyanines in different ways. A mechanism for multiple chlorine adsorption appears to exist for CoPc and MgPc, which suggests that the chlorine is associated with the organic part of the molecule. In contrast, chlorine adsorption on PbPc seems to involve the metal atom.[28]

Although surface science studies of organic surfaces are still in their infancy, the use of SSIMS along with the electron and vibrational spectroscopies should ensure that the tools necessary to make significant progress are there.

6.9. Conclusions

The information which has been shown to be accessible on the 'ideal model' systems described has important implications for SSIMS analysis. First, they show that the technique does not destroy information on very delicate structures; if adsorbate structures are stable then there is a good prospect for all other surfaces. Second, we have very strong evidence that SSIMS is sensitive to the adsorbate chemical structure and will therefore be sensitive to the surface chemical structure of materials in general. That this is true will become very clear from the examples presented in Chapter 7. Third, although it is necessary to analyse the data with care, the various contributions to

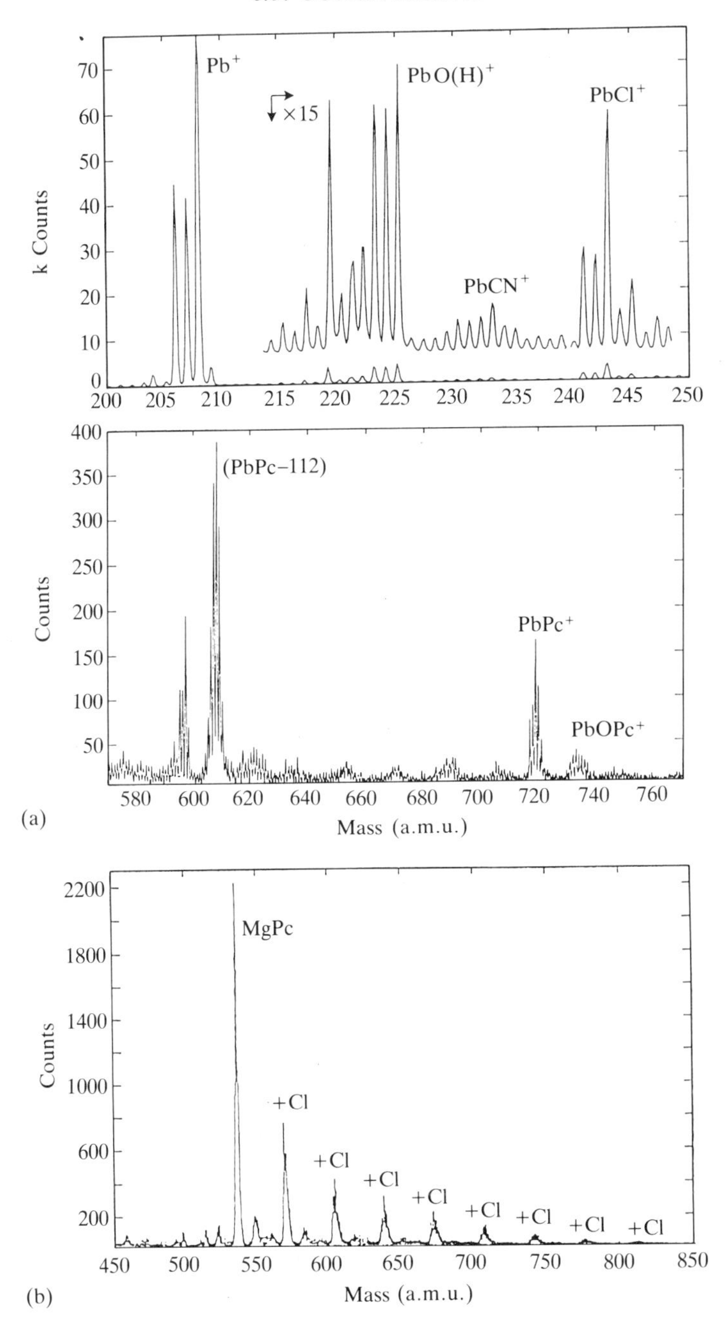

FIG 6.20. (a) Part of the positive ion ToF-SIMS spectrum of lead phthalocyanine obtained with a primary Ga^+ particle dose of 6×10^9. (b) Part of the positive ion ToF-SIMS spectrum of magnesium phthalocyanine showing multiple chlorine addition. Reproduced with permission from Refs 13 and 29.

secondary ion intensities can be dealt with to yield valuable quantitative or semi-quantitative information on the coverage and state of the surface layer. The combination of SSIMS with other surface spectroscopies—XPS and HREELS—and with TDMS make a powerful combination in the study of surface chemical interactions.

References

1. Bauer, E. (1984). In *The Chemical Physics of Solid Surfaces and Heterogeneous Catalysis*, Vol 3B (ed. D.A. King and D.P. Woodruff), p.1. Elsevier, Amsterdam; (1980). In *Phase Transitions in Surface Films* (ed. J.G. Dash and J. Ruvalds), p. 267. Plenum Press, New York.
2. Biberian, J.P. and Somorjai, G.A. (1979). *Appl. Surf. Sci.*, **2**, 352.
3. Brown, A. and Vickerman, J.C. (1984) *Surf. Sci.*, **140**, 261.
4. Müller, A. and Benninghoven, A. (1973). *Surf. Sci.*, **39**, 427; (1974). *ibid.* **41**, 493.
5. Benninghoven, A. (1973). *Surf. Sci.*, **35**, 427.
6. Aggarwal, H., Hirschwald, W., and Vickerman, J.C. (1989). *Surf. Interface Anal.*, in press.
7. Barber, M. Vickerman, J.C., and Wolstenholme, J. (1976). *J. Chem. Soc. Faraday I*, **72**, 40 and (1977). *Surf. Sci.*, **68**, 130.
8. Kishi, K. and Roberts, M.W. (1975). *J. Chem. Soc. Faraday I*, **71**, 1715.
9. DeLouise, L.A. and Winograd, N. (1985). *Surf. Sci.*, **159**, 199.
10. Lange, W., Jirikowsky, M., and Benninghoven, A. (1984). *Surf. Sci.*, **136**, 419.
11. Jirikowsky, M., Holtkamp, D., Klüsner, P., Kempken, M., and Benninghoven, A. (1987). *Surf. Sci.*, **182**, 576.
12. Bordoli, R.S., Vickerman J.C., and Wolstenholme, J. (1979). *Surf. Sci.*, **85**, 244.
13. Vickerman, J.C. (1987). *Surf. Sci.*, **189/190**, 7.
14. Sakakini, B., Dunhill, N., Harendt, C., Steeples B., and Vickerman J.C. (1987). *Surf. Sci.*, **189/190**, 211.
15. Benninghoven, A., Beckmann, P., Greifendorf, D., Müller K.H., and Schemmer, M. (1981). *Surf. Sci.*, **107**, 148.
16. Brown, A. and Vickerman, J.C. (1983). *Surf. Sci.*, **124**, 267.
17. Brown, A., van den Berg, J.A., and Vickerman, J.C. (1984). *Proc. 8th Int. Congr. Catalysis* (Berlin, 1984), p. IV–35.
18. Harendt, C., Sakakini, B., van den Berg, J.A., and Vickerman, J.C. (1986). *J. Electron Spectrosc. Rel. Phenom.*, **39**, 35.
19. Foley, K.E., Winograd N., and Garrison, B.J. (1984). *J. Chem. Phys.*, **80**, 5254.
20. Hills, M.M. Parmeter, J.E., Mullins, C.B., and Weinberg, W.H. (1986). *J. Am. Chem. Soc.*, **108**, 3554.
21. Surman, D., Vickerman, J.c., and Wolstenholme, J. (1980). *Proc. 4th Int. Congr. Surface Sci.* (Cannes, 1980) 525; Sakakini, B., Swift, A.J., Vickerman,

J. C., Harendt, C., and Christmann, K. (1987). *J. Chem. Soc., Faraday Trans. 1*, **83**, 1975.
22. Matolin, V., Gillet E., and Gillet, M. (1985). *Surf. Sci.*, **162**, 354.
23. Ogle, K.M. and White, J.M. (1986). *Surf. Sci.*, **165**, 234.
24. Lauderback, L. L. and Delgass, W. N. (1985). *Solid State Chemistry in Catalysis, Am. Chem. Soc. Symp. Ser.*, **279**, 239.
25. Kaminsky, M.P., Winograd, N., and Geoffroy, G.L. (1986). *J. Am. Chem. Soc.*, **108**, 1315.
26. Delouise, L.A. and Winograd, N. (1985). *Surf. Sci.*, **154**, 79.
27. Tindall, I.F. and Vickerman, J.C. (1985). *Surf. Sci.*, **149**, 577.
28. Eccles, A.J., Humphrey, P., Jones, T.A., and Vickerman, J.C. (1988). *Sixth Int. Conf. on SIMS* (Versailles 1987), p. 1059. John Wiley, Chichester / New York.

7

STATIC SIMS FOR APPLIED SURFACE ANALYSIS

7.1. Introduction

A mass spectral analysis is capable of providing a wealth of information regarding the chemistry of the surface layers of a material. For a true surface analysis it is desirable that only the first monolayer is sampled, since the chemical and physical properties of a surface are determined by the uppermost layers of the sample. SSIMS provides us with the capability to perform such detailed studies. It was in the early 1970s that true surface analysis by SIMS was shown to be possible. Benninghoven[1] defined the conditions for the SIMS technique to create SSIMS, which uses a low-level primary ion beam current density. In this mode a very low erosion rate results, so in order to obtain enough ions for analysis and to minimize fluctuations in the ion yields, a large target area is sampled. Since the ion dose in SSIMS is very small the individual primary ions statistically each hit a different spot on the target and thus damage is minimal. In order to prevent adsorption on the sample surface superseding the erosion rate, the pressure must be maintained in the UHV region.

The analytical conditions required for such an analysis are discussed in detail in Section 2.1. Typically, a beam of primary ion current density of the order of 10^{-9} A cm^{-2} bombards a relatively large sample area (0.1 cm^2). In this manner a sputtering time of several hours may be achieved for a single monolayer. Since the primary beam energy has an effect on the sputter rate of the surface, SSIMS generally uses low primary ion beam energies, of the order of 100 eV–10 keV, hitting the surface at an angle of about 70° to the surface normal in order to maintain surface specificity. With these conditions it is possible to generate fingerprint spectra defining the chemical state of the samples by employing high-sensitivity mass spectrometers, of which there are several types available (see Chapter 4).

SSIMS is a rapidly expanding technique in the surface analysis of many materials; it has wide applications in the fields of semiconductors, catalysis, polymers, oxides, drugs, adhesion, and lubrication, to name but a few. SSIMS is not without its problems, however, and as with many other surface analysis techniques, the most common problem is that concerning the role of surface potential.

7.2. Surface potential

As indicated in Section 2.4, when attempting to analyse insulating materials using standard ion bombardment techniques, a rapid build-up of charge occurs on the surface, which is often severe enough to result in the total loss of the spectrum.[2] Charging may also influence the composition of the sample itself.[3] The diffusion of alkali metals away from the surface is also a result of charging which will result in an inaccurate analysis.[4]

Under positive ion bombardment an insulating sample will charge-up positively. Assuming that there is no leakage of current between the surface and earth then the surface will charge to the energy of the beam. Furthermore, as the voltage increases it may interfere electrostatically with the incoming primary particles, deflecting them away. In practice, of course, most samples are less than perfect insulators under such a voltage, so the charge built up will be less than the energy of the incoming beam. It will, however, still be sufficient to greatly degrade the spectrum obtained. This is the situation illustrated in Fig. 7.1. We see that as the surface potential on an insulating sample increases, the intensity of the copper secondary ion signal increases at first then decreases rapidly.

The consequences of this are illustrated schematically in Fig. 7.2. The analyser system will have a restricted energy acceptance window. If an insulator charges by a voltage ΔV, then the KE distribution shifts to ($E_0 + \Delta V$) for positive ions and ($E_0 - \Delta V$) for negative ions. The energy of the positive ions will be too high to be detected whilst negative ion emission may be suppressed. Quadrupole instruments are particularly vulnerable to surface

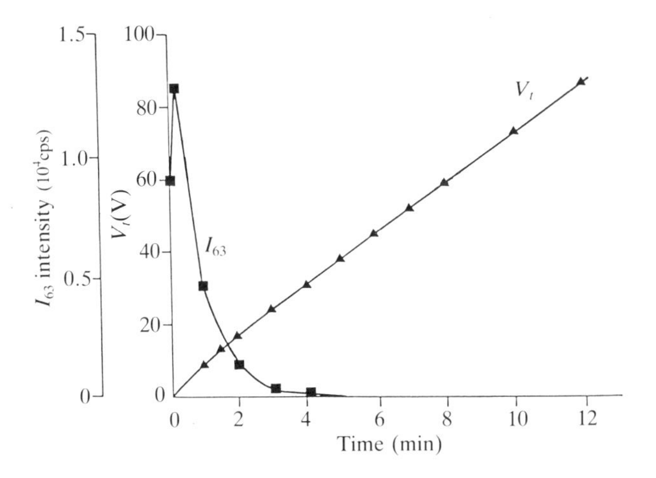

FIG 7.1. Change in Cu intensity as target voltage increases with time.

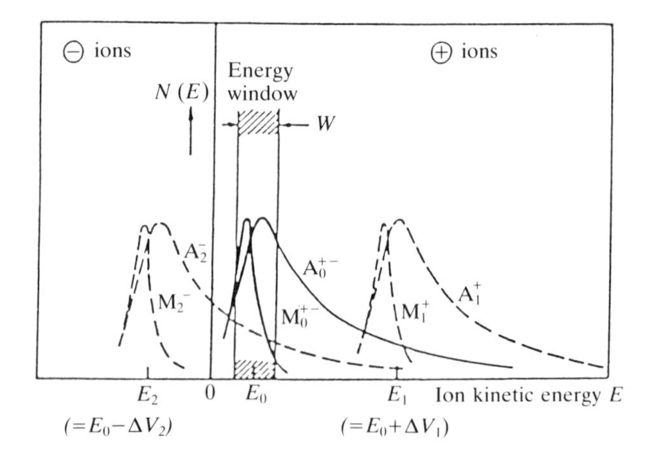

FIG 7.2. Schematic representation of the kinetic energy distributions of positive and negative atomic (A) and molecular (M) secondary ions in relation to the acceptance window of the energy analyser. Solid—no charging; dashed line—charging. Reproduced with permission of Pergamon Press Ltd from van den Berg, J. A. (1986). *Vacuum*, **36**(11/12), 985, Fig.5.

potential changes since the energy acceptance window to obtain a well-resolved spectrum is between 5 and 10 eV. ToF and magnetic sector instruments are less susceptible to this changing surface potential for positive ions since they employ a high extraction field and have a wider energy pass band.

7.2.1. Control of surface potential—electron beams

There are various methods used to overcome these 'charging problems', which include mixing silver or graphite with the sample or placing a conducting grid over the surface and applying a variable bias.[5] These methods do not always provide a uniform surface potential across a sample, which may lead to ion yields that are not truly representative of the surface composition. The most common method employed to compensate for such problems is to use a low-energy electron flood gun.[3,6]

A carefully balanced electron beam will compensate for some of the excess positive surface potential, allowing this potential to be lowered back into the energy acceptance window of the analyser. However, excess electrons will make the surface potential become negative, and the positive signal will be lost. Although this is often a suitable method of charge neutralization it can sometimes produce problems such as electron-stimulated ion emission

(ESIE) which is particularly common when attempting to analyse polymers.[7] Furthermore, when endeavouring to analyse powders in this manner it is found that the different particles in the sample may have different neutralization parameters, leading to sudden discharges from the surface due to interparticle resistances.

For 500 eV electrons the coefficient of secondary electron emission from metals is near unity; however,for oxides and other insulators this figure can be much greater than one,[8] so the available electron current to effect charge neutralization for a positive ion beam would be greatly reduced for these materials. In practice, it appears that a negative surface potential can be obtained for the emission of negative secondary ions if the electron beam current is several times that of the primary beam current. This is in agreement with some of the observations made in the analysis of polymers.[9,10]

7.2.2. Control of surface potential—neutral primary beams

Surface charging is very successfully curtailed by minimizing the charge being put into the surface initially by the use of a neutral beam, i.e. FABMS.[11,12] Bombardment by a neutral beam immediately reduces the problem of sample charging to that associated with particle emission.[13] Under atom bombardment the sample will usually charge slightly positively due to the emission of secondary electrons, although this is not usually so severe as to disrupt positive ion emission. The emission of electrons can be reduced by applying a positive potential directly to the sample. Any electrons emitted, as a result of uncollected positive ions and neutral species colliding with the interior of the vacuum system, will serve to counterbalance any residual charge as a result of emitted positive ions. However, for optimum negative ion emission the surface potential must be made negative by an input of electrons. This surface potential situation will result in the suppression of the positive secondary ions.

Since under atom bombardment the samples tend to charge slightly positively it has been suggested that a negative primary ion beam would alleviate the problems. Negative ions are formed in the discharge region of plasma ionization sources such as duoplasmatrons in dynamic SIMS instruments. They are generally formed less efficiently than positive ions but typically by reversing the extraction voltages appropriately a negative ion beam can be formed at ~10 per cent of the flux of the positive beam. Although such beams are commonly found in ion microscopes, the difficulty with negative primary ion bombardment is that the surface potential tends to stabilize at a negative potential that is too high for a quadrupole analyser.

7.2.3. Electron neutralization vs. ESIE

There is a delicate balance between the electron flux needed to neutralize surface charging and that which gives rise to ESIE. This has been studied by Humphrey[14] for a variety of polymers. The variation of secondary ion emission as a function of electron beam flux is very similar for all the polymers and is typified by the behaviour of the F^- signal from 0.75 mm thick poly (tetrafluoroethylene) (PTFE) film plotted in Fig. 7.3. The graph shows the F^- signal generated by electron bombardment alone (ESIE) and by simultaneous atom and electron bombardment.

The ESIE yield becomes significant at higher electron currents where it then becomes a major effect. Atom bombardment alone produces a steady though rather low secondary ion signal. When the electron flux is increased while the atom flux remains constant a level is reached at which the secondary ion signal has increased by a factor of ten while it can be seen from the other curve that ESIE alone would not be a major contributing factor at this point.

It must be concluded that, at this stage, the departing secondary electron flux due to atom bombardment is being matched by the incoming primary electrons, and the surface potential is stabilized close to zero or somewhat negative. As the primary electron flux is increased further the surface potential is made more negative and negative secondary ions will be accelerated beyond the energy acceptance of the analyser, so the signal falls.

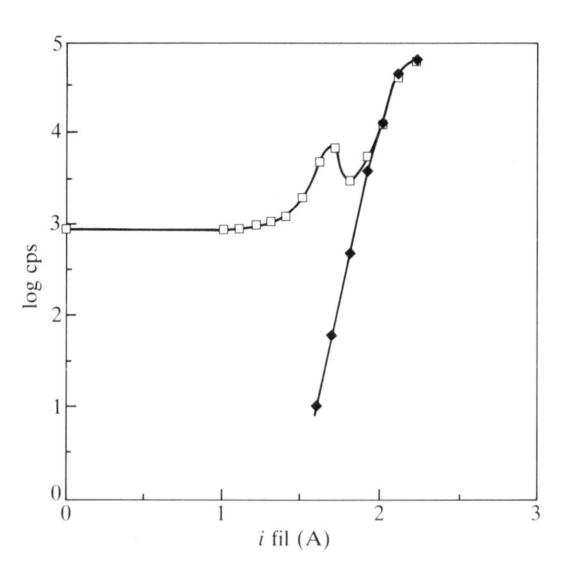

FIG 7.3. Variation of secondary ion emission as a function of electron beam flux. □, atoms and e^-; ♦, e^- only.

Increasing the electron beam flux yet further rapidly increases the contribution of ESIE and this becomes the dominant effect, as can be seen by the precise match with the ESIE curve.

It is clear that the conditions for electron-neutralized SIMS (SIMS-EN) without ESIE are a very narrow set of values which have to be carefully defined for each material analysed. It is interesting to note that the primary electron flux arriving at the sample at the conditions for SIMS-EN is often considerably higher than the secondary electron current lost as a consequence of primary atom bombardment. The reasons for this cannot be stated with certainty. The mechanism of electron-induced neutralization involves many parameters. As was illustrated earlier, the degree of sample charging experienced under atom bombardment was +10 to 30 eV. This will not be sufficient to trap a 500 eV primary electron, but electrons of this energy flooding a broad area will also produce lower-energy secondary electrons from the sample and from the surrounding hardware. These electrons may then be attracted into the charging sites on the sample to neutralize the charge build-up. Providing a metal plate around the sample or a grid over it from which electrons can be formed seems to allow easier establishment of neutralized conditions.

The arguments here applied to neutralization under atom beam bombardment also apply also to ionic primary beams. However, when an ion beam is used, the charges developed are not only greater but they change more rapidly. This serves to make the achieving of SIMS-EN more difficult in practice but no less impractical in theory. (See the Appendix to this chapter for a systematic protocol which enables the correct conditions for SSIMS analysis to be obtained quickly and reproducibly.)

7.2.4. Consequence of incomplete control of surface potential

The different coefficients of secondary electron emission from different materials in the same sample will lead to differential charging which makes spectral acquisition and interpretation more difficult. This problem reveals itself in several ways, for example,[15] in a study of an alumina–copper system, under Ar^+ bombardment no secondary ion signal could be obtained for the aluminium from the alumina, although copper was readily detected. The application of a primary atom beam, however, resulted in a high yield of positive secondary ions from the copper and the alumina even without a compensating electron beam.

The absence of any detailed information in the negative spectrum is also often a consequence of a lack of adequate charge neutralization, as is indicated in Fig. 7.4. Fig. 7.4(a) shows the negative ion spectrum of poly (ethylene terephthalate) with insufficient charge neutralization. There is no useful information observed above $m/z = 25$. If we now compare this with

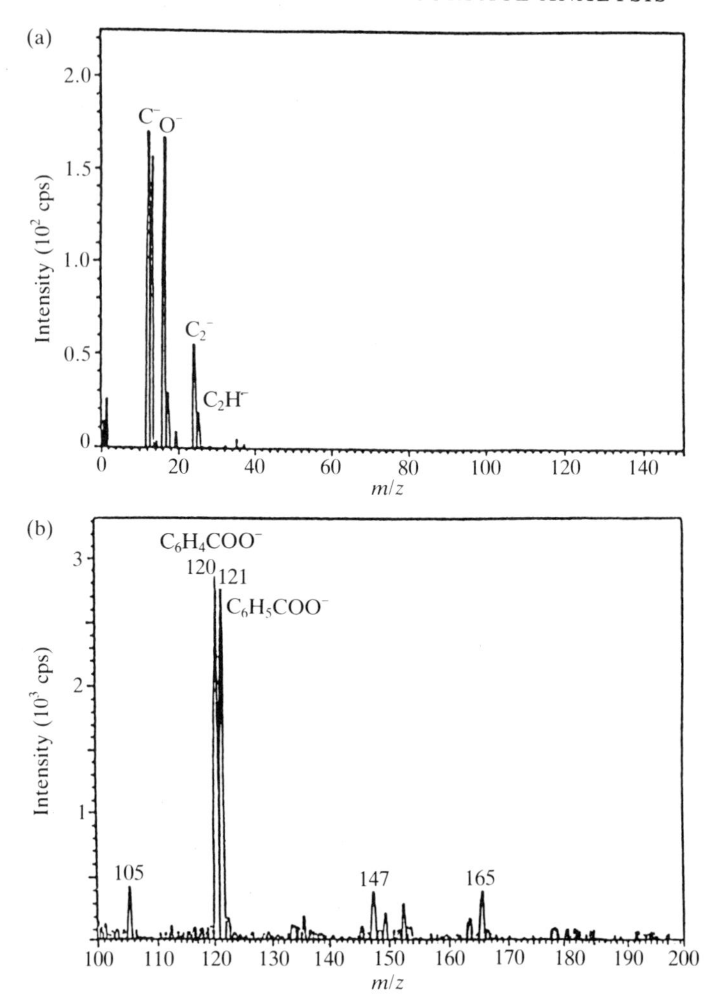

FIG 7.4. Effect of optimum charge neutralization of polyethylene terephthalate. (a) With insufficient charge neutralization; (b) with charge neutralization optimized. Reproduced with permission from Ref. 9.

Fig. 7.4(b), where the charge neutralization has been optimized, there are now intense signals in the m/z = 100–200 region, which was previously uninformative. This clearly demonstrates the importance of overcoming the problem of sample charging to avoid loss of useful information, which frequently is found at the higher mass region of the spectrum.

Using FAB or electron neutralization enables SSIMS mass spectra to be obtained which define the elemental composition and detailed chemical states of the surfaces of both conducting and insulating materials, for all elements, to a detection limit in the p.p.m. region. This makes SIMS a tremendously useful technique in many fields of surface technology.

7.3. Comparison of ToF and quadrupole instruments

A number of different analysers are used in SSIMS; perhaps the most common is the quadrupole, although the time-of-flight (ToF) is becoming more widely used. The quadrupole has several points in its favour, namely, compactness and the fact that it is relatively inexpensive. However, compared with a ToF instrument, the transmission of ions through the quadrupole is low, of the order of 1 in 10^3. Furthermore, a quadrupole must scan the mass range to produce a spectrum so only one mass at any given time is being transmitted, thus 'wasting' many of the particles sputtered from the surface. Taking all these points into account, possibly as few as 1 ion in 10^6 is actually detected, which imposes great limits on the sensitivity of the quadrupole technique. A further limitation which is difficult to overcome is that many polymer and pharmaceutical problems demand good resolution over increasingly high mass ranges. Depending on the electronic configuration of the quadrupole it will have an upper mass limit, commonly around $m/z=$ 1200–2000. At high masses the transmission is reduced (transmission $\propto M^{-1}$) so the higher mass signals are attenuated, which is a problem since the higher mass peaks are frequently the weaker signals in the spectrum. With the advent of liquid metal guns, small area analysis of features down to 100 nm diameter are increasingly being demanded; with such a small area there will only be 10^5 ions per monolayer available for sampling so that several monolayers may have to be consumed for one analysis, thus destroying any possibility of SSIMS analysis. As a result of these problems the ToF instrument is becoming increasingly popular, since it allows a high and uniform transmission for all masses of up to 10 per cent, parallel detection of all species, so no ions produced are 'wasted', and, theoretically, an unlimited mass range. The mass resolution may be around 10 000 for a reflectron type ToF analyser; resolutions up to 13 000 have even been reported.[16] A discussion of the ToF system is presented in Chapter 4.

ToF instruments do suffer from some drawbacks. The first important problem is that of analysis time; compared with a quadrupole this may be 10 times longer (see Section 4.4.3). Insulators also pose a problem when analysed by ToF since charging does occur even though the primary beam fluence is much smaller than for quadrupole systems. Experiments using similar charge compensation methods as in conventional quadrupole analyses, i.e. use of continuous low-energy electron flooding, have been tried for ToF systems, but with little success. The main effect of such efforts has been to produce noise in the spectrum, possibly due to ESIE. A more successful method has been the use of a pulsed electron beam system. This neutralizes the surface between primary ion pulses. The electron beam must operate in the pulsed mode in order to prevent electrons from being accelerated on to the sample in the positive ion mode when the extraction voltage is on. Such electrons could acquire energies up to 3 keV, enough to sputter particles from the surface, i.e. to create ESIE. However, this method also requires pulsing of the extraction field between zero, when the electron beam is on, to keV energies, when the electron beam is off, for ion extraction.[17,18]

Generally the spectra produced by the ToF and quadrupole systems are very similar. (Fig. A5.1) The relative intensity of the high mass ions is usually higher in the ToF spectra and the primary ion beam fluence required for an equivalent spectrum is about 10^4 times less for the ToF than for the quadrupole analyser. In some cases there are significant differences because in the ToF system the secondary ions are extracted very rapidly (in ns); consequently, metastable ions have less time to decay and be detected in the ToF than in the quadrupole system.

7.4. Primary beam effects

The parameters of the primary beam are important in the generation of a useful spectrum, i.e. one which is not characteristic of a *damaged* surface or a result of a charging sample. It is important to try to understand these effects in order to be able to optimize the conditions for a SIMS experiment. For example, in the analysis of an organic surface it is useful to be able to maximize the yield of the higher mass clusters, since they make the fingerprint of a particular sample unique. It might be expected that the use of a heavier primary particle would assist in this optimization of the larger clusters. Similarly, increasing the primary beam energy might be expected to result in increased sample damage, which is undesirable. The parameters with which we are mainly concerned here are fluence, particle mass and energy.

7.4.1. Fluence

This is perhaps the most important parameter governing the sensitivity and the rate of generation of structural damage. As discussed in the section on delicate materials, we see that a change in the relative intensity of clusters is indicative of damage. Too high a fluence will result in a damaged sample, thus giving misleading analyses. The variation of PMMA major peaks with changing primary ion dose is shown in Fig. 7.5. At low fluences damage is minimal; there is no major decline in the characteristic ions. At higher doses the intensity of these peaks decreases and is accompanied by an increase in the

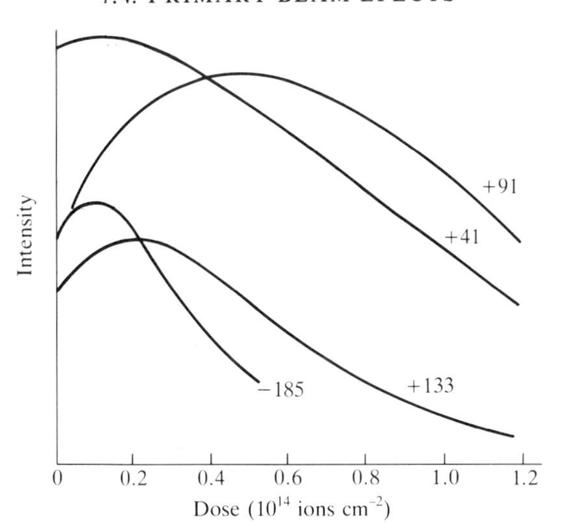

FIG 7.5. Change in cluster intensities of the PMMA major peaks with changing primary ion dose. Reproduced with permission of Pergamon Press Ltd from Briggs, D. and Hearn, M. J. (1986). *Vacuum*, **36**, 1007, Fig.5.

clusters characteristic of a damaged surface, i.e. a $m/z = 133$ cyclic structure which forms as the polymer becomes unsaturated. Eventually aromatic units are formed, such as the $m/z = 91$ cluster. This peak goes through a maximum at a higher fluence than the other peaks, and still further damage occurs at higher doses, revealed by the fact that the overall cluster intensities decrease as a consequence of cross-linking and the production of a carbon-rich surface. Polymer damage can be followed by monitoring oxygen loss because the CH^-/O^- ratio rises with ion dose.

It is concluded from these studies that 10^{13} primary particle impacts cm^{-2} is the maximum dose which will generate a SIMS spectrum which reflects the chemistry of an undamaged surface. In Section 2.1.1 it was pointed out that each primary impact influences an area of 10 nm in diameter. Thus it is estimated that 10^{13} impacts cm^{-2} would result in the whole surface being affected. The experimental data clearly accord with this estimate. Sensitivity to bombardment-induced damage does vary. Organic materials show evidence of this at the 10^{13} impacts cm^{-2} level, whereas many inorganic materials can sustain significantly higher doses before damage is evident.

Taking these primary beam requirements together with charge neutralization conditions, a typical set of parameters can be specified.[19] For a 4 keV primary Ar^+ beam, rastered over a 5 mm × 5 mm area, where the ion current is typically 0.3–0.6 nA, to effect neutralization a 700 eV electron beam is defocused in order to flood a greater area than that of the impinging primary ion beam. The precise current density is typically 1–5 nA cm^{-2}, approximately three times the primary ion beam current. It is frequently found that if the electron beam is directed onto the edge of the sample then neutralization is most effective, probably due to low-energy secondary electrons generated outside the sample area from the edge of the sample-holder or surrounding components.

7.4.2. Primary particle mass and energy transfer

The influence of the primary beam energy has been discussed in Section 2.5.4. The effect of different energies and masses on cluster yield can be seen by looking at a series of cluster intensities recorded under different conditions.[20] The polymers LDPE (low density polyethylene), P^tBuMA poly(*t*-butyl methacrylate), and PEO, poly(ethylene oxide), were studied for the purpose of this investigation. Looking at Table 7.1 we can see that there is a general trend to increasing intensity in the higher mass clusters with increasing primary ion mass and energy. Considering first increasing only the mass of the bombarding particles, whilst maintaining the energy constant, the total ion yield is increased (Xe $>$ Ar $\approx$ Ne $\gg$ He) concomitant with an increase in the partial ion yield of the higher mass clusters. The variation of relative cluster intensities of P^tBuMA vs. the mass of the primary ion is shown in Fig. 7.6. In the case of Xe, where ion yields have been measured for varying energies, the same increases in total and partial ion yields are observed with increasing energy (see Fig. 7.7). Thus the trend observed is that

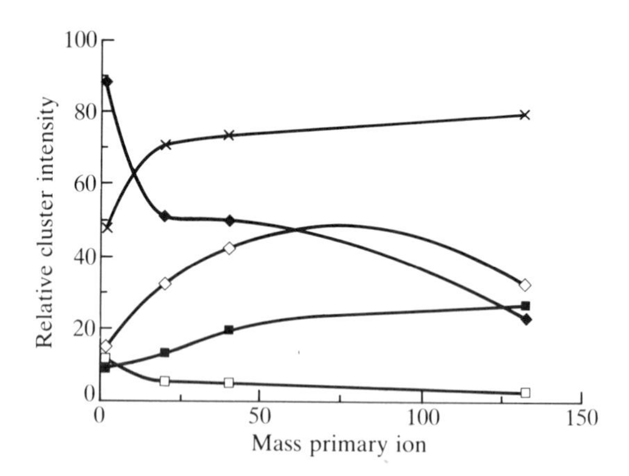

FIG 7.6. Relative cluster intensities of P^tBuMa as a function of changing primary ion mass. □, 15; ◆, 29; x, 57; ◇, 71; ■, 85.

TABLE 7.1

(a) Positive secondary ion relative intensities for LDPE

Ion	Energy	Cluster m/z/relative intensity								$\Sigma I^{\dagger}$
	(keV)	15	29	43	57	71	85	99	113	(Counts s^{-1})
(e^-)	0.7	6.6	111	100	47	16	5.9	0.4		
He^+	4	12	88	100	48	15	9.1	2.1		9.0×10^3
Ne^+	4	4.8	51	100	71	32	13	2.9	0.3	8.5×10^4
Ar^+	4	4.7	50	100	73	42	19	3.1	1.0	5.6×10^4
Xe^+	4	2.1	23	100	79	32	26	3.2	1.2	1.5×10^5
Xe^+	1	3.1	35	100	88	35	20	2.1	1.0	8.2×10^4
Ga^+	10	6.9	34	100	77	43	18	5.4	2.0	1.1×10^5

(b) Positive secondary ion relative intensities for P^{t}BuMA

Ion	Energy	Cluster m/z/relative intensity										$\Sigma I^{\dagger}$
	(keV)	15	29	41	57	69	81	95	109	121	141	(Counts s^{-1})
(e^-)	0.7	0.9	47	76	100	1.3	0.5	0.4	0.4	0.5		
He^+	4	7.0	40	75	100	1.8	0.2					1.3×10^5
Ne^+	4	6.5	47	66	100	6.5	2.2	2.2	3.4	2.3		5.4×10^5
Ar^+	4	7.7	46	63	100	7.7	1.5	1.2	2.4	1.8	0.8	4.6×10^5
Xe^+	4	4.9	41	63	100	22	3.3	4.1	7.3	4.9	3.3	9.4×10^5
Xe^+	2	5.0	31	56	100	8.7	2.5	2.5	4.4	2.9	1.9	5.2×10^5
Xe^+	1	3.8	31	46	100	6.9	2.1	2.3	3.8	2.7	1.5	1.3×10^5
Ga^+	10	5.5	28	50	100	14	5.5	7.8	10	8.3	2.8	1.1×10^6

Table 7.1(c) Negative secondary ion relative intensities for PEO

Ion	Energy	Cluster m/z/relative intensity								ΣI †
	(keV)	13	16	25	43	61	85	133	219	(Counts s^{-1})
He^+	4	80	100	47	13					1.3×10^4
Ne^+	4	74	100	56	6.5	2.5				7.6×10^4
Ar^+	4	84	100	51	3.0	0.5				6.0×10^4
Xe^+	4	67	100	91	64	32	5.4	4.6		2.4×10^5
Xe^+	2	48	37	82	100	49	15	18	5.2	8.5×10^5
Xe^+	1	80	100	84	39	21	5.9	14		8.9×10^4
Ga^+	10	100	95	46	0.4					1.1×10^5

†Summation of absolute intensities of tabulated peaks, using a constant ion primary beam current density of 6.7×10^{-10} A cm^{-2}

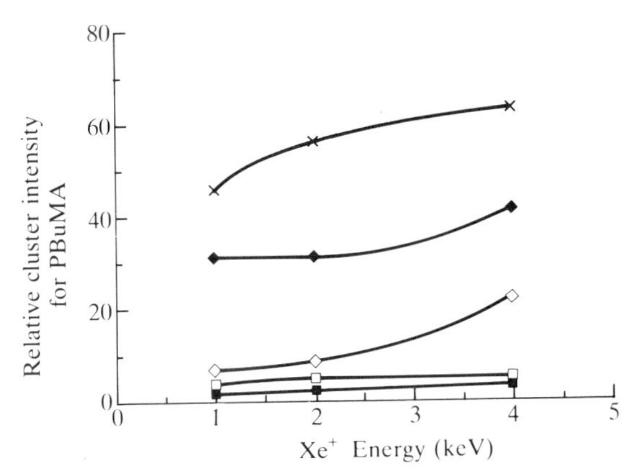

FIG 7.7. Relative cluster intensities of P^{t}BuMa as a function of primary ion energy. □, 15; ♦, 29; ×, 41; ◇, 69; ■, 81.

increasing the primary beam mass and/or energy tends to result in an increased yield of the higher mass clusters.

This conclusion was supported in an analysis of spin cast polystyrene films on a silver substrate.[21] A comparison of primary bombarding particles also suggested that a more intense spectrum resulted from xenon primary beams as opposed to argon and it was proposed that this is due to a greater momentum transfer in the case of the Xe bombarding particles (see Fig. 7.8).

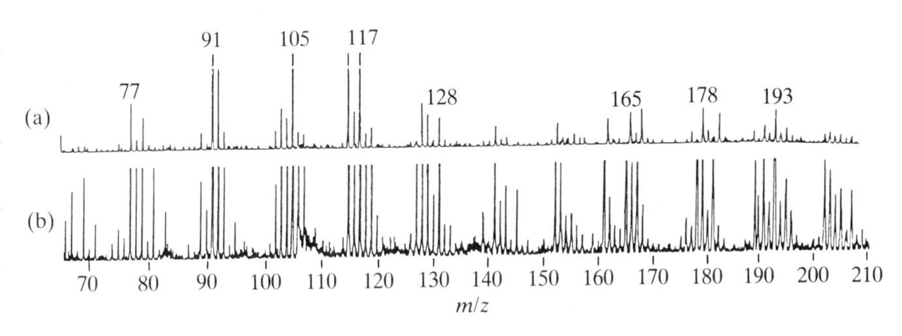

FIG 7.8. (b) shows the more intense polystyrene spectrum obtained using the higher mass primary bombarding particles Xe compared with (a) which was acquired using Ar. Reproduced with permission, from Ref. 21.

7.4.3. *Atom vs. ion beam damage*

Sputtering as a result of ion bombardment is generally partially explained in terms of the incoming ion beam being neutralized just before entering the sample surface. If a charged particle is implanted in the surface of an insulating material then to maintain electrical neutrality there must be some redistribution of charge which will inevitably result in some bond breaking and molecular fragmentation. If we look at the primary ion and primary atom spectra of polystyrene (Fig. 7.9) recorded with the same current density, two significant differences can be seen.[22] First, the overall cluster intensity is much greater for the neutral primary beam spectrum and, second, the low mass clusters are more intense in the ion-induced spectrum than in the neutral beam spectrum. The rate of decay of high mass clusters with beam fluence has been shown to be over four times slower under atom bombardment than under ion bombardment.

A further comparison of the different primary beams is indicated in the investigation of the decay profiles of niobium pentoxide. Under Ar^0 bombardment the NbO^+ increases initially, over time, and then levels out.

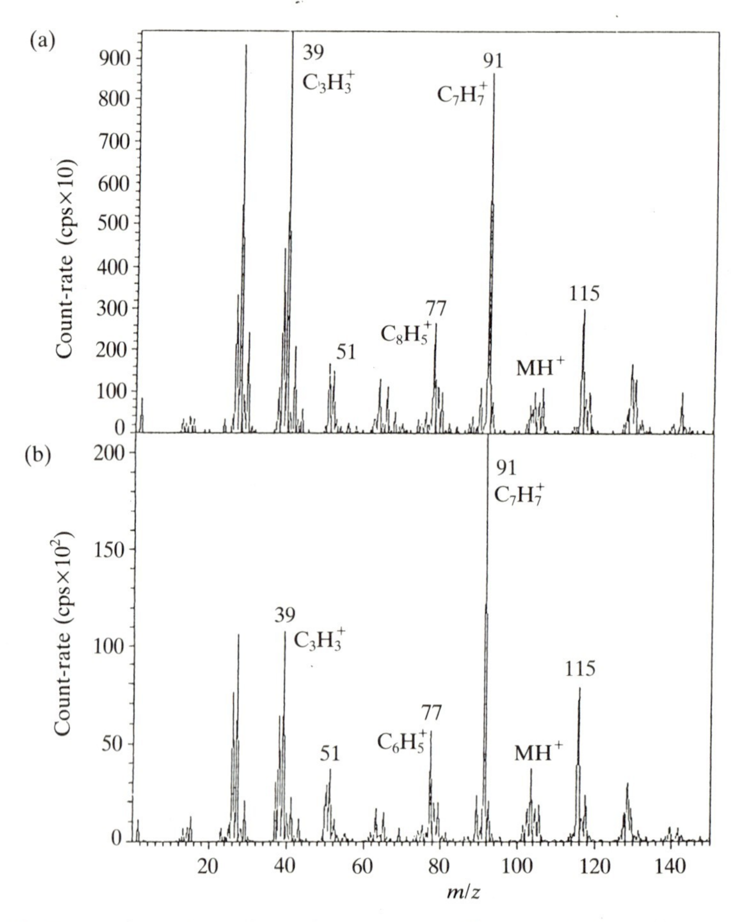

FIG 7.9. Comparison of (a) primary ion spectra and (b) neutral primary beam spectra of polystyrene. (a) SIMS; 2 keV Ar^+; 3×10^9 particles $cm^{-2}s^{-1}$. (b) FABMS; 2 keV Ar^o; 3×10^9 particles $cm^{-2}s^1$. Reproduced with permission from Ref. 22.

Comparing this with the ion bombardment spectra, after the initial rise in intensity there is a sharp decrease for both NbO^+ and NbO_2^+. If we consider a positive particle being implanted in the oxide lattice then to effect neutralization there could be a loss of an electron from the metal oxide bond resulting in the loss of oxygen from the surface; hence the decrease in the oxygen-containing clusters under ion bombardment.

7.5. Spectral interpretation

The primary aim in SSIMS is to relate the mass spectra produced to the surface chemical structure. For a simple compound, spectral interpretation is usually not a serious problem since the number of possibilities for a given ion is limited; this is particularly true for inorganic systems. In the analysis of complex mixtures, particularly of organic materials, spectral assignments are facilitated by first obtaining a spectrum of the pure substances in the mixture. Such an approach would benefit from the provision of a SIMS spectral library such as is currently available in organic mass spectrometry. The first edition of which has recently been publisbed.[23]

In some cases, peak assignment may require verification if there is a choice as to the possible structure of the cluster. For example, in the case of the cluster ion at $m/z = 69$ in the positive ion mass spectrum of PMMA, the first assignment was made as $C_5H_9^+$, a cyclic dimethyl propyllium ion. Recently an alternative structure corresponding to $CH_3{\cdot}C{\cdot}\,CH_2CO^+$ has been proposed. Accurate mass measurement could be employed to discern between the two possible structures if a high-resolution magnetic sector instrument was available. Otherwise another method is clearly required to clarify the nature of such a cluster. A SSIMS analysis was carried out on *perdeuterated* PMMA;[24] the peak at $m/z = 69$ now appeared at $m/z = 74$, i.e. five mass units higher, indicating that there were only five hydrogen atoms in the cluster, which is good evidence for assigning the structure of the cluster to that of $CH_3{\cdot}C{\cdot}CH_2CO^+$. The method of isotopic labelling has great potential in the verification of organic polymer clusters.

It is apparent that tandem mass spectrometers able to perform parent/daughter ion analyses will also contribute to such spectral interpretation problems.

7.5.1. *MS–MS*

To facilitate spectral interpretation tandem mass spectrometery is increasingly being used in SIMS. Magnetic sector based systems are very expensive consequently triple quadrupole systems have been investigated. The triple quadrupole system is an extension of the single quadrupole instrument to incorporate two extra quadrupoles into the path of the ions. Q2 is not operated as a mass filter but operates in r.f. mode only to contain and focus

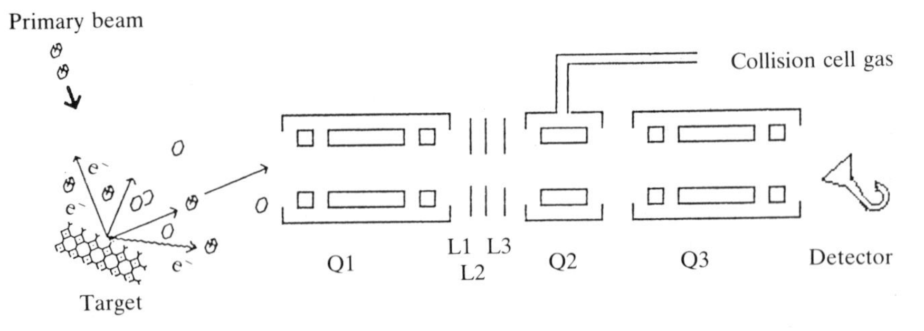

FIG 7.10. Schematic diagram of the triple quadrupole analyser system.

any ions transmitted from Q1 through Q2 into Q3 and then to the detector. The second quadrupole is filled with collision gas, generally an inert gas such as argon. In this collision chamber, fragmentation of the parent produces daughter ions which are transmitted into Q3. See Fig. 7.10 for a schematic diagram of the triple quadrupole system. There are several modes of operation of this type of instrument.

Daughter scans. This is the most commonly used mode of operation. Here Q1 selects the chosen parent ion; only this ion will be transmitted by Q1 into Q2. The third quadrupole is scanned over a selected mass range to detect any daughter ions produced by fragmentation of the parent ion in Q2, which are transmitted into Q3.

Parent scans. This is a reverse of the first process. Q3 is set to transmit only one preselected daughter ion. Q2 is filled with gas as in the daughter scan mode, and Q1 is scanned through the mass range to detect any parent ion which gives rise to the selected daughter ion.

Neutral loss scans. In this case Q1 and Q3 are scanned simultaneously with a constant defined mass difference corresponding to the mass of the neutral molecule lost.

Collision-induced dissociation (CID) efficiency is commonly between 5 and 15 per cent and can be as high as 80 per cent as opposed to that obtained in a multisector instrument which may only be $\approx$ 1 per cent. The daughter ion intensities will be of lower intensities than the parent ion so it is necessary to have a significant parent ion signal before a satisfactory daughter ion spectrum can be collected. The mass of the collision gas molecules in the second quadrupole has an effect on the parent ion fragmentation; in general, the heavier the collision the greater the degree of fragmentation (see Table 7.2).

The triple quadrupole instrument is a valuable tool from both a research and an industrial viewpoint as it may provide an insight into the fragmentation processes of secondary ions and an aid to the assignment of structures

TABLE 7.2

Effect of collision gas on parent ion fragmentation

Parent	Daughter	% fragmentation		
		He	Ar	Xe
VO^+	V^+	0.1	0.5	0.2
VO_2^+	VO^+	0.75	4	1.4
V_2^+	V^+	2	4	10
V_2O^+	V^+	0.1	1.3	2.5
	VO^+	0.05	0.6	1.25
$V_2O_2^+$	VO^+	0.7	2	4
$V_2O_3^+$	VO^+	0.3	3	10
$V_2O_5^+$	V^+		3	5
	V_2O^+		1	3
	$V_2O_2^+$		2.5	5

whose compositions are unknown. As an example, consider the analysis of a series of $Al_{2-x}Cr_xO_3$ solid solutions mounted in indium. It was unclear whether the parent peak detected at $m/z = 131$ ought to be attributed to

$AlCr_2$ $52 + 52 + 27 = 131$
or InO $115 + 16 = 131$

or a contribution from both clusters. An MS–MS analysis was performed which revealed only one daughter ion at $m/z = 115$, thus the peak was attributed solely to InO^+ as there was no evidence of Al or Cr in the daughter ion spectrum.

A further example is shown in the analysis of two materials being investigated for use in the drug industry. Hydroxypropylmethylcellulose (HPMC) and hydroxypropylcellulose (HPC) both show a peak at $m/z = 59$. Superficially this would suggest that the same fragmentation process had occurred in both samples to produce the same cluster; however, on further analysis it was discovered that the two parent ions gave different daughter ion spectra.

Sample	Parent	Daughters	Loss
HPMC	59 ⟶	25, 41, 43	34, 18, 16
HPC	59 ⟶	31, 41, 43	28, 18, 16

On this basis two different structures have been proposed for the fragments (see Fig. 7.11).

The loss of $m/z = 34$ in HPMC is attributed to two OH groupings, to leave C_2H. Both structures can conceivably lose 18, equivalent to HOH, to give the observed $m/z = 41$ structures; similarly, a loss of 16 will produce the

```
(a)                              (b)   H
      H   OH  H                        |     //O
      |    |  |                     +C—C
  H—C—C—C+                            |     \OH
      |    |  |                        H
      H   H   H
```

FIG 7.11. Proposed structures of the 59 m/z clusters: (a) 59 fragment HPC; (b) 59 fragment HPMC.

43 peak, but only HPC shows a signal at $m/z = 31$, i.e. a loss of $m/z = 28$, indicating C_2H_4. This suggests that the $m/z = 59$ peak in HPMC originates from splitting of the ring whereas the HPC fragment is a result of cleavage of the side-chain 2-hydroxypropyl ether substituent.

The lack of complexity in performing an analysis using the triple quadrupole mass spectrometer renders it compatible with computers both for data collection and handling and instrument control.

7.5.2. Fingerprint spectra—chemical structure—organic materials

The ability of SSIMS to provide precise surface chemistry information on organic materials is illustrated particularly well by a number of studies on polymer surfaces. An early study in the analysis of *n*-butyl, *sec*-butyl and *tert*-butyl polymers[25] pointed the way. SSIMS highlighted clear differences in the spectra of the isomers. The butyl cation $C_4H_9^+$ showed varying relative intensities ranging from 0.12 for the *n*-butyl, 0.41 for the *sec*-butyl and 0.22 for the *tert*-butyl; this is in accordance with the predicted stability series for the carbonium ions.

The fingerprinting abilities of the technique are demonstrated in a series of systematic studies of polymers. Consider the positive and negative ion spectra obtained from PET (Fig. 7.12). The positive spectrum of PET in the region $m/z = 15$–91 consists of non-diagnostic hydrocarbon fragments. Above this region the spectrum is simply interpreted in terms of bond cleavage to release stable fragments of OH, CO, COOH. This leads to proposals for the structures of the larger fragment ions (see Fig. 7.13). The ions detected at $m/z = 191$ and 193 are characteristic of the monomer repeat unit ± 1.

In the negative ion spectrum the ions observed between $m/z = 12$ and 25 provide no direct structural information but the CH/O ratio for polymers may correlate with the ratio for the monomer repeat unit.[26] The peaks at $m/z = 40$ and 41, C_2O and C_2OH respectively, are directly related to the polymer structure; they are derived from the OCH_2CH_2 function. The higher mass clusters at $m/z = 48$ (C_4), 49 (C_4H), 65 (C_5H_5), and 76 (C_6H_4) are thought to result from aromatic ring fragmentation.

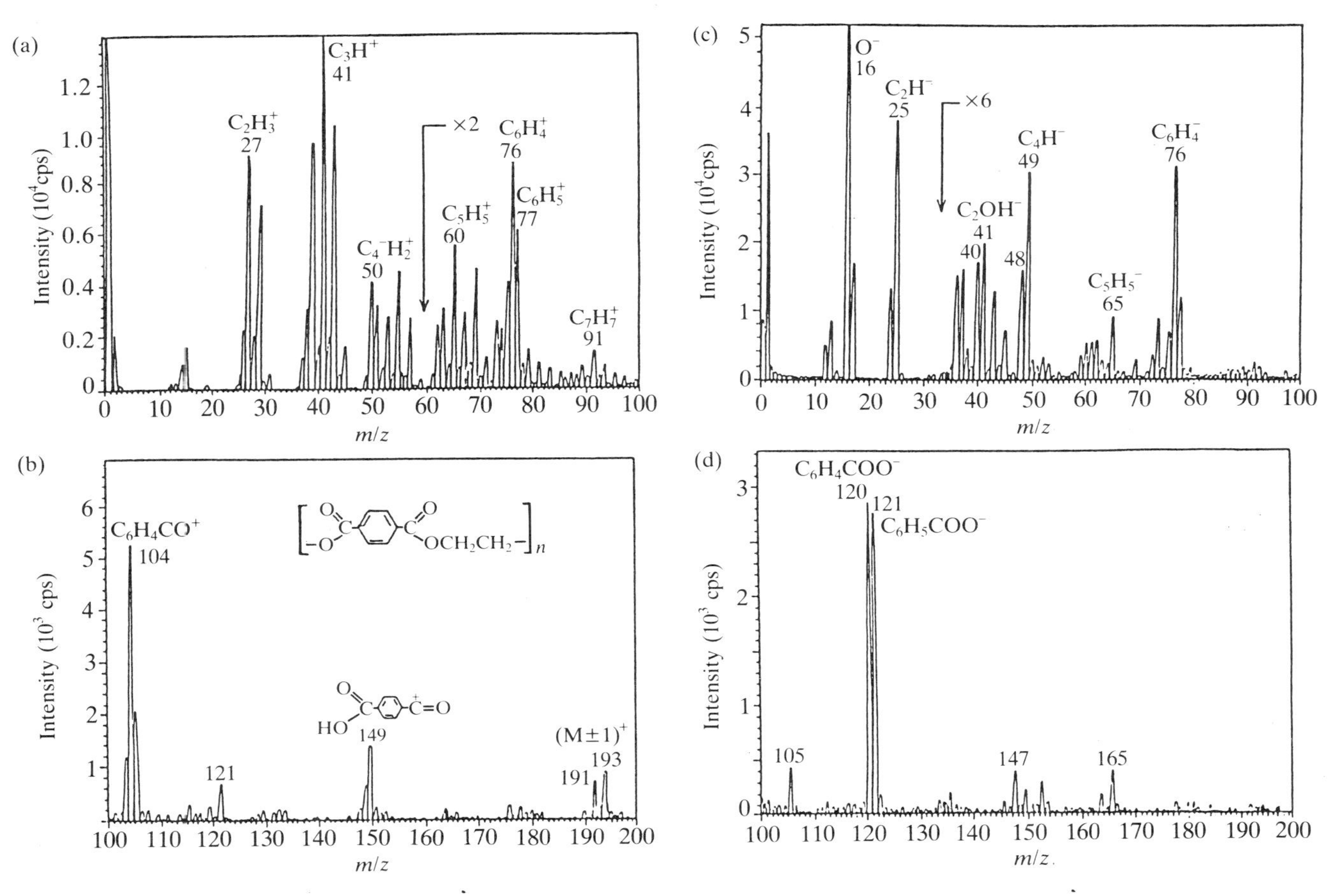

FIG 7.12. Positive (a,b) and negative (c,d) ion spectra of PET. Reproduced with permission from Ref. 9.

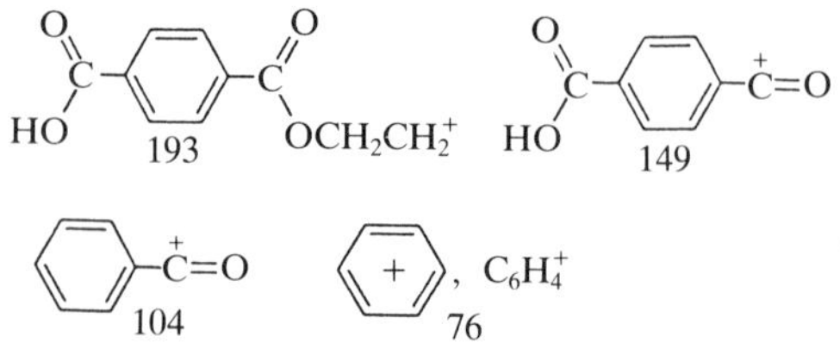

FIG 7.13. Structures proposed for the larger mass ions from the PET spectra. Reproduced with permission from Ref. 9.

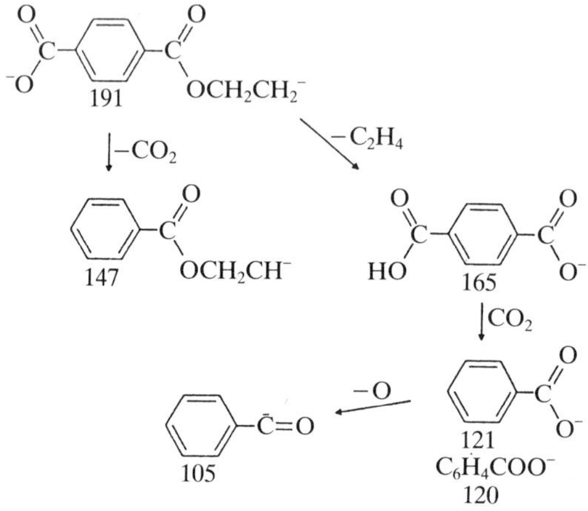

FIG 7.14. Fragmentation pathways of the large mass clusters observed in the PET spectra. Reproduced with permission from Ref. 9.

These data obtained from the negative spectrum of PET lead to the proposition of a fragmentation pathway for the larger clusters (see Fig. 7.14).

Briggs has reviewed the relationship between the structure of the polymers and their SSIMS spectra.[27] A series of detailed investigations of methacrylate polymers[26,28] has assisted in the assignment of the main ions (see Fig. 7.15) in the positive spectrum of PMMA and PHEMA.

The negative spectra of the methacrylate polymers are far more informative regarding structure than the positive, although some spectral differences in the positive spectra do exist. In the case of PMMA the CH_3^+ group is the most intense cluster, originating from the methoxy side-chain, whereas for PHEMA $CH_2CH_2OH^+$ is the predominant ion. In polymers of this type the positive ion clusters in the m/z range 15–57 typify a hydrocarbon backbone, so they are not useful diagnostic fragments of a specific polymer by themselves, although the relative intensities of these smaller ions

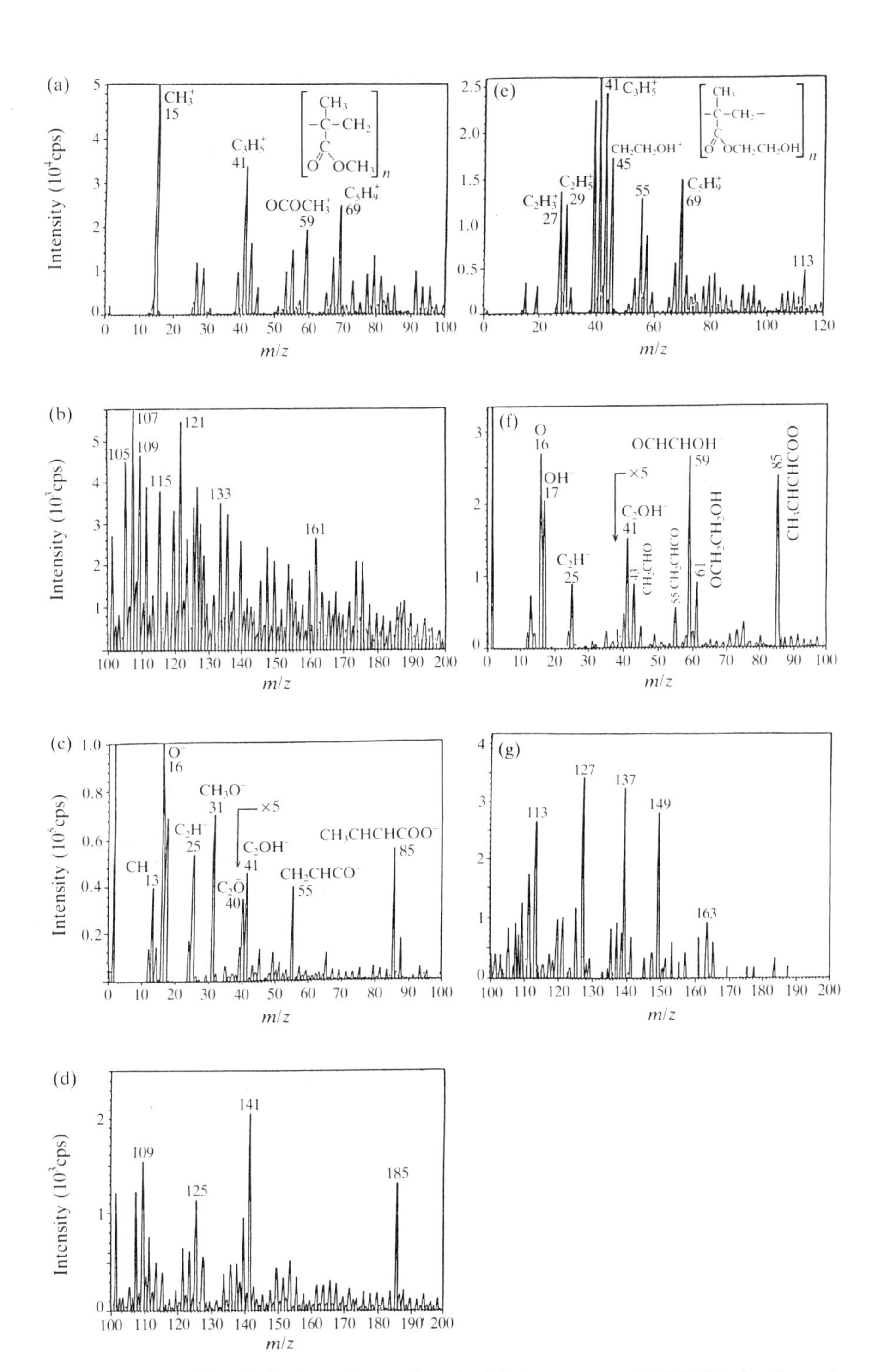

FIG 7.15. Positive (a,b,e) and negative (c,d,f,g) spectra of PMMA (a–d) and PHEMA (e–g). Reproduced with permission from Ref. 9.

may be supporting evidence in structure elucidation.

The advantage of the negative spectra is that they provide a great deal of structural information without the hindrance of the intense C_xH_y fragments.[29] The low mass negative ion clusters originate mainly from the methacrylate base unit, although they also provide some useful structural information in the methacrylate polymer systems. Comparing the PHEMA and PMMA spectra, we see that the cluster at $m/z = 43$ is more intense in the PHEMA system, which suggests that it originates from the side-chain group, i.e. CH_2CHO^-. Similarly, the strong $m/z = 31$ peak evident in the PMMA spectrum which is due to cleavage of the methoxy side-chain, CH_3O^-, is replaced by peaks at $m/z = 59$ and 61 which are due to $OCHCHOH^-$ and $OCH_2CH_2OH^-$ respectively. These differences observed in the spectra are a consequence solely of the different groups substituted in the side-chain.

Since many of the observed results are the same as those known to occur in electron impact (EI) mass spectrometry of aliphatic esters, we can often begin to interpret the results obtained from SIMS using the same rules as those applied to electron bombardment ionization mass spectrometry. The ions at $m/z = 55$ and 85 are common to all the reported spectra of alkyl methacrylate polymers, originating as fragments of the backbone system.

With increasing chain length, there is a tendency for the resonance stabilized $R\text{-}CH{=}CHO^-$ ion to predominate in the spectra. In a recent detailed study of the methacrylate homopolymers,[30] intense fragments containing several polymer units were observed, which are attributed to the basically linear structures occurring as a result of scission of the main polymer chain, induced by elimination of the side-chain groups. Table 7.3 summarizes the negative ion data obtained from a range of alkyl methacrylates. The ions observed exactly mirror the changing identity of the R group and so can be used diagnostically in an analytical situation. The powerful fingerprinting approach in SSIMS has been further demonstrated in a number of papers, using a wide range of polymers,[31-34] including low-density polyethylene (LDPE), polypropylene (PP), polystyrene (PS), polymethylmethacrylate (PMMA), polymethacrylic acid (PMAA), nylon-6 and polyethylene terephthalate (PET). These were investigated with experimental parameters which enabled spectra to be obtained, showing several polymer repeat units, from undamaged surfaces.

7.5.3. *Fingerprint spectra—inorganic materials*

The relationship between the SSIMS spectra and chemical structure is well illustrated for inorganic materials by the metal oxides. We saw in Chapter 6 that the oxidation of metal surfaces has been studied by SSIMS for a number

TABLE 7.3

Negative ions formed in the alkyl methacrylate spectra

POLYMER	R Group
PMMA	CH_3
PEMA	C_2H_5
PPMA	C_3H_7
PBMA	C_4H_9
PHMA	C_6H_{11}

POLYMER FRAGMENT (negative ions)		MASS				
	R =	**CH_3**	**C_2H_5**	**C_3H_7**	**C_4H_9**	**C_6H_{11}**
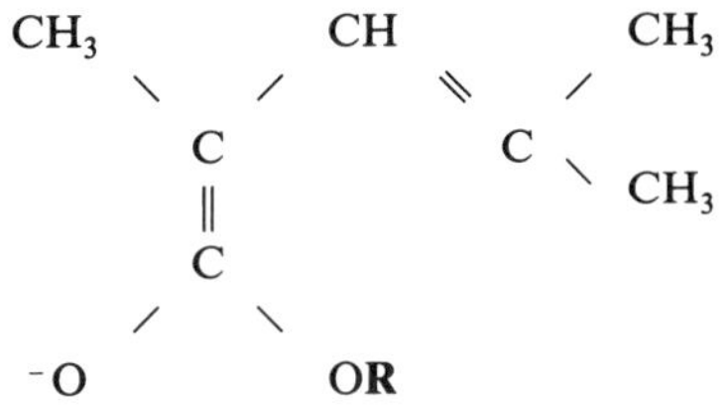 Structure 1 (below)		141	155	169	183	209
Similarly	M + 55	155	169	83	197	223
	M + 67	167	181	195	209	235
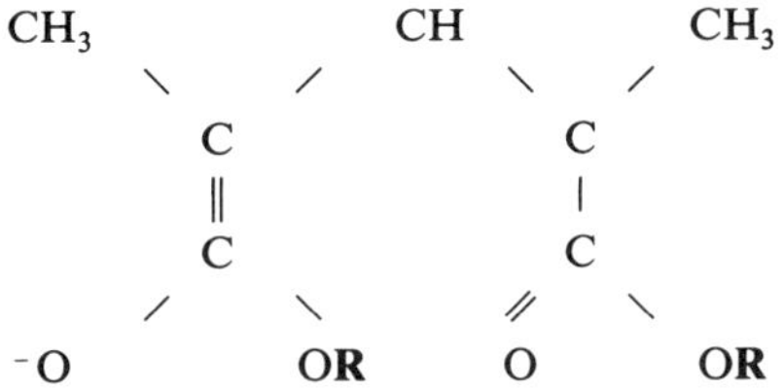 Structure 2 (below)		185	213	241	269	321
Similarly	2M + 25	225	253	no clusters seen		
		241	269	no clusters seen		
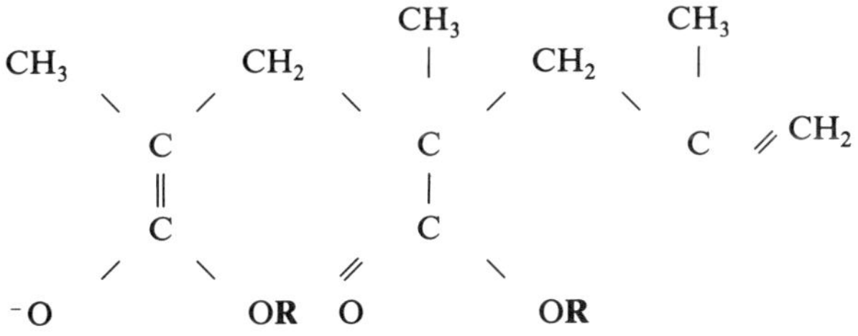 Structure 3 (below)						
Similarly	2M + 55	255	283	no clusters seen		
	2M + 67	267	295			

```
CH3          CH          CH3
   \        /    \\     /
     C             C
     ||              \  CH3
     C
    /   \
 -O      OR
```

```
CH3            CH            CH3
   \          /    \        /
     C               C
     ||              |
     C               C
    /   \          //   \
 -O      OR       O      OR
```

```
                        CH3            CH3
CH3          CH2        |      CH2     |
   \        /    \           /      \
     C               C                C  // CH2
     ||              |
     C               C
    /   \          //   \
 -O      OR      O       OR
```

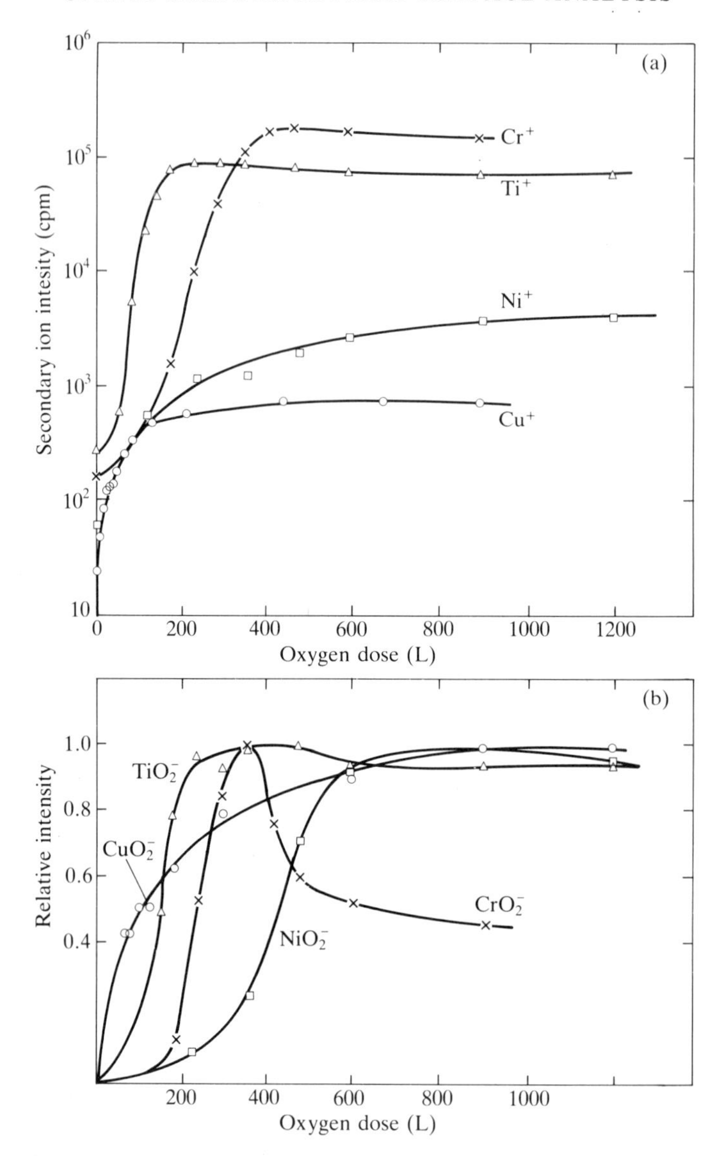

FIG 7.16. (a) Increase in Me^+ emission during oxygen exposure of different clean metal surfaces. Primary ion current intensity: 4×10^{-10} A cm^{-2}. (b) Change in relative intensity of the MO_2^- clusters with increasing oxygen dose. Reproduced with permission from Ref. 35.

of metals.[35] The typical oxidation behaviour of some metals is illustrated by Fig. 7.16. In general, oxide formation on metals occurs at high oxygen doses,[36] it proceeds in two steps. The first step is characterized by the appearance of MeO_2^- and MeO_3^- ions, and the second step is shown by the detection of MeO^+ fragments.

The relationship between the relative yields of $MeO_x^\pm$ ions and chemical structure will be influenced by the mechanism of their formation. Benninghoven and Werner *et al.*[36,37] have suggested that the secondary particles were originally formed as ions and subsequently reneutralized, whereas Plog proposed that the particles were emitted as neutrals which then broke into ions. Benninghoven *et al.*[38] have investigated the relative yields of the $MeO_x^\pm$ ions and proposed an empirically based valence model which indicates that the oxidation state of the metal ions plays a significant role in determining secondary ion cluster yield. They have developed these ideas and proposed a formula for the calculation of secondary ion yields from oxides and oxidized metal surfaces. They studied 15 oxidized metal surfaces before arriving at the formula, which was developed for mono-metallic oxides. The fragment yields from the oxidized surfaces of polycrystalline Mg, Al, Ti, V, Cr, Mn, Fe, Ni, Cu, Sr, Nb, Mo, Ba, Ta, and W metals were recorded. The metal samples were exposed to an oxygen dose of 100–1000 L, after being cleaned by ion bombardment. At these doses an oxide structure of several monolayers is formed which does not undergo any subsequent change under increasing oxygen exposure.

The fragment valence K was defined in order to introduce a uniform parameter for emitted fragments of a different charge. This value K is the formal valence number of the metal in the emitted fragment, and is simply calculated by the formula $K = q + 2n$ (providing there is only one metal atom in the fragment), where q is the fragment charge and n is the number of oxygen atoms. A plot of the relative ion yields of MeO_x^- and MeO_x^+ as a function of the fragment valence K allows a Gaussian curve to be fitted through the data, which is demonstrated by the chromium plots (see Fig. 7.17). Similar curves can be constructed for any of the investigated metals provided there are at least three data points available for a given class of ions. In order to compare the curves there must be common parameters. These are G^+ and G^-, which are the maxima of the positive ion and negative ion plots respectively. The maximum yields at these points are S^+ and S^-. G^0 is the lattice valence; this is the mean of the lattice parameters G^+ and G^-. The difference between the lattice valence G^0 and G^+ or G^- is called the displacement parameter α (see Fig. 7.18). The results for some of the metals are summarized in Table 7.4.

These systematic relationships are also found to hold for bulk oxides. There is thus a clear relationship between SSIMS spectra and the oxidation

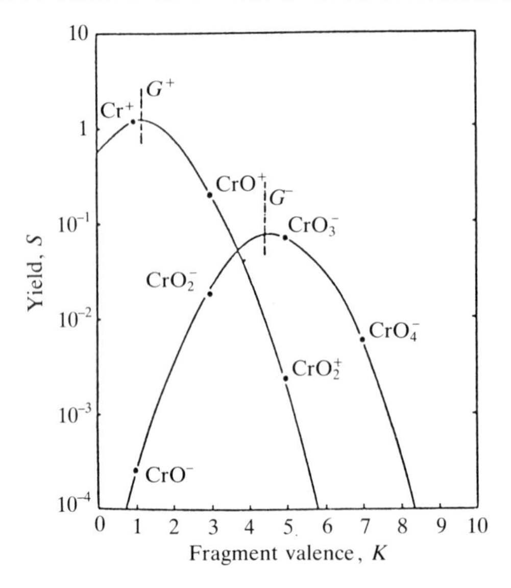

FIG 7.17. Yield of the fragment ions $CrO_n^{\pm}$ as a function of the fragment valence K. Reproduced with permission from Ref. 38.

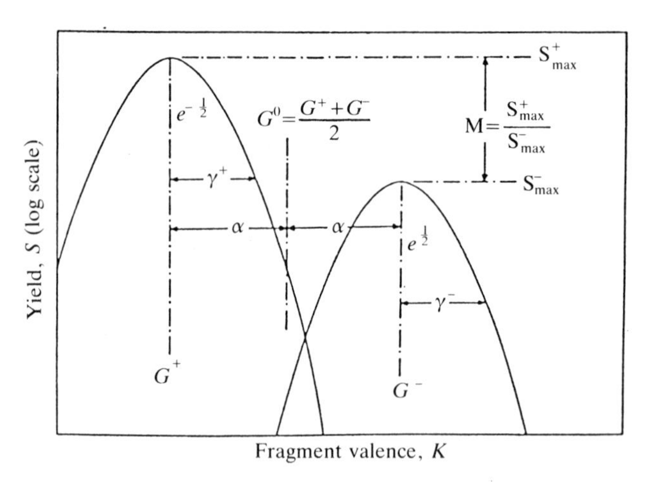

FIG 7.18. Definition of the maximum yield S, the lattice valence parameter G^0 and the displacement parameter α. Reproduced with permission from Ref. 38.

TABLE 7.4

(a) Parameters of some of the investigated metals (determined directly by the best fit)

Metal	S^+_{max}	G^+	γ^+	S^-_{max}	G^-	γ^-
Ti	8×10^{-1}	2.1	0.92			
V	8×10^{-1}	2.3	0.87	1.4×10^{-2}	4.2	0.84
Cr	1.2×10^{0}	1.0	1.15	7.5×10^{-2}	4.6	1.05
Mn				3×10^{-2}	3.0	1.02
Fe				1×10^{-2}	3.45	1.05
Nb	3×10^{-1}	3.05	1.09			
Mo	5×10^{-1}	1.8	1.23	9.5×10^{-2}	5.35	0.86
Ta	2×10^{-2}	3.2	1.05	8.5×10^{-3}	4.7	0.86
W	1.6×10^{-1}	2.7	1.02	1.4×10^{-1}	5.3	0.76

(b) Parameters of the 15 investigated metals (determined by assuming constant γ)

Metal	G^+	S^+_{max}	G^-	S^-_{max}	G^0	α
Mg	−0.9	6×10^{0}	1.3	1.1×10^{-2}	0.2	1.1
Al	−1.0	4.7×10^{0}	2.0	3.4×10^{-2}	0.5	1.5
Ti	2.05	7.5×10^{-1}	4.3	2.2×10^{-2}	3.2	1.1
V	2.25	7×10^{-1}	4.3	1.2×10^{-2}	3.3	1.0
Cr	1.35	1.3×10^{0}	4.45	7.5×10^{-2}	2.9	1.55
Mn	0.1	5×10^{-1}	3.0	3×10^{-2}	1.55	1.45
Fe	0.45	4×10^{-1}	3.4	9.5×10^{-3}	1.9	1.5
Ni	−1.3	6.5×10^{-1}	2.5	6.5×10^{-2}	0.6	1.9
Cu	−1.9	5.5×10^{-1}	2.5	1.7×10^{-2}	0.3	2.2
Sr	1.25	1.7×10^{-1}	1.6	1.6×10^{-2}	1.4	0.2
Nb	3.1	3×10^{-1}	5.0	2×10^{-2}	4.05	0.95
Mo	2.0	6.5×10^{-1}	5.35	8.5×10^{-2}	3.7	1.7
Ba	1.7	4×10^{-2}	2.5	8×10^{-3}	2.1	0.4
Ta	3.2	2.2×10^{0}	4.7	8.5×10^{-3}	3.95	0.75
W	2.7	1.6×10^{-1}	5.4	1.4×10^{-1}	4.05	1.35

state of the oxide. The relative intensities of the $Me_yO_x^{\pm}$ ions can be used to monitor the surface oxidation state.

7.6. Applied polymer analysis

The properties of polymer surfaces and interfaces are of great technological importance and scientific interest; consequently, much effort has been directed towards gaining an understanding of the chemical and physical nature of the surfaces. The inability of the other widely used technique, XPS, to provide *full* molecular information for polymers, since the chemical states of common polymer elements such as carbon, oxygen and silicon do not produce easily resolvable chemical shifts, led a number of workers to explore the possibility of employing SSIMS.[23,39,40] SSIMS has a great potential for polymer analysis since it offers mass spectral data with a high degree of surface specificity, with a good signal to noise ratio, monolayer sampling depths, and (of particular importance) the ability to detect hydrogen. The technique has routine areas of application for materials of great technological importance, especially in the field of polyester film surface analysis. PET, as an example, is used in a wide range of products, including packaging films, photographic films, reprographic materials, magnetic, video and audio tapes, credit cards, and electronic components.

7.6.1. Polymer surface—end group analysis

The acquisition of information regarding polymer surface structure is illustrated in a study which seeks to identify the end groups of a poly(methyl-methacrylate) latex by combining the techniques of XPS and SSIMS. This gives us a powerful combination of analytical techniques for the determination of the surface chemistry.

The PMMA latex colloid was prepared by adding purified methyl methacrylate to distilled water in the presence of potassium persulphate.[41] The resulting latex particles were 156 nm in size. The negative ion spectra are shown for the PMMA latex and pure PMMA for comparative purposes (see Fig. 7.19). There is a large reduction in the intensity of the PMMA specific ions in the low m/z region at 45, 55, and 85. Relatively intense signals appear at m/z = 64, 80, and 96; these are attributed to ions from the sulphated chain ends of the polymer, i.e. SO_2^-, SO_3^-, and SO_4^-. These also agree with the XPS data from the S 2p peak. According to theoretical considerations, the presence of these charged groups at the surface should result in a stable surface. The high intensity of the SSIMS ions arising from the sulphated polymer chains as compared with the low intensity of the XPS peaks is an indication of the greater surface specificity of SSIMS. More detailed information regarding the positioning of the surface groups could be obtained using angle resolved XPS. The combination of these two techniques finds an important application in the surface chemical analysis of polymer colloids.

7.6.2. Surface composition—copolymer analysis

Information about the sequencing of copolymers is potentially available from SSIMS.[42] To demonstrate this capability, ethyl methacrylate–hydroxy-

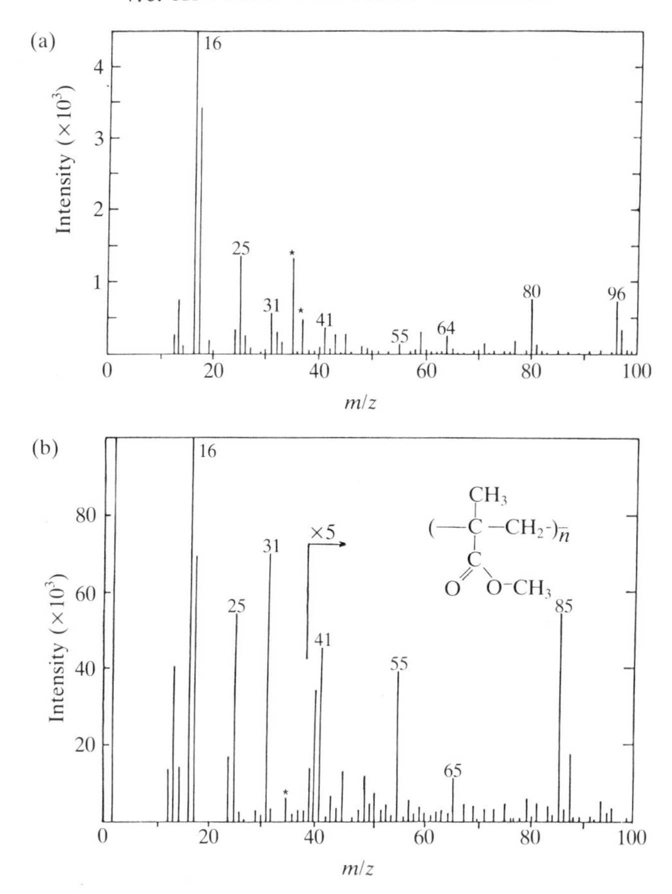

FIG 7.19. Negative ion spectra of PMMA latex, (a) and pure PMMA (b).

ethyl methacrylate (EMA–HEMA) was spin cast onto cleaned glass discs. The important peaks in the negative analysis are found at m/z = 127, 141, 155, 185, and 213, for which the following structures (see Fig. 7.20) were assigned.

The m/z = 155 and 213 structures are characteristic of PEMA, the m/z = 127 fragment is characteristic of the PHEMA system, since the loss of ROH^+ is common for hydroxylalkyl ester groups, and the m/z = 185 peak represents a EMA–HEMA fragment from the copolymer. The fragment at m/z = 141 is also representative of the EMA–HEMA system as a result of olefin elimination from the alkyl methacrylate. If the m/z = 127 and 155 peaks are taken as representative of the HEMA and the EMA monomers respectively then plots of $A/(A+B)$ and $B/(A+B)$, where $A+B$ = intensity

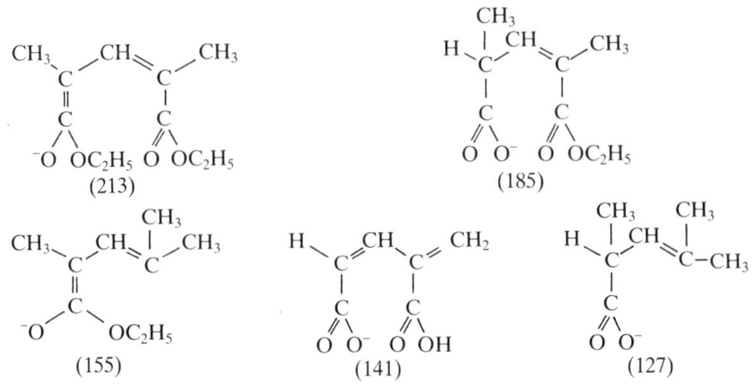

FIG 7.20. Structure assignment of the important characteristic peaks in the EMA–HEMA copolymer system. Reproduced with permission from Ref. 42.

of m/z = 127 peak + intensity of m/z = 155 peak will show smooth trends. The two intensities representative of the linked units are plotted normalized to the sum of 127 + 155. As predicted, the curves pass through their maxima at the mid-range bulk composition (see Fig. 7.21). Graph (a) indicates that the relative yield of a given cluster is not totally matrix independent since there is a curvature in the plots, which would otherwise be straight lines. This type of analysis indicates that the relative intensities of a small number of ions may provide detailed, quantitative information on the surface composition of copolymers.

7.7. Adhesion

Adhesion is vital in such areas of technology as surface coatings, composite materials, and corrosion protection. It is important to understand the mechanisms by which interactions between materials occur. Clearly, it is necessary to define the properties of the interfacing materials which give rise to effective surface bonding and if possible to characterize the chemistry of the interfaces.

7.7.1. Optical coatings

Polycarbonate (PC) is used in the manufacture of optical lenses. In some cases where it may be desirable to attach a coating to the PC surface, it is found that this is possible for pure PC but is not viable for the commercially produced material, suggesting that the surface compositions are different. In a SSIMS analysis of the commercial polymer, which contains 0.5 per cent of

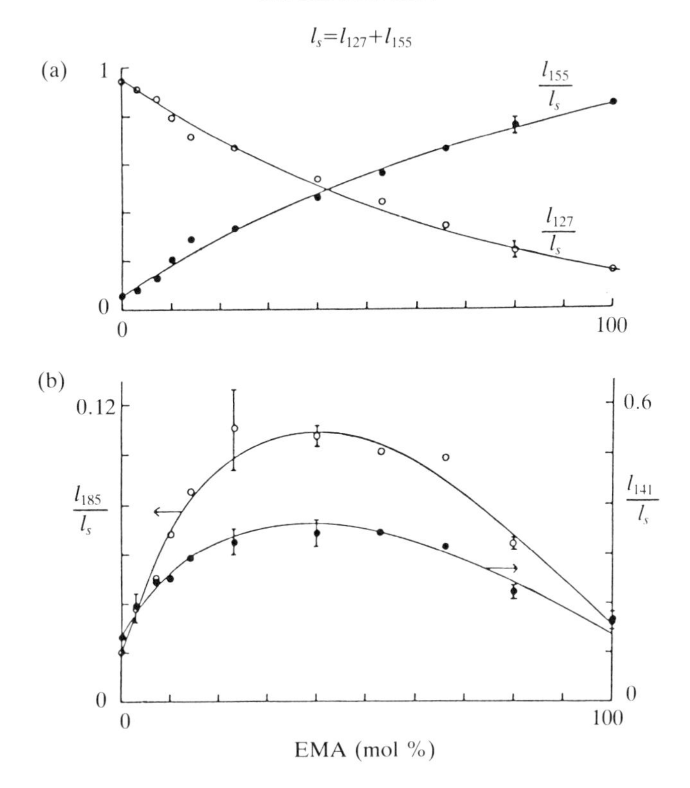

FIG 7.21. Relative intensities of characteristic ion fragment peaks from EMA–HEMA copolymers plotted as a function of bulk composition. Reproduced with permission from Ref. 42.

a release agent (pentacrythritol esterified by palmic and stearic acids), the typical PC peaks were observed. Additional signals were also observed at m/z = 205, 255, and 283 which are not present in the spectrum of pure PC (see Fig. 7.22). The m/z = 205 peak is assigned to the *i*-octylphenolate anion, which originates from the end group of the polymer. It is likely that this group may influence the adhesive properties of the PC. The palmitate and stearate anions manifest themselves as the m/z = 255 and 283 peaks respectively; thus the release agent is present at the surface, which clearly has serious consequences for adhesion.

7.7.2. Composite materials

Another problem of adhesion is demonstrated by the analysis of composite materials. These materials are usually a polymeric matrix with an inorganic filler. In this instance, the interface between the matrix and the filler is

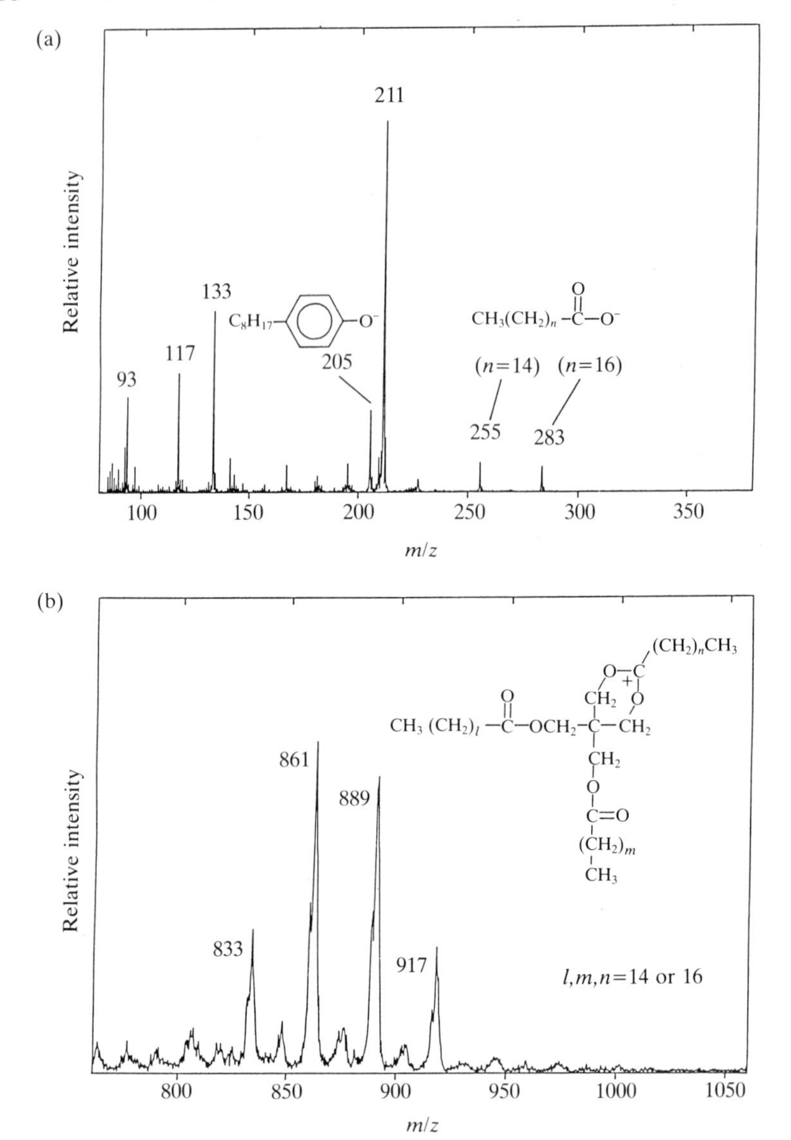

FIG 7.22. Spectrum of commercially produced polycarbonate (PC). Reproduced with permission from Ref. 44.

critically important, in order to obtain good adhesion to maintain stress transfer, weather resistance, etc. This particular study involved 'E-glass', alumina and quartz, treated with 3-methacryloxypropyltrimethoxysilane (3-MPS).[43]

SSIMS spectra were recorded for both the silyated samples and the untreated substrate. Positive spectra from both the clean and silyated 'E-glass' spectra both show hydrocarbon fragmentation patterns, part of which may be attributed to the organic silane side-chain in the case of the treated sample. In the negative spectra, ions at m/z = 41 and 85 were evident from the silyated sample but not from the clean sample. These were assigned to the silane side-chain. The m/z = 41 ion is particularly intense and this was explained by the formation of an allylic carbanion, which is particularly stable. The m/z = 85 peak is the methacryloxy fragment which is also seen in the methacrylate spectrum. The alumina sample gave very similar spectra. It was deduced that the 3-MPS on 'E-glass' and alumina maintain the methacryl unsaturation. In the analysis of quartz, the positive ion spectrum shows a prominent peak at m/z = 69; this was assigned as the stable dimethylcyclopropyl carbocation, which is present in polypropylene and polymethacrylates. The negative ion spectrum confirms the hypothesis for the formation of a polymethacrylate at the quartz surface, which was proposed on the basis that methacryloxypropyl silane is particularly susceptible to polymerization. The polymerization prevents the methacryl unsaturation which explains the reduction in intensity of the m/z = 41 and 85 peaks compared with the previous samples.

7.7.3. Polymer surface treatment

The effects of surface treatments on polymers can also be studied.[44] In order to read video information on an optical disc (which is fabricated from PMMA) using a laser, a silver film must be deposited on the surface of the disc. In order to deposit such a film the polymer surface has to be sensitized with stannous ions. A PMMA surface treated with a tannic acid solution will bind to the stannous ions more readily than one which is untreated. SSIMS has been used to study the effect of tannic acid on the chemistry of the PMMA surfaces. Figure 7.23 shows the spectra obtained from the treated surfaces. The positive ion spectrum of the treated polymer (a) shows a strong peak at m/z = 153 which is absent from the spectrum of pure PMMA. This is a stable ion of tannic acid, the galloyl ion. The lower mass spectrum shows peaks originating from PMMA and tannic acid. Similarly, in the negative spectrum the peak at m/z = 169 is due to the gallate anion and those at m/z = 123, 124, and 125 were assigned to pyrogaloate-type structures. These peaks suppress most of the characteristic PMMA signals. Thus tannic acid

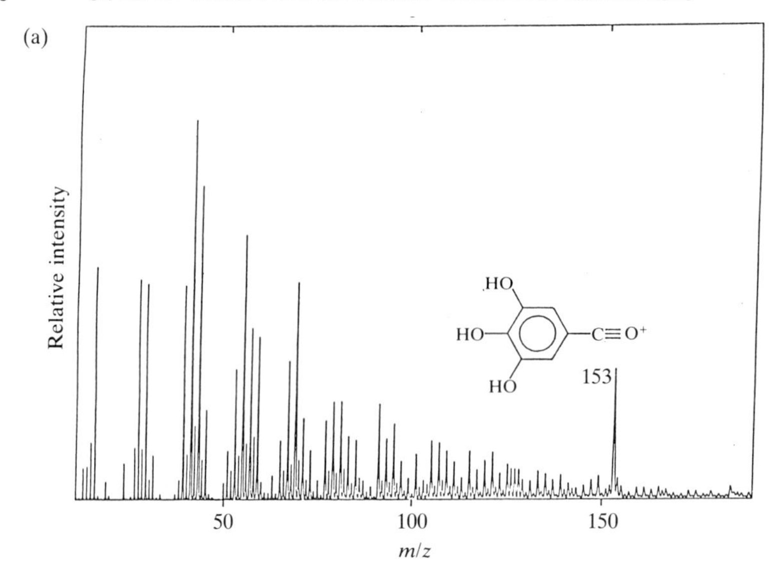

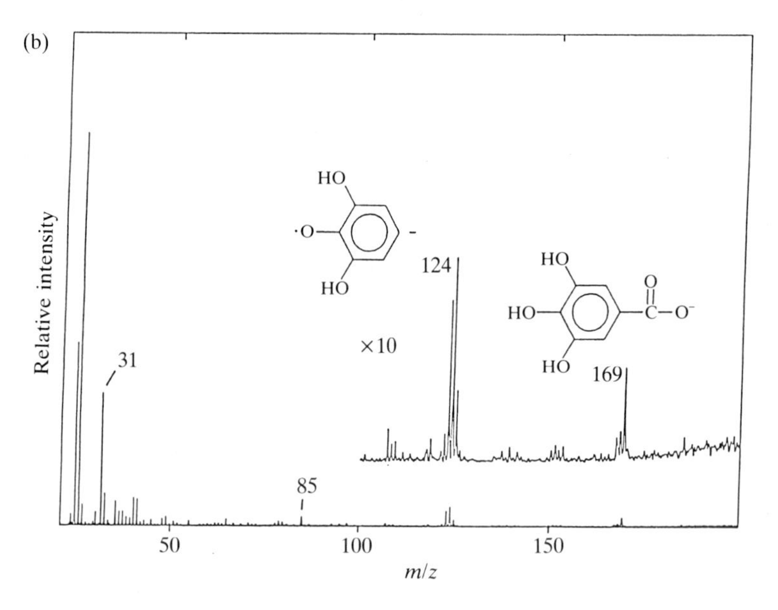

FIG 7.23. Positive (a) and negative (b) ion spectra of pure PMMA treated with tannic acid. Reproduced with permission from Ref. 44.

does modify the polymer surface, encouraging subsequent binding of the stannous ions to the PMMA surface.

7.8. Pharmaceutical applications

The surface analysis of drug delivery systems has been neglected to date. This may seem surprising as a knowledge of the surface composition is important in the characterization of pharmaceuticals, from the aspects of biocompatibility and the rate of drug release. The ability of SSIMS to provide surface specific information led to its application to several particular pharmaceutical problems. The application requirements are very similar to those of polymers. To illustrate the approach a study of two typical polymers employed in drug delivery—hydroxypropylcellulose (HPC) and hydroxypropylmethylcellulose (HPMC)—is described. They have been surface characterized and the surface orientations of the drugs used in a polymer bead formulations have been studied.

7.8.1. Characterization of polymer delivery systems

It is first necessary to understand the spectra of the basic polymers used. The positive ion spectrum of HPC is shown in Fig. 7.24. As with most organic samples it is dominated by $C_xH_y^+$ fragments; however, there are characteristic peaks from the cellulose backbone and from the side-chain molecules. The ions at m/z = 99 and 117 arising from $C_5O_2H_7^+$ and $C_5O_3H_9^+$ were assigned as fragments of the repeating unit of the cellulosic molecule. The ions at m/z = 59, 73, and 87 are characteristic of the 2-hydroxy-propyl substituent ether group ($^+CH_2CHOHCH_3$, $OCHCOHCH_3^+$, and $^+CH_2OCHCOHCH_3$). Previous analyses of ether cellulose materials have shown only small intensities for the m/z = 59 peak, thus corroborating evidence that in this spectrum this peak is derived from fragmentation of the 2-hydroxypropyl ion. This has been verified by MS–MS analysis (see Section 7.5.1). The negative ion information in this case was particularly relevant as it provided much structural information without the confusing interference of the intense organic species which are often present, particularly at low masses, in the positive spectra.

7.8.2. Drug release

An important aspect of drug design and development is the distribution of molecules within the polymer structure. A preferential orientation of the drug at the polymer surface may lead to a rapid release of the drug into the

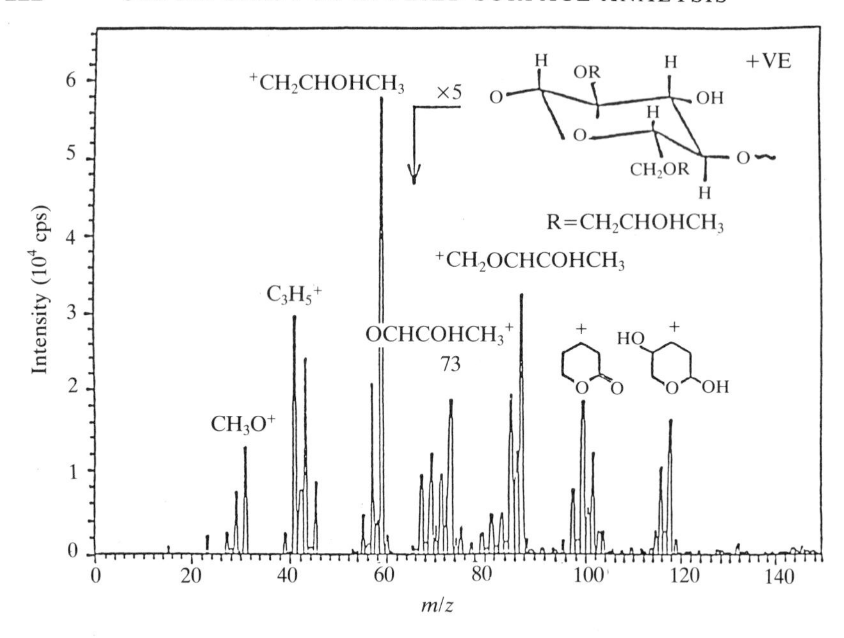

FIG 7.24. SIMS positive ion spectrum of pure hydroxypropyl cellulose (HPC). Reproduced with permission from Ref. 48.

body, which could have drastic toxicological consequences. Two studies illustrate the use of SSIMS to study these problems.

Theophylline is a drug used to dilate the bronchioles, often used in the treatment of asthma patients and to assist premature babies with breathing problems. A simulated drug release system was prepared by mixing theophylline, microcrystalline cellulose, and lactose with water, the mixture was extruded, and formed into beads which were then dried. SSIMS spectra of the beads were collected along with spectra of the polymer beads without the drug. Both the positive and the negative spectra show distinctly the presence of the molecular drug ion superimposed on the spectrum of the cellulose beads. At the lower mass end of the spectrum, CN^- and CNO^- ions are observed which arise from the drug alone. Analysis of the drug on the polymer surface was further studied by collecting a SIMS image of the beads which showed the drug to be evenly distributed over the surface.[45]

Ethylcellulose is widely used in the drug industry as a coating to prolong drug release in sustained release tablets. It is essential to maintain this coating intact in order to prevent any potentially toxic effects on the patient as a result of an increased drug release rate. To be effective the plasticizer must reside between the polymer chains, reducing the cohesive forces in the matrix and thereby increasing the free volume of the system.

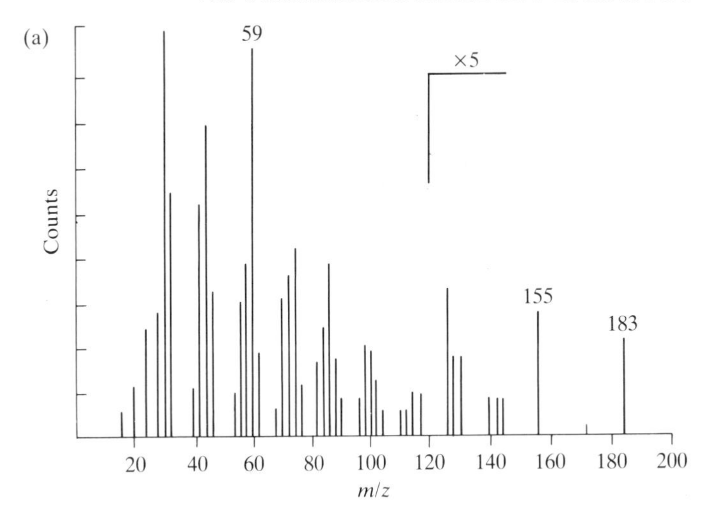

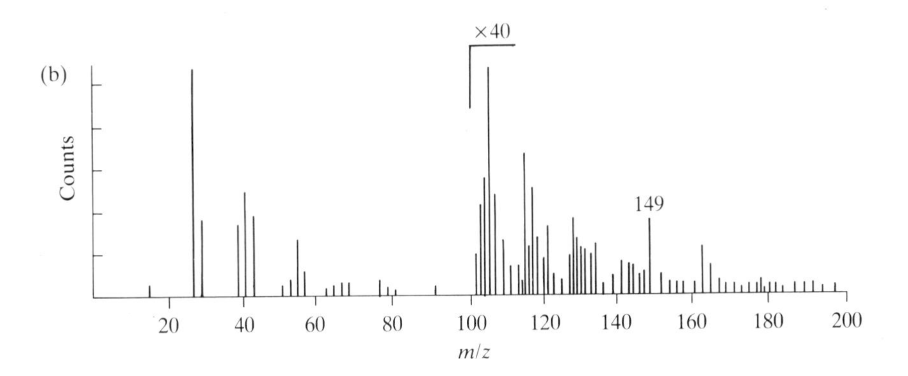

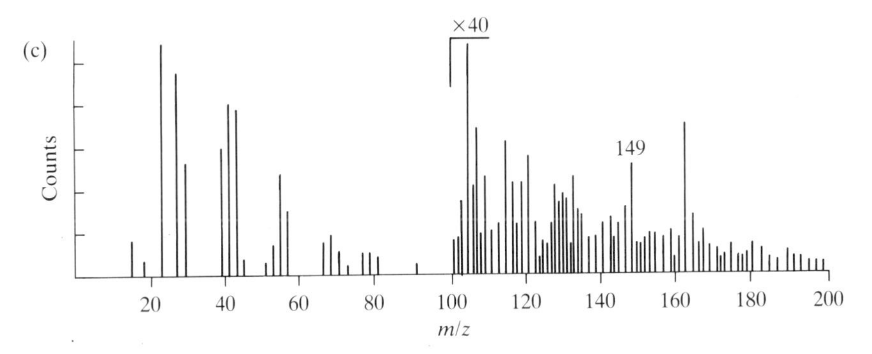

FIG 7.25. Surface chemical composition of ethyl cellulose films plasticized with a range of dialkyl phthalates. (a) Ethyl cellulose; (b) dimethyl phthalate; (*cont.*)

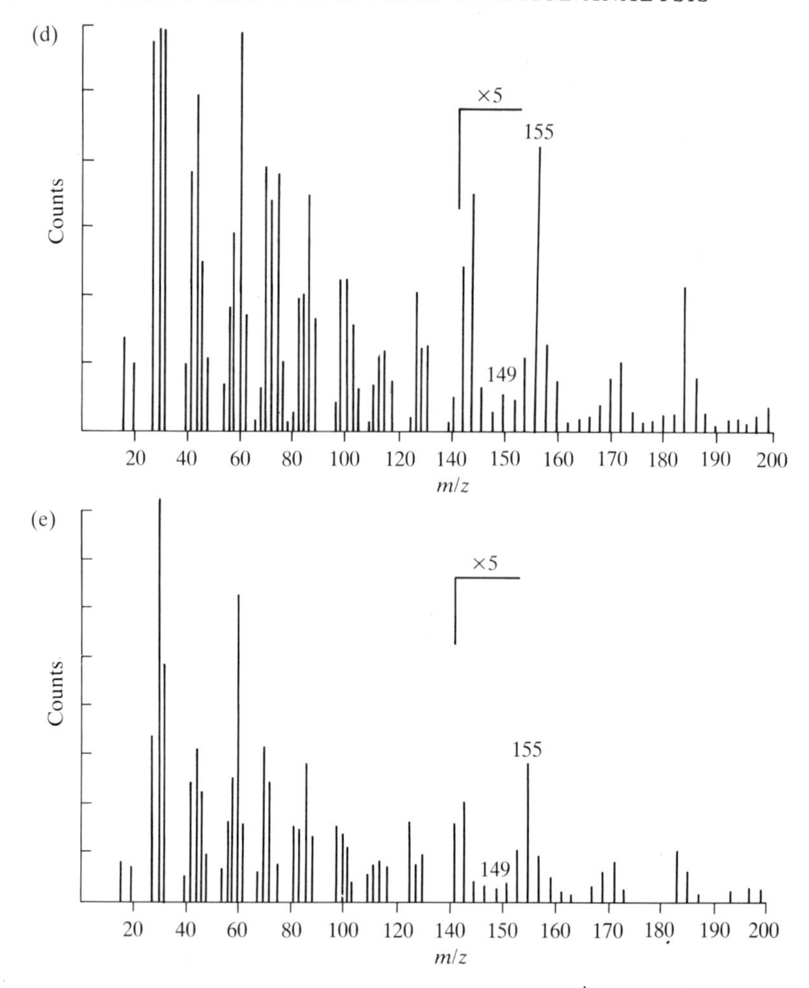

FIG 7.25. (c) diethyl phthalate; (d) ethyl cellulose–diethyl phthalate; ethyl cellulose–dimethyl phthalate. (Reproduced with permission from Ref. 46.)

The surface presence of these plasticizers has been examined by SSIMS, by looking at the surface chemical composition of ethylcellulose films plasticized with a range of dialkyl phthalates[46] (see Fig. 7.25). The low molecular weight fragments $C_xH_y^+$ again dominate the positive spectrum, but the peak at $m/z = 59$ is characteristic of ethoxy substitution at one of the hydroxyl sites of the anhydroglucose unit. The major diagnostic ion in the range of phthalate dialkyl plasticizers studied is a peak at $m/z = 149$ which is attributed to a phthalate anhydride peak. The surface concentration of plasticizer was monitored by following the intensity of the $m/z = 149$

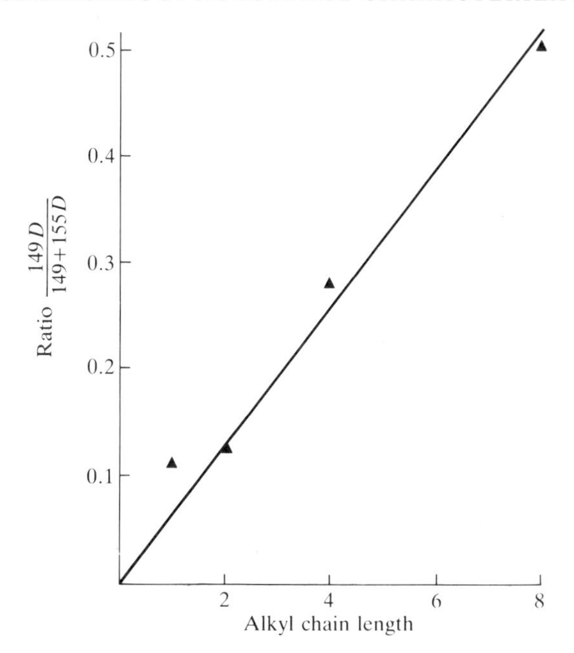

FIG 7.26. Linear relationship between the alkyl chain length and the ratio of ion intensities for different plasticizers Reproduced with permission from Ref. 46.

peak, relative to that of the $m/z = 155$ ion attributed to the polymer. A linear relationship was found to exist between the alkyl chain length and the ratio of ion intensities for the different plasticizers (see Fig. 7.26). The diagnostic fragment of the dimethyl and the diethyl phthalate systems is barely detectable, demonstrating that there is a low surface concentration of these plasticizers, whilst the dibutyl and the dioctyl derivatives show a much greater surface presence.

On the basis of these studies, there is evidence to suggest that SSIMS could play a significant role in the analysis and development of drug delivery systems. It is capable of providing a detailed characterization of drug and matrix interactions in the surface. Any additional surface contaminants,[47,48] e.g. plasticizers, surfactants, production process by-products, etc., will be observed, which may have an influence on the drug release profile and the resulting biocompatibility.

7.9. Semiconductor surface characterization

Electrical performance of semiconductors is so sensitive to purity levels that the ability to measure elemental concentrations to the p.p.b. levels is essential for the semiconductor industry. Although dynamic SIMS is an invaluable technique in the area of electronic materials, because depth profiling provides an analysis of the distribution of elements through the layers of the semiconductor, SSIMS is of considerable value in assessing the surface purity. A wide-ranging review concerning use of SIMS in the development and analysis of microelectronic devices is given by Grasserbauer *et al.*[49]

7.9.1. Surface cleanliness

Surface cleanliness is of prime importance to the successful growth of a good semiconductor. Before the growth of these layers the surface has to be meticulously cleaned. The method commonly employed is to degrease the surface with a detergent solution, then to create a passivating oxide layer by etching in an oxidizing solution. This process is carried out in solution, and various procedures and etchants have been studied for semiconductor growth.

The chemical treatments reported for GaAs include sulphuric acid–hydrogen peroxide solutions and hydrochloric acid.[50,51] Figure 7.27 shows the GaAs sample before and after the cleaning process. Before treatment the surface shows intense peaks from organic contamination, CH^- and C_2H^-, some halide contamination, and a comparatively low AsO_2^- signal. After surface degreasing and etching in H_2SO_4–H_2O_2–H_2O the spectrum shows clear differences. The previously intense organic fragments are replaced by O^- and OH^-, and the gallium and arsenic fragments are more apparent, indicating that a native oxide has been formed on the surface.

A similar analysis was carried out on InAs[52] which had been degreased by boiling in trichloroethylene for 5 min, then isopropanol followed by methanol for 5 min each.[53] The etching procedure was carried out in a 2 per cent solution of bromine in methanol for 5 min; after rinsing in methanol, a second etching was performed in concentrated HF also for 5 min, then the sample was rinsed in distilled water.[54] As with the previous example, the untreated sample was badly contaminated with hydrocarbons. The negative ion spectrum showed phosphate ions, which may have been introduced at the polishing stage (see Fig. 7.28).

After the first stage of degreasing, the indium signal is absent and the surface appears to be covered in a thin layer of hydrocarbon. The second degreasing procedure results in less hydrocarbon contamination and an improved indium signal. The etching with bromine and methanol shows a much-improved surface: the barium contamination, also from the polishing process, has been removed, and an intense In_2^+ and the appearance of such cluster ions as $InAs^+$, $InAsO^+$, and In_2O^+ indicate an increasingly clean surface. The formation of this oxide film is due to the use of oxidizing agents.

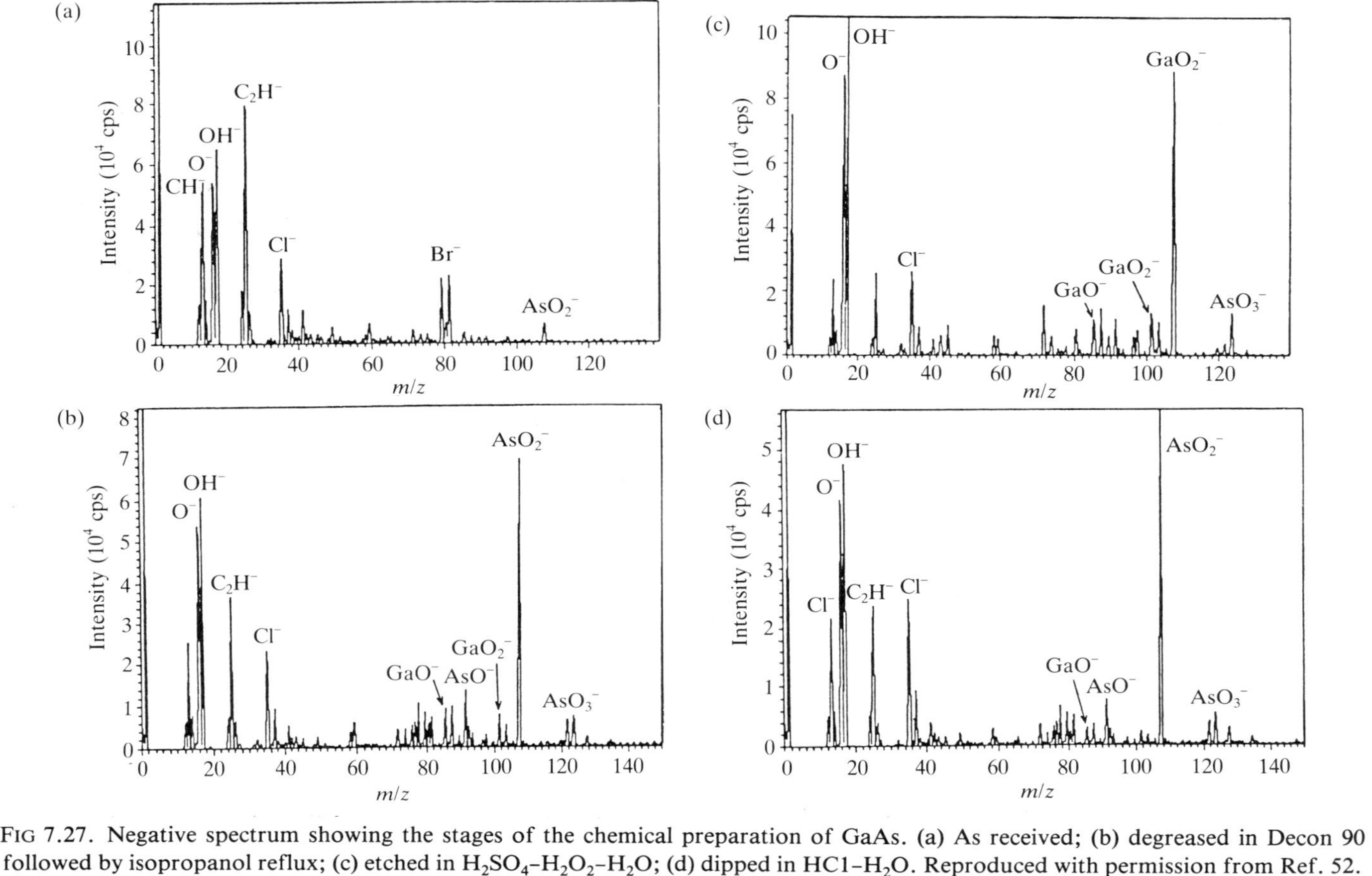

FIG 7.27. Negative spectrum showing the stages of the chemical preparation of GaAs. (a) As received; (b) degreased in Decon 90 followed by isopropanol reflux; (c) etched in H_2SO_4–H_2O_2–H_2O; (d) dipped in HC1–H_2O. Reproduced with permission from Ref. 52.

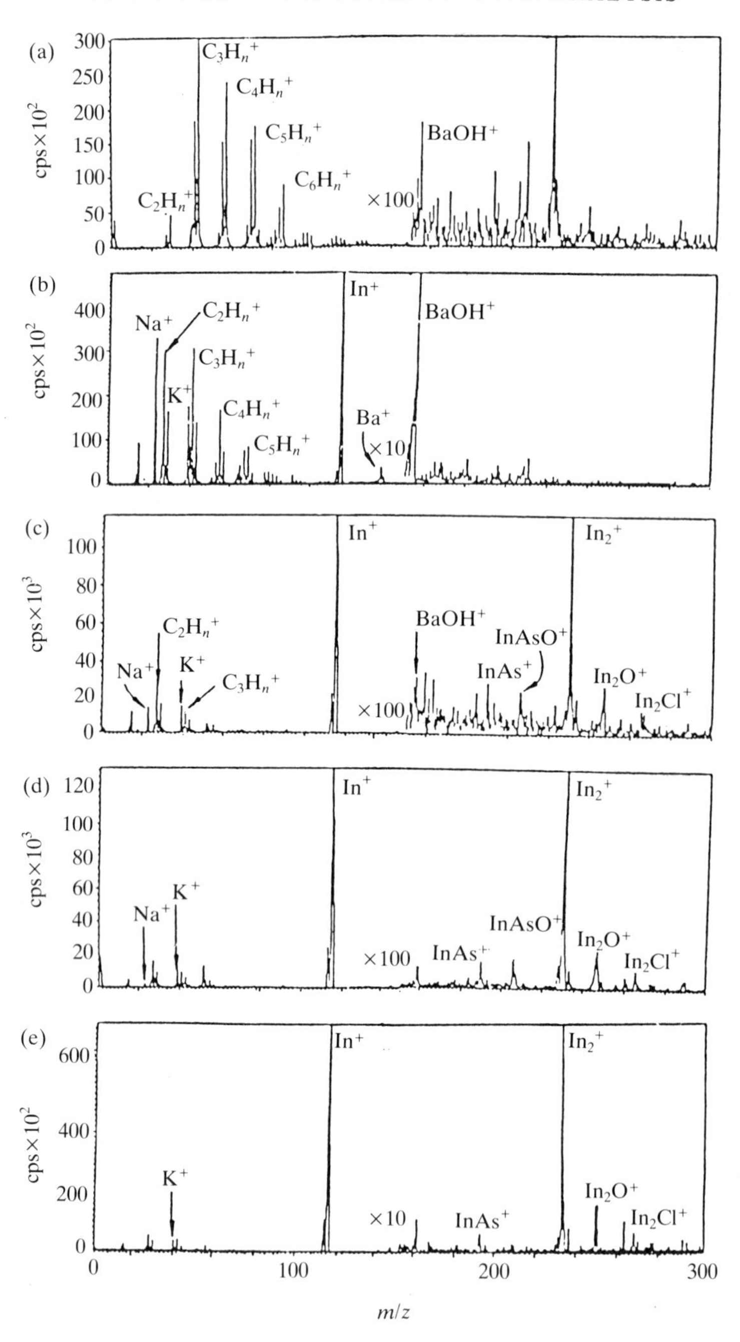

(a)
(b)
(c)
(d)
(e)
cps×10²
cps×10³
$C_2H_n^+$
$C_3H_n^+$
$C_4H_n^+$
$C_5H_n^+$
$C_6H_n^+$
$BaOH^+$
Na^+
K^+
Ba^+
In^+
In_2^+
InAs⁺
InAsO⁺
In_2O^+
In_2Cl^+
×100
×10
m/z

Contamination is still apparent, as is indicated by the abundance of halide ions in the negative spectrum. The final treatment results in an even cleaner surface, ready for semiconductor growth.

By SIMS analysis the contaminants introduced to a semiconductor surface due to packaging and storage may be investigated. For example, the positive ion spectrum of an as-received InP sample was found to be dominated by organic fragments and peaks at $m/z = 73, 147, 207, 221$, and 281. These are characteristic fragments of poly-dimethylsiloxane, which probably originated from the packaging material. The negative ion spectrum confirmed the organic contamination and also showed the presence of surface oxide and Cl^-. The higher mass fragments CH_3SiO^-, SiO_2^-, and $CH_3SiO_2^-$, confirmed the siloxane presence, and the clusters at m/z 63 and 79 (PO_2^- and PO_3^-) showed that the surface phosphorus was present as a phosphate.[55]

SSIMS is also applied in the field of organic semiconductors, in studies on the metal phthalocyanines.[56] These organic semiconductors undergo conductivity changes as they adsorb strongly electrophilic gases; consequently, they are potential sensors for hazardous gases such as Cl_2 and NO_2. The variations in conductivity are found to be dependent on the central metal species, so a SSIMS study can be performed to assess any change in the surface chemistry and hence the adsorption capabilities of the phthalocyanines (see Chapter 6).

7.9.2. *Microelectronics analysis*

SSIMS not only has an application to semiconductors within the electronics industry but also in the fields of solar cells, solders (PbSn), and electroless plating.[57]

PdSn colloidal particle alloys are used as a catalyst in the deposition of Cu on epoxy circuit boards, a process known as seeding. It is known that the surface segregation of tin and the formation of a tin oxide outer layer on the colloidal particles are important steps in the effectiveness of this seeding process.[58] A combination of XPS and SSIMS have been used to analyse the Pd_3Sn alloys at different temperatures with a view to understanding the process and discovering the optimum conditions for effective catalysis.

At the start of the experiment a general idea of the saturation coverages of

---

FIG 7.28. Positive ion spectra of (a) InAs (100) degreased in trichloroethylene, (b) InAs (100) degreased in trichloroethylene and exposed to IPA and methanol, (c) InAs (100) degreased in trichloroethylene and etched in 2 per cent Br_2–methanol. (d) InAs (100) degreased in trichloroethylene and etched in (1) 2 per cent Br_2–methanol, (2) HF and rinsed in distilled water, (e) treatment (d) using 12 MΩ water. Reproduced with permission from Ref. 52.

oxygen at varying temperatures was measured using XPS. The apparent coverages were 0.7 monolayers at 200°C and 1.6 monolayers at 650°C. Above this temperature the oxygen dissolves into and desorbs from the alloy. Pd remains in a metallic state throughout the exposures, as is demonstrated by the Pd $3d_{3/2,\,5/2}$ peaks, which do not shift significantly under any oxidation conditions. The oxygen peak does develop a shoulder at lower binding energy. By comparing these oxygen shifts with the Sn XPS signals it was concluded that the different oxygen peaks are associated with the different Sn peaks. At 650°C, about 60 per cent of the metallic Sn signal at 493.3 eV has been converted to a higher BE, 495.0 eV, which is characteristic of tin oxide. The preferential oxidation of tin at elevated temperatures is accompanied by a surface segregation of tin from the bulk. The SSIMS data were obtained under equivalent conditions to allow interpretation of known surface situations from XPS (see Fig. 7.29). At 500°C all the positive ions show a linear increase as a function of oxygen coverage up to 1.2 monolayers. At high exposures, the Pd^+ and the Sn^+ intensities increase significantly but the oxygen-containing clusters stay constant or decrease. The SnO_2^- behaves differently, showing two linear segments up to high coverage. The XPS data

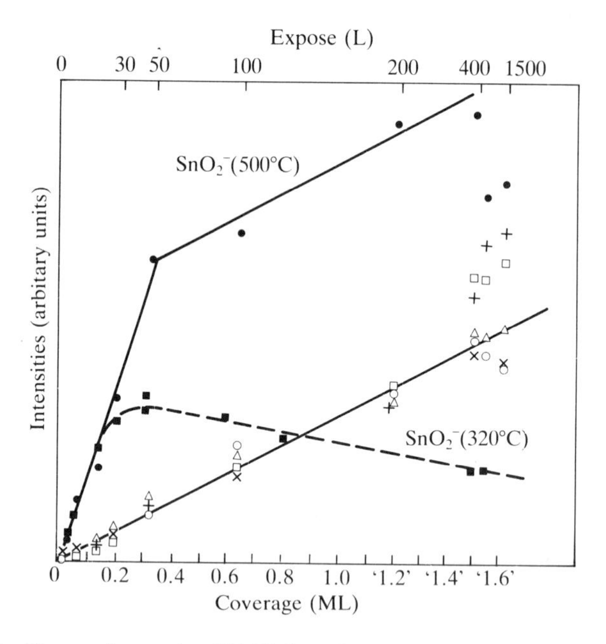

FIG 7.29. Positive and negative SIMS data for Pd_3Sn vs. apparent coverage and exposure: × Sn oxide signal taken from XPS, + Pd^+, □ Sn^+, △ $PdSnO^+$, ● SnO_2^+; all signals after dosing with oxygen at 500°C, ■ SnO_2^- after dosing with oxygen at 320°C. Reproduced with permission from Ref. 57.

add credence to the proposition that oxide nucleates strongly at the onset of adsorption at 500°C.

Since the SnO_2^- and the SnO^- intensities and the Sn oxide growth from XPS data show a linear relationship, these ion intensities must be a quantitative measure of the two-dimensional oxidation up to exposures of 0.3 ML. Above this, the break in the SnO_2^- and the SnO^- curves suggests three dimensional oxide growth, as SSIMS is sensitive mainly to the top monolayer. At lower temperatures (320°C), SnO_2^- shows the same increase as at 500°C; this increase slows down after a coverage of 0.15 ML has been attained then decreases after 0.3 ML coverage. This would imply that two-dimensional oxide growth occurs at the same rate at both temperatures at lower coverages. The growth slows down as the coverage increases for low temperatures and finally ceases at coverage above 0.3 ML for the low temperatures.

The use of a cluster ion $PdSnO^+$ to follow the formation of the quasimetallic state is illustrated by following the ratio of $PdSnO^+/SnO_2^-$, which provides information concerning the development of this state over the exposure range at varying temperatures. At high coverages this ratio is eight times higher for the lower temperatures than for 500°C. This indicates, in agreement with XPS data, that the quasimetallic state is dominant at saturation for low temperatures. The study indicates that SSIMS can be reliably used to investigate the reactions in the outer atomic layers, with respect to oxide nucleation and the generation of quasimetallic states.

The main criticism of SSIMS in this type of application is the non-quantitative nature of the results; however, relative intensities may be studied to gain quantitative comparisons. Althought XPS can monitor different oxidation states to provide information on contamination, it has the problem of limited spatial resolution.

7.10. Oxide analysis

By comparison with the amount of knowledge concerning the segregation effects in metals and alloys, there is relatively little information in the same field for metal oxides. This may seem surprising since a number of technological processes would benefit from a clearer understanding of the oxide structures and reactions: catalytic reactions, semiconductor production, reduction of metal ores, and many more. The study of oxides and oxide solid solutions has developed steadily over the past 15 years. Much of the interest stems from these technological problems, but originally the studies were to provide an insight into the fundamentals of catalytic reaction. A comprehensive review concerning the present understanding of geometric, electronic and chemisorption properties of metal oxide surfaces has been written by

Henrich.[59] Many of the reviews are restricted to single crystal oxide surfaces as it is a more straightforward task to correlate surface properties with a characterized, known geometry. The similarities in the various metal oxides may be correlated with their surface geometry and cation electronic configuration. Sufficient work has been carried out for trends to begin to develop regarding the basic surface properties of the metal oxides.

7.10.1. Surface state and reactivity of oxides

As in the case of metals the surface cleanliness of the oxides can be monitored by SSIMS, but of more interest is the ability of the technique to investigate surface chemical structure. Here the relative yields of the cluster ions $M_xO_y^{\pm}$ are important parameters. As we have seen, the emission of these cluster ions is very sensitive to the structure of the oxide, the structure of the exposed plane, the stoichiometry, lattice energy and degree of covalency of the oxide. Since all these parameters are interrelated it is difficult to establish the precise roles of each one[60] although an oxide surface can be clearly characterized from a SSIMS spectrum. To illustrate the point, Table 7.5 compares the yields of positive and negative cluster ions from V_2O_5, NiO, Ga_2O_3, PbO and

TABLE 7.5

(a) Positive ion intensities

Fragment	V_2O_5	NiO	Ga_2O_3	SnO_2	PbO
M^+	1	1	1	1	1
MO^+	0.1	0.2	0.2	0.5	0.07
MO_2^+	2×10^{-4}			0.03	
M_2^+	2×10^{-4}	0.05	1.7		0.004
M_2O^+	1×10^{-4}	0.1	0.01	0.6	1.6
$M_2O_2^+$	4×10^{-5}	0.01		0.14	0.15
$M_2O_3^+$	2×10^{-6}				
$M_2O_4^+$	7×10^{-5}				
$M_2O_5^+$	2×10^{-4}				
M_3^+					0.003
$M_3O_2^+$					0.25
$M_3O_3^+$	7×10^{-4}			0.02	0.03
$M_3O_4^+$	4×10^{-4}			0.006	
$M_3O_5^+$	1×10^{-5}				
$M_4O_3^+$	2×10^{-5}				
$M_4O_4^+$	1×10^{-5}				0.03
$M_4O_5^+$	1×10^{-5}				
M^+ Yield/nA	1×10^7	1.5×10^4	1.5×10^4	1×10^4	4×10^2

(b) Negative ion intensities

Fragment	V_2O_5	NiO	Ga_2O_3	SnO_2
M^+	1	1	1	1
MO^-	4.6×10^{-5}	0.06	0.5	0.5
MO_2^-	4.6×10^{-4}	0.186	1.25	0.05
MO_3	2.76×10^{-4}	3.6×10^{-3}		1.0
MO_4				0.1
M_2O^-	2.3×10^{-5}			
$M_2O_2^-$	1.84×10^{-5}	3.0×10^{-3}		0.5
$M_2O_3^-$	1.38×10^{-5}	0.024		2
$M_2O_4^-$	1.38×10^{-5}			0.45
$M_2O_5^-$	4.6×10^{-5}			
$M_2O_6^-$	9.2×10^{-6}			
$M_3O_3^-$	1.38×10^{-5}	5.4×10^{-3}	0.85	
$M_3O_4^-$	2.3×10^{-5}	1.2×10^{-3}		
$M_3O_5^-$	1.84×10^{-5}			

TABLE 7.6

Reduction in intensity of the oxygen-rich clusters as a result of primary beam bombardment

Peak	Before etching	After etching
Ti^+	100	100
TiO^+	131	53
TiO_2^+	2.6	0.2
Ti_2^+	0.1	0.2
$Ti_2O_2^+$	0.6	0.25
$Ti_2O_3^+$	0.5	0.05
TiO_2^-	0.3	0.15
TiO_3^-	0.5	0.13

SnO_2. The information obtained can be used to monitor the surface stoichiometry under reduction conditions. Thus when an oxide is subjected to ion bombardment there may be a preferential loss of oxygen, as reflected by the relative yields of the cluster ions. In Table 7.6 it can be seen that there is a reduction in the intensity of the oxygen-rich clusters accompanied by an increase in the metal-rich clusters. This can be correlated with XPS measurements, bearing in mind that SSIMS is more surface sensitive than XPS.

This type of measurement has been used to monitor the reduction of TiO_2 surfaces followed by the reaction of water vapour.[61] $TiO_2(100)$ surfaces were reduced either by argon etching or by hydrogen for several hours at 1300 K.

TABLE 7.7

Extent of reduction shown by SSIMS and AES ratios

	TiO^+/Ti^+ SIMS	O/Ti AES
White oxidized	0.7	1.3
sputtered	0.3	0.6
Blue H_2 reduced sample	0.3	0.3

The extent of reduction was measured by the AES O/Ti ratio and the TiO^+/Ti^+ SIMS ratio. Table 7.7 shows the results.

It was found that the hydrogen reduction process produced the same surface stoichiometry as the sputtering process, but the greater sampling depth of AES showed that the reduction penetrated further into the bulk of the sample. The reactivity of the surfaces towards adsorption and reaction of water vapour on the surface can also be monitored by SSIMS using the $TiOH^+/TiO^+$ and the OH^-/O^- ratios, to follow the hydroxyl surface coverage. Langmuir adsorption isotherms result if the two ratios are plotted as a function of exposure; however, we see that if the ratios are plotted against each other for surfaces of different starting stoichiometries, linear plots with a constant gradient result (see Fig. 7.30). The intercept increases with increasing non-stoichiometry. Thus

$$OH^-/O^- = S_H + 1.4TiOH^+/TiO^+.$$

It is suggested that the presence of two types of surface hydroxyl give rise to this behaviour on the TiO_2 surface. One hydroxyl group is strongly dependent on the surface stoichiometry; it is the more strongly bound to the titanium surface of the two hydroxyl groups. This hydroxyl group is reflected in the behaviour of the $TiOH^+$ and OH^- ions, and the other by OH^- only. By monitoring the two ratios as a hydrated sample is heated, it was shown that the second hydroxyl has the greater stability and is stable to above 500°C, whereas the first hydroxyl constantly decreases with temperature. Hydrogen-pretreated samples show a lower propensity to form hydroxyl surface species, especially the more stable hydroxyl.

7.10.2. Surface segregation

SSIMS can also be used in the analysis of mixed oxides to assess the surface segregation of the components.[37] An example is the mixed spinel system

$$Mg\,Y_{2-x}V_xO_4 \text{ where } Y = \text{Cr or Al}.$$

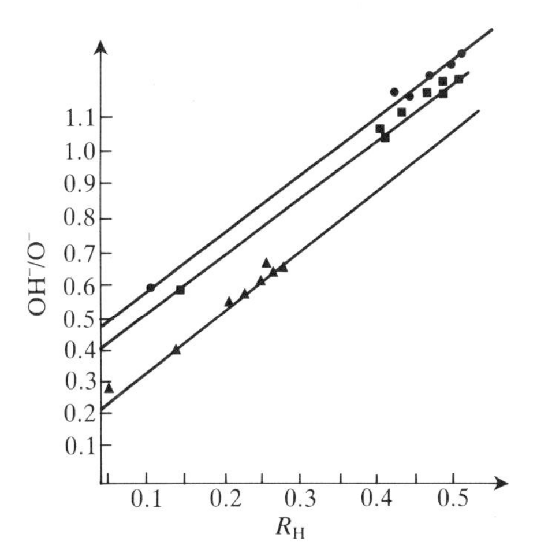

FIG 7.30. Linear dependence of OH^-/O^- on R_H ($TiOH^+/TiO^+$) for different starting stoichiometries TiO^+/Ti^+(SIMS): ▲, 0.69; ■, 0.39; ●, 0.30. Slope = 1.4 ± 0.1. Reproduced with permission from Ref. 61.

The oxides were prepared by mixing stoichiometric quantities of the samples, pressing the material into pellets and heating to temperatures between 1250 and 1350°C.

The secondary ion ratios $(V^+/Mg^+)/(V/Mg)_{bulk}$ and $(Cr^+/Mg^+)/(Cr/Mg)_{bulk}$ vs. composition were plotted for the $MgCr_{2-x}V_xO_4$ series. The resulting plots showed a nearly horizontal line in both cases, which indicates that the surface concentrations reflect the bulk concentrations (see Fig. 7.31); there is no surface segregation. The plot of VO^+/V^+ as a function of composition showed essentially a straight line, which indicated no change in the oxygenated species; this suggests that the V-O bond strength did not alter very much with the level of doping. A similar plot for the chromium species showed an increase in this ratio with increasing Cr concentration, implying that the Cr–surface oxygen bond strength and covalency increases.

For the series $MgAl_{2-x}V_xO_4$ the ratios $(V^+/Al^+)/(V/Al)_{bulk}$ and $(V^+/Mg^+)/(V/Mg)_{bulk}$ were plotted as a function of the composition, as for the chromium oxide series. From the plotted results it was apparent that the surface composition did not reflect the bulk composition. There was a steady fall in the ratio as the concentration of vanadium increased; this shows that there is a surface enrichment of vanadium at low concentrations, a phenomenon which is known in some spinels.[62] As the concentration of vanadium increases, there is a fall in the relative intensity of the signal from the guest ion V^{3+}, suggesting a surface depletion of these ions in the corresponding matrices.

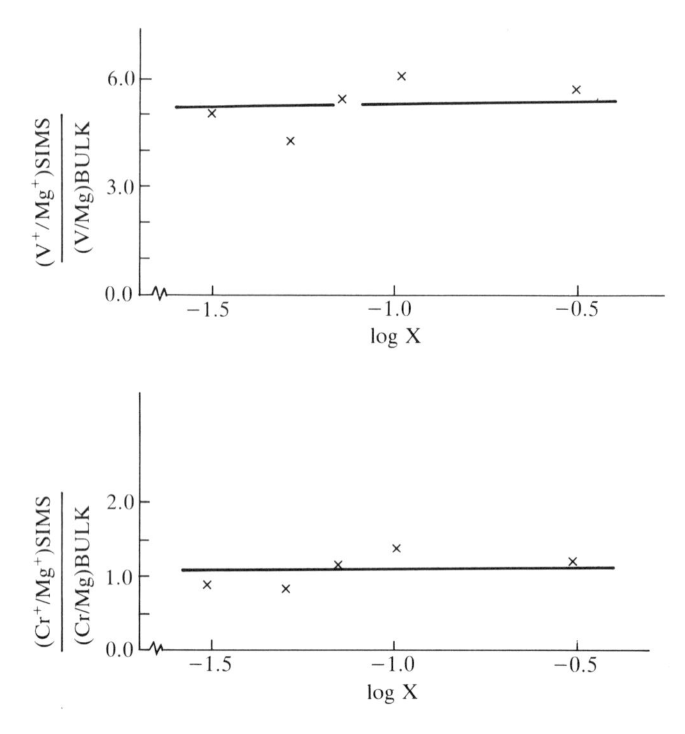

FIG 7.31. Variation of the ratios of the secondary ion intensity to the bulk ratio as a function of the composition of $MgCr_{2-x}V_xO_4$. Reproduced with permission from Ref. 37.

This may be explained from a thermodynamic viewpoint: in order to minimize the surface free energy, the surface will be enriched by the component which has the lowest free energy. In multicomponent systems this may result in a gross difference between the analysed surface and the bulk composition.

7.11. Catalyst analysis

There are two main areas of interest in the analysis of catalysts, the elemental surface analysis and the surface chemical structure before and after catalysis.

7.11.1 Catalyst preparation

Zeolites form an important class of cracking catalysts, and the effects of different preparation and pretreatment procedures on the surface Si/Al ratios have been investigated. This is important since the surface composition and framework composition will clearly influence the catalytic behaviour of the zeolites.[63] A calibration plot for the zeolites was first prepared. This ranged from a Si/Al ratio of almost unity (Faujastite X) to a ratio of ≈ 18 (ZSM 5). The Si^+/Al^+ ratio is recorded after etching for 5 min and again after 20 min. The results were as shown in Fig. 7.32. Using this calibration plot it is possible to analyse a number of unknown zeolites. After treatment with an acid solution, the zeolite mordenite was analysed. It was observed that considerable dealumination of the surface layers occurred. Surface dealumination by acids is an easy process so it is necessary to control the acid leaching step carefully. A sample zeolite treated with $SiCl_4$ analysed by SIMS showed a surface rich in aluminium and a framework rich in silica.[64] Steam treatment is often used for zeolite activation although little is known as to the effect of the treatment on the framework of the zeolite. A FABMS analysis of a steam-treated sample showed a decrease in the Si^+/Al^+ ratio, suggesting that the aluminium moves from the framework of the zeolite and the surface becomes rich in aluminium.

Since the silica–aluminas are such an important class of compounds they have been extensively studied.[65] They are frequently prepared by precipitation, followed by thermal treatment. The most significant features in the SIMS spectra of the precipitates were the M^+ (M = Si or Al) and the MOH^+

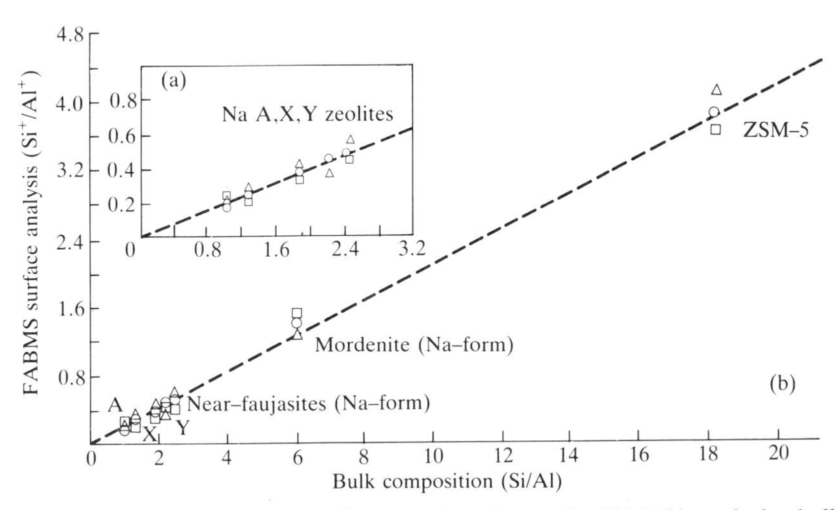

FIG 7.32. Correlation between the secondary ion ratio Si^+/Al^+ and the bulk composition Si/A1, for zeolites. △ First layer; ○ after first bombardment; □ after second bombardment. The low Si/Al region is expanded in inset (a). Reproduced with permission from Ref. 62.

ions. Evidence of some surface hydroxylation came from the pure alumina and silica spectra as the MOH^+ peak was always more intense than the MO^+ peak; this was partially removed by etching.

First, the surface silica content was measured for a simple physical mixture of silica and alumina (10 per cent SiO_2). Similar spectra were then recorded for a calcination series, after a calibration plot had been constructed. Calcination of this material resulted in an increase in the surface silicon concentration (SSC) as corundum formed. A further large increase occurred at the formation of cristobalite. These observations follow the established rules that, for mixed oxides, the more surface active component (the one with the lower melting point) will coat the other oxide particles.

The behaviour for the coprecipitate of the same composition was different; here there was an increase in the SSC as $\delta-$ and $\epsilon-Al_2O_3$ were formed. Similar events were described for the 5 per cent SiO_2; the rise in the SSC was attributed to the formation of alumina structures with less tolerance for combined silica. This rise reaches its maximum at the point where corundum is formed as no tetrahedral cations can be accommodated in this structure. For a 30 per cent composition, an increase in the SSC is first recorded as tegragonal mullite forms; this then decreases, contrary to what is expected from the bulk, as silica-rich particles are transformed into a homogeneous mullite. Although these results are only semiquantitative they do give important information on the thermal chemistry of the silica-modified aluminas.

7.11.2. Cr–silica catalysts

SIMS has been used to obtain detailed information on the surface chemistry of Cr–silica catalysts.[66] The catalysts were prepared by aqueous impregnation of silica with chromium oxide and calcined at temperatures up to 525°C. Krauss and co-workers confirmed that after reduction in CO or H_2, isolated Cr species were responsible for the catalytic activity.[67] There are contrasting views concerning the dispersion of the active sites.

Recent SSIMS studies have shown that aggregation of supported Cr occurs at even relatively low Cr loadings. In the SSIMS fragmentation pattern many of the Cr clusters overlapped with those of the Si-based fragments due to the relative abundances of the natural isotopes, and the presence of hydrogen or hydroxyl species. Computer analysis was used to calculate the ion count from a specific fragment. The prediction of a particular isotope pattern was deduced based on a combinatorial calculation. This is discussed in detail in Ref. 66. These analyses yielded useful data for the investigation of species such as $Cr_2SiO_2^+$ and $Cr_2O_4^+$. From the results obtained, it was evident that the $Cr_xSi_yO_z^+$ clusters are dominant features in the spectrum irrespective of the Cr loading on the support; this indicates that significant amounts of Cr must be in intimate contact with the Si. The multiple chromium clusters were more abundant in the higher Cr loadings. Increasing dispersion of the chromium is observed as the loading is reduced, since the $Cr_xSi_2O_z^+$ fragments are consistently more abundant at lower ion yields. By monitoring the variation of these ions with Cr content, it was concluded that significant clustering of Cr species occurred at low concentrations of Cr on SiO_2, possibly during the sample preparation stage. These clusters may be redistributed to an extent during calcination of the sample but they were still observed after they were calcined. This example illustrates the sensitivity of SSIMS, in producing cluster ions which are indicative of detailed changes in the surface chemistry.

7.11.3. Surface reactivity

SSIMS also finds an application in the study of real catalyst surface reactivity, since unlike some methods of surface analysis there is no requirement for a smooth single crystal surface. The CO on Pd system was discussed in detail in Chapter 6. This model CO adsorption onto Pd has been extended to the study of adsorption and oxidation of CO over small Pd particles supported on mica (a model of a real supported catalyst).[68] The particle size ranged from 2 to 12 nm. The same linear relationship was observed for θ_{CO} vs. $\Sigma_n(Pd_nCO^+/Pd_n{}^+)$ for the Pd particles as for the Pd single crystal. The particles above 6 nm in diameter are stable; they are triangular with their (111) plane parallel to the mica substrate. When they are exposed to CO the SSIMS analysis, monitored by ($PdCO/\Sigma Pd_xCO$), indicates that the bonding is initially linear; as coverage increases there is evidence for the development of bridge bonding. This is similar to the results obtained from the Pd (111) single crystal. The particles below 5 nm in diameter are unstable; they tend to become spherical when exposed to gas, and faceting may even occur under more severe conditions.

In an investigation of the influence of CO on Pd, the authors show that after 11 cycles of CO adsorption–desorption on fresh Pd particles, the total adsorbed CO falls. However, the total possible coverage on exposed Pd stays constant, indicating that particle coalescence has taken place. The molecules are predominantly in a linear co-ordination and there is no evidence of surface carbon (monitored by PdC^+ and Pd_2C^+). After 1 h annealing in an O_2–CO atmosphere at 570 K, to simulate an extended period of CO and O_2 reaction, there is a large rise in the total surface covered by adsorbates. A proportion of the CO molecules have moved into bridged sites and some carbon has appeared. This correlates with the development of new surface edges and corner sites due to faceting. After a further 11 CO adsorption–desorption cycles, there was a significant rise in the surface carbon, which results in a loss of CO adsorption sites; the CO which does adsorb moves back to predominantly linear bonding.

Another system of great interest which demonstrates the ability of SSIMS to monitor surface reactivity is the interaction of alcohols with metals and metal oxide surfaces.[69] CH_3OH, CD_3OH, and CD_3OD were used in the study. On initial exposure of clean aluminium to MeOH, the most prominent changes occurred in the negative ion spectrum where the major ions detected were those expected from oxidized aluminium or a partially hydroxylated surface. The AlO^-, AlO_2^-, and the O^- peaks rose swiftly to a plateau, and there was no evidence of hydrocarbon ions. The deuterated equivalents of methanol showed more clearly the relationship of the formation of OH^- and the appearance of carbon-containing ions. Upon further exposures of methanol, the second stage of reaction is uncovered. The SSIMS spectra show the production of OH^- concurrent with OCH_3^-; the absence of a molecular ion indicates that there is no physisorption of methanol, but there is evidence for the formation of surface hydroxide species. The positive spectrum shows the $AlOCH_3^+$ peak, suggesting that the OCH_3^- fragment adsorbs as methoxide. Combining the data collected from SSIMS and XPS, it is apparent that clean aluminium at room temperature promotes the decomposition of methanol to methane and surface oxide, until all the metallic surface states have reacted. Once this surface oxide has formed the methanol begins to form a surface methoxide and hydroxide. When the surface was heated the methoxide decomposed to methane leaving an oxidized surface.

7.12. Conclusions

Properly used, SSIMS does not destroy delicate surface structures; consequently, it can be usefully employed in all areas of technology. It is difficult to imagine how any other technique could provide the detailed surface chemical information demonstrated in the above examples. The range of potential applications in this area is considerable. However, SSIMS is not the universal panacea for surface analysis. Other techniques are frequently required to confirm, enhance or help to quantify the analysis. The combination of SSIMS with other surface spectroscopies will be immensely powerful for the study of surface chemical interactions.

Appendix

General protocol for optimization of SIMS parameters (see Appendix 5)

1. Move sample to estimated optimum position, according to the internal geometry of the instrument.
2. Switch on ion/atom beam with static conditions, i.e. a beam current density in the order of 10^{-9} A cm^{-2}.

3. The quadrupole analyser should be set to standard operating conditions initially (according to the instrument manual). If there is no charging, i.e. a stable, measurable signal is observed, then optimization can continue. This necessitates a series of gradual adjustments to the set of standard conditions.
4. The sample position may also be moved in order to maximize the signal; care must be taken not to saturate the detector. This is recognized by 'split peaks'. In this event the primary beam current must be reduced in order to limit the number of secondary ions produced.
5. If no signal can be found even with a high target bias then charging (or an instrumental fault) is indicated. In the case of charging it will be necessary to apply a compensating electron beam.
6. The current of the electron beam should be gradually increased until a signal is detectable. Since optimum charge neutralization lies within a small range, too great a flux will not effect complete neutralization, but may result in electron-stimulated ion emission (ESIE). The optimum position for the electron beam is often found to be impinging on the edge of the sample rather than directed at the same spot as the incoming primary beam.
7. The target position and operating conditions of the analyser can now be re-optimized.
8. It is a worthwhile task to ensure that there is no ESIE by blanking off the primary ion/atom beam, when the signal should disappear. Any residual signal will be a consequence of ESIE.
9. Spectral acquisition can now proceed.
10. Finally, familiarity with a particular instrument leads to a better knowledge of the specific operating conditions for different types of sample.

References

1. Benninghoven, A. (1970). *Z. Physik*, **230**, 403.
2. Surman, D. and Vickerman, J.C. (1981). *Appl. of Surf. Sci.* **9**, 108–21.
3. Gardella, J.A. and Hercules, D.H. (1980). *Anal. Chem.*, **52**, 226.
4. Fontaine, J.M., Durand, J.P. and Gressus, C.L. (1979). *Surf. and Int. Anal.* **1**, 196.
5. Morgan, A.E. and Werner, H.W. (1979). *Anal. Chem.* **49**, 927.
6. Briggs, D. and Wooton, A.B. (1982) *Surf. and Int. Anal.*, **4**, 109.
7. Hunt, C.P., Stoddart, C.T.H., and Seah M.P. (1981). *Surf. and Int. Anal.*, **3**, 157.
8. *CRC Handbook of Physics and Chemistry*, 62nd edn (1981–2), pp. E373–4.
9. Brown, A. and Vickerman, J.C. (1986) *Surf. and Int. Anal.* **8**, 75.
10. Hearn, M.J. and Briggs, D. (1988). *Surf. and Int. Anal.*, **11**, 198–213
11. Eccles, A.J., van den Berg, J.A., Brown, A. and Vickerman J.C., (1986). *J. Vac. Sci. and Technol.*, **A4**, 1888.
12. van den Berg, J.A. 1986 *Vacuum* **36**, 11/12, 981–9.
13. Surman, D. and Vickerman, J.C. (1981). *Appl. of Surf. Sci.*, 9, 108–21.
14. Humphrey, P. (1988). Ph.D. thesis, UMIST.
15. Lai, S.Y., Briggs, D., and Vickerman, J.C. (1989) *Surf. and Int. Anal.*, in press.

16. Niehuis, E., Heller, T., and Benninghoven, A. (1985) *Ion Formation from Organic Solids*, Vol. 3, Springer Proc. in Phys. 9. Springer, Berlin.
17. Hagenhoff, B., Niehuis, E., van Leyen, D., and Benninghoven A., (1987) 235 *Proc. VIth Int. Conf. on SIMS*, Versailles, France (ed. A. Benninghoven, A.M. Huber, and H.W. Werner,). John Wiley, Chichester.
18. Lub, J., van Velzen, P.N.T., van Leyen, D., Hagenhoff, B., and Benninghoven A. (1989). *Surf. and Int. Anal.*, **12**, 53.
19. Briggs, D. (1984) *Polymer*, **25**, 1379.
20. Briggs, D. and Hearn, M.J. (1985). *Int. J. Mass. Spec. and Ion Proc.*, **67**, 47–56.
21. Campana, J.E., Ross, M.M., Rose, S.L., Wyatt, J.L. and Colton R.J. (1983) *Ion Formation from Organic Solids*, Vol. 25, Springer Series in Chemical Physics, p. 144. Springer Verlag, Berlin.
22. Brown, A., van den Berg, J.A., and Vickerman, J.C., (1985). *Spectrochimica Acta*, **40B** (Nos. 5/6), 871–7.
23. Briggs, D., Brown, A., and Vickerman, J.C. (1989) *Handbook of Static SIMS*, John Willy, Chichester.
24. Van Ooij W., private communication.
25. Gardella, J.A., and Hercules, D.M. (1971). *Anal. Chem.* **53**, 1879.
26. Briggs, D., Hearn, M.J., and Ratner, B.D. (1984) *Surf. and Int. Anal.*, **6**, 184.
27. Briggs, D. (1984) *Polymer*, 25, 1379.
28. Briggs, D. (1982) *Surf. and Int. Anal.*, **4**, 151.
29. Brown, A. and Vickerman, J.C. (1986), *Surf. and Int. Anal.*, **8**, 75–81.
30. Briggs, D. and Hearn, M.J. (1989) *Surf. and Int. Anal.*, in press.
31. Briggs, D. and Wootton, A.B. (1982) *Surf. and Int. Anal.*, **4**, 109.
32. Briggs, D. (1982). *Surf. and Int. Anal.*, **4**, 151.
33. Briggs, D. (1983). *Surf. and Int. Anal.*, **4**, 113.
34. Briggs, D. (1986). *Surf. and Int. Anal.*, **9**, 391.
35. Benninghoven, A. (1973). *Surf. Sci.*, **35**, 427–57.
36. Benninghoven, A. and Mueller, A. (1973) *Thin Solid Films*, **12**, 439.
37. Sakakini, B. (1987) Ph.D. thesis UMIST.
38. Plog, C., Wiedman, L., and Benninghoven, A. (1977). *Surf. Sci.*, **67**, 565–80.
39. Gardella, J.A. and Hercules, D.M. (1971) *Anal. Chem.*, **53**, 1879.
40. Gardella, J.A., Novak, F., and Hercules, D.M. (1984) *Anal. Chem.*, **56**, 1371.
41. Lynn, R.A.P., Davis, S.S., Short, R.D., Davies, M.C., Vickerman, J.C., Humphrey P., and Hearn, M.J. (1988). *Polymer Commun.*, **29**, 365–7
42. Briggs, D. and Ratner, B.D. (1988). *Polymer Commun.*, **29**, 6–8,
43. Garbassi, F., Ochiello, E., Bastioli, C., Romano, G., and Brown A. (1986). *Proc. of 1st Int. Conf. on Comp. Interfaces*, May 1986.
44. Lub, J., van Velzen, P.N.T., van Leyen, D., Hagenhoff B., and Benninghoven A. (1988). *Surf. and Int. Anal.*, **12**, 53.
45. Davies, M.C., Brown, A., Newton, J.M., and Chapman S.R. (1988). *Surf. and Int. Anal.*, **11**, 591.
46. Wilding, I., Davies, M.C., Melia, C.D., Brown, A., Humphrey, P., and Rowe R.C. (1988). Submitted to *J. Appl. Polymer Sci.*
47. Peppas, N.A. and Buri, P.A. (1985). *J. Contr. Rel.*, **2**, 266.
48. Davies, M.C. and Brown, A. (1986) *Proc. 13th Int. Contr. Rel. Soc. Symp.*, 194–5.
49. Grasserbauer, M., Stungeder, G., Potzl, H., and Cruerrero E. (1986). *Fresenius Z. Anal. Chem.*, **323**, 421–49.

50. Brown, A., Humphrey, P., Hunt, N., Patterson, A.M., Vickerman, J.C., and Williams J.O. (1986). *Chemtronics*, **1**, 11.
51. Brown, A., Gerrard, N.D., Humphrey, P., Patterson, A.M., Vickerman, J.C., and Williams, J.O. (1986). *Chemtronics*, **1**, 64.
52. Brown, A., van den Berg, J.A., and Vickerman, J.C. (1986) *Surf. and Int. Anal.*, **9**, 309–17.
53. Astles, M.G., Dosser, O.D., McLean, A.S., and Wright P.J. (1981). *J. Cryst. Growth*, **54**, 485.
54. Hancock, B.R. and Kroemer H. (1984). *J. Appl. Phys.*, **55**, 4239.
55. Brown, A., Gerrard, N.D., Humphrey, P., Patterson, A.M., Vickerman, J.C., and J.O. Williams. (1986). *Chemtronics*, **1**, 64.
56. Eccles, A.J., Humphrey, P., Jones, T.A., and Vickerman J.C. (1987) *Conf. Proc., SIMS VI*, Versailles, France (ed. A. Benninghoven, A.M. Huber, and H.W. Werner) p. 27. John Wiley, Chichester.
57. Rotermund, H.H., Penka, V., Delouise, L.A., and Brundle C.R. (1987) *J. Vac. Sci. Technol.*, **A5**, (4, Jul/Aug), p. 198–202.
58. Delouise, L.A., Miller, D.C., Auerbach, D.J., and Brundle C.R. (1984, 1985): IBM internal publications.
59. Henrich, V.E. (1985). *Rep. Prog. Phys.*, **48**, 1481-541.
60. Aggarwal, H., Reed, N.M., and Vickerman J.C. (1988). To be published.
61. Bourgeois, S., Gimenez, C., and Perdereau M. (1985). *Mater. Sci. Monog.*, **23B**, 931.
62. Surman, D.J., van den Berg, J.A., and Vickerman, J.C. (1982) *Surf. and Int. Anal.*, **4**, 160.
63. Dwyer, J., Fitch, F.R., Machado, F., Qin, G., Smyth, S.M., and Vickerman, J.C. (1981). *J. Chem. Soc. Chem. Commun.*, 422–4.
64. Dwyer, J., Fitch, Qin and Vickerman, J.C. (1982). *J. Phys. Chem.* **86**, 4574–8.
65. Espie, A.W. and Vickerman, J.C. (1984). *J. Phys. Chem.*, **80** 1903–13.
66. Ellison, A., Moulyn, J.A., Scheffer, B., Brown, A., Herbert, B., Humphrey, P., Diakun, G., Worthington, P., Mabbs, F.E., and Collinson D. (1987). *Advances in Polyolefins* (ed. R.B. Seymour and T. Cheng), p. 111. Plenum press, New York.
67. Morys, P., Gorges, U., and Krauss, H.L. (1984). *Z. Naturforsch.*, **398**, 458.
68. Gillet E., Channakhone S., Matolin, V., and Gillet M. (1985) *Surf. Sci.*, 152/153. p. 603–14.
69. Tindall, I.F. and Vickerman, J.C. (1985). *Surf. Sci.*, **149**, 577.

8

SIMS IMAGING

8.1. Introduction

SIMS imaging is a method of performing the SIMS experiment which provides two- and three-dimensional information on elemental and molecular distributions with sensitivities higher than other imaging methods such as SEM, EDAX and SAM.

Scanning electron microscopy (SEM) has for many years been used to obtain images of topographic features at high magnifications. More recent adaptations of the technique, such as Energy Dispersive X-ray Analysis (EDX or EDAX) have also allowed a degree of chemical mapping. There are, however, problems with this technique in terms of the range of elements that can be analysed (below m/z ~20 it becomes very difficult), the surface specificity, and the ultimate sensitivities achievable. Similarly, with the scanning Auger microprobe (SAM), although the lateral resolution is comparable with SIMS the sensitivity is much lower (0.5 per cent for SAM compared with p.p.m. for SIMS).

SIMS imaging provides a *direct* method of chemically mapping the distributions of materials with the high sensitivity, extreme surface specificity, and full range of elemental and molecular information available from the SIMS technique.

It is worth considering at this stage the nomenclature of SIMS imaging. Werner[1] defined two imaging variations of the SIMS technique:

(1) **microprobe SIMS**, in which a small spot beam (~1 μm diameter) is scanned across the sample of interest, [2,3] or

(2) **imaging SIMS**, in which the beam is broader (~300 μm diameter) and stationary while the secondary ion optics are scanned to produce the image.[4]

When the development of the liquid metal ion source revitalized microprobe SIMS by allowing sub-micron imaging, the technique became known as SIMS imaging. This is potentially the source of some confusion with the already common imaging SIMS and so there has recently been a move towards the new name of **scanning SIMS**. This brings the nomenclature more into line with scanning electron microscopy (SEM) and scanning Auger microprobe (SAM). In this work the names 'scanning SIMS' and 'imaging

SIMS' will be used for the two variants and the term 'SIMS imaging' will be used to refer to both.

Alternative terms for the instrumentation are **ion microprobe** for the scanning beam instrument and **ion microscope** for the scanned optics system. The spatial resolution limits are currently ~20 nm for microprobes[5] and 0.2 μm for ion microscopes.[6]

8.2. Experimental

The equipment for SIMS imaging has already been described in Chapter 4 from the point of view of the ion sources and mass analysers used so it will not be reconsidered here in any detail. The aim of this section is to provide a general overview of SIMS imaging by both experimental methods and also to describe the data systems which are so important to the efficient use of the technique.

8.2.1. Ion microprobe

In scanning SIMS, a microfocused primary beam is scanned across the sample while the mass spectrometer is set to the mass (line) of interest. If the secondary ion signal is displayed synchronized to the beam scan then an image can be built up of the distribution of that particular species over a given area of surface. This is shown schematically in Chapter 4 (Fig. 4.2). The time-scale for acquisition of these images is typically 20–80 s and the standard method used is to display the image on a oscilloscope and expose a photographic film to the screen as the image forms. Nowadays, dedicated data systems are also available for image acquisition (see Section 8.2.4).

The advent of the liquid metal ion source allowed high-resolution scanning SIMS images with a sub-micron probe to be acquired for the first time.[7] Since then, several groups have produced images of a variety of materials ranging from metals, polymer fibres, and semiconductor devices to insects.[5, 8–10]

The field of view is variable between several millimetres and a few microns.

An additional feature of scanning SIMS is that the incoming primary beam also forms secondary electrons in sufficient numbers for secondary electron images of the surface under analysis to be generated by scanning the beam at TV rate. These provide a useful guide when moving about the sample, allowing accurate positioning on small features or imperfections.[11,12] These ion-induced secondary electron images are comparable with those produced in a scanning electron microscope, but due to the much smaller escape depth of the secondary electrons (1–2 nm compared with 10–20 nm in SEM) the contrast effects observed are often much more surface specific.[13]

The ion microprobe can, of course, also be used to produce SIMS spectra from small areas (down to a few square microns) by scanning the ion beam at a fixed rate while the mass spectrometer scans the mass range. Spectra can even be obtained from points by using a stationary beam and taking a spectrum, although in this mode it must be borne in mind that the primary beam flux density at the point of analysis is usually well above the SSIMS regime. For example, a 1 pA beam of 0.2 μm diameter has a flux density of 3 mA cm^{-2}, i.e. 10^6 times greater than Werner's[1] level for the SSIMS regime (damage rates are considered further in Section 8.3).

8.2.2. Ion microscope

Ion microscopes such as the Cameca IMS series have been available for many years. The sample is held at a fixed potential of 4.5 kV while being bombarded with high-energy ions (5–17.5 keV) from a variety of sources (O_2^+, Ar^+, O^-, Cs^+). The beam diameter is variable from 3 to 200 μm and provides a maximum current density on target in excess of 50 mA cm^{-2} into a 50 μm spot.

As the lateral resolution of the image is determined by the secondary ion optics it is independent of the diameter of the primary beam. The transfer optics allow three preset circular image fields with diameters of 400, 150, and 25 μm.

The secondary ions produced are passed into a magnetic sector mass spectrometer offering high sensitivity and superior mass resolution to the quadrupole commonly used in ion microprobes. The mass-resolved ion image is magnified and translated into an equivalent electron image. This is then displayed on a fluorescent screen, where it can be viewed directly or photographed (see Chapter 4, Fig. 4.1). A recent alternative is the option of transferring the image directly into a computer. This is discussed further in the next section.

Ion imaging is often a secondary feature of these instruments, the primary use of which is in rapid-profiling dynamic SIMS. The ability to produce a secondary ion image during simultaneous acquisition of a depth profile is of great importance to accurate profile acquisition. Any lateral inhomogeneities in the distribution of the species studied will introduce errors into the final profile (see Chapter 5).

8.2.3. Data systems

Several SIMS imaging computer systems are currently available on the market. In this section we shall briefly consider two types, one for scanning SIMS and one for imaging SIMS.

Vacuum Generators have produced a computer data acquisition system which is fairly typical of the scanning SIMS approach. The VGX 7000 series data system is based on a DEC Micro PDP-11/23 Minicomputer. The software has been written under DEC's Micro-RSX operating system. This allows a degree of multi-user operation, which is essential for the efficient operation of such a system. Data can be acquired very quickly whereas the time to process and output data can be lengthy due to the amount of data available.

The system is able to acquire and manipulate secondary ion and secondary electron images as well as conventional mass spectra and depth profiles.

A 512 × 512 8-bit framestore image buffer allows four 256 × 256 images to be held in memory at the same time. These can be stored subsequently on fixed or removable Winchester Discs. Other storage systems, such as tape streamers or optical discs, can be added if necessary.

The scanning of the beam and the mass settings of the quadrupole are controlled by the computer. Minimum image acquisition times are just under 20 s per ion per pass (256 × 256 pixels with 0.3 ms per pixel). Up to four ions can be acquired at the same time by rapid switching of the quadrupole. A version is available to operate a scanning ToF-SIMS system where up to eight images can be acquired simultaneously in no extra time and each image can contain up to twenty peaks or groups of peaks. This gives a tremendous advantage over the quadrupole system, which is limited to collecting the single mass/charge signal that is being transmitted at any one time.

Many of the examples in Section 8.6 were acquired on this type of data system.

For imaging SIMS Charles Evans & Associates have developed the PC-1 data system to run a Cameca ion microscope. Based on the IBM PC - AT, the system provides the usual mass spectral and depth profiling software and also some imaging capability. The abilities to store and recall images, adjust the intensity scaling, produce retrospective linescans, and perform multi-dimensional image acquisition and manipulation (Section 8.6), are all available as software options from this package.[14]

As has been described previously, in the ion microscope the specimen is bombarded with a 'broad' primary beam (150 μm), and the secondary ions from different areas, though they may be emitted at the same time, maintain their relative spatial relationship during the extraction and mass separation. For computer acquisition the secondary ions are allowed to impinge on a position-sensitive, pulse counting detector (Charles Evans & Associates Resistive Anode Encoder, Model RAE-5PC) which enables the computer to assign a position to the arriving ion, thus building up an elemental distribution map without the need to scan the primary beam. An example of results from this system is given in Section 8.7.

As was mentioned previously, before the advent of computer image acquisition systems, SIMS images were generally displayed on an oscilloscope screen which was then photographed. In the 1970s Fassett and Morrison[15] devised a computerized system for imaging SIMS in which the photographed image was then scanned by a photodensitometer which digitized the brightness information and fed it into a computer. This was clearly a limited system and the clarity of the computer image depended on the quality of the photograph. The acquisition of SIMS images directly onto computer has several advantages:

1. The rapidity of acquisition is of great importance when the sample is prone to damage or when rapid analysis is required.
2. The fact that the computer stores all the information and contrast scales automatically means that time wasted with under- or over-exposed photographs need no longer be a problem.
3. The ease of storage of images on magnetic or optical discs or tape is a great advantage and any computer can then be programmed to access the data and display it on the screen. By using a computer camera or an appropriate printer the image can be transferred to different media quickly and easily.
4. The single major benefit of image computerization is the ability to manipulate the data after acquisition to gain the maximum information. Some examples of the post-processing and manipulation of images are given later in this chapter.

8.3. Sensitivity vs. damage

8.3.1. Sensitivity

As has been mentioned previously in the section on SSIMS, it is one of the basic tenets of SIMS analysis that the secondary ions detected are representative of the initial state of the surface. With the exception of multi-dimensional SIMS discussed later it is frequently the aim of scanning SIMS to provide surface images under SSIMS conditions. This is less so with imaging SIMS, where the images are usually generated during a depth profile under dynamic SIMS acquisition conditions. Even these are still generally regarded as being representative, as far as elemental distribution is concerned (although not necessarily chemical state), of a given depth in the profile, although they are obviously formed over a depth range of many monolayers.

The advantage of imaging SIMS is that the lateral resolution is independent of the beam spot size and hence of the beam current. Thus the damage rate can be increased without decreasing the image quality. In scanning SIMS, however, there is a strong link between damage rate and lateral resolution.

Thus for scanning SIMS it is not always possible to obtain an image under truly 'static' conditions, especially at high magnification. The lifetime for a monolayer under ion bombardment can be estimated by the following equation:

$$t_m = N_s / Y(I_p/e) \qquad (8.3.1)$$

where t_m is the lifetime (s), I_p is the primary flux density (A cm^{-2}), e is the electronic charge (1:6 $\times$ 10^{-19} C per particle). N_s is the number of atoms in a 1 cm^2 monolayer, and Y is the number of atoms removed per incoming ion (sputter yield—atoms per ion).

For SSIMS of a 1 cm^2 sample, N_s is estimated at 10^{15} particles cm^{-2} and Y is assumed to be 1. Then for an I_p of 1 nA cm^{-2} the lifetime t_m is 1.6 $\times$ 10^5 s or 44 h. Given that the SIMS experiment can typically be performed in tens of minutes the amount of material removed is small ($<$ 1 per cent of a monolayer in 1000 s).

The damage rate in scanning mode can be calculated using a modification of this equation. The primary flux density can be split into two components; the primary current and the area analysed ($I_p = i_p/a$). The primary current can be measured directly by connecting a meter between a conducting sample and earth. The sample must be biased + 15 V to suppress secondary electrons which would otherwise appear as a contribution to the incoming current. The area analysed can be calculated if the magnification factors are known accurately.

If the number of atoms per monolayer and the sputter yield are again assumed to be 10^{15} cm^{-2} and 1 respectively then the number of monolayers removed per 100 s can be calculated as a function of current and analysis area according to the equation

$$M_r = \frac{i_p \, Y \times 100}{N_s \, a \, e} \qquad (8.3.2)$$

where M_r is monolayers removed (in 100 s), i_p is primary current (A), an a is area being etched (cm^2).

The results of such a calculation for a range of currents and magnifications appear in Table 8.1.

It can be seen that only at low currents and low magnifications can static conditions (i.e. sub-monolayer removal per image) be achieved.

These figures are only intended as a guide, however. Some materials sputter at much higher rates than others. A selection of sputter yields under a range of bombardment conditions is given in Appendix 7.

8.3.2. *Damage*

There is an important difference between damage and sputtering. Material that has been sputtered is removed from the surface. Some material, however, may be chemically or physically altered by the primary particle

TABLE 8.1

Monolayers removed as a function of ion current and magnification

Nominal magnification	Current (nA)						
	0.01	0.05	0.1	0.5	1.0	5.0	10.0
100	0.0005	0.0025	0.005	0.025	0.05	0.25	0.5
200	0.002	0.01	0.02	0.1	0.2	1.0	2.0
500	0.015	0.075	0.15	0.75	1.5	7.5	15
1000	0.05	0.25	0.5	2.5	5.0	25	50
2000	0.2	1.0	2.0	10	20	100	200
5000	1.2	6.2	12.5	62	125	625	1250
10,000	5	25	50	250	500	2500	5000
20,000	20	100	200	1000	2000	10,000	20,000
50,000	125	625	1250	6250	12,500	62,500	125,000

Assumes: sputter yield $Y = 1$; frame time = 100 s; numbers of frames = 1 (256×256 points).

collision but not be removed. This material is 'damaged'. It is often assumed that, for ordered systems, subsequent surface layers are removed quite neatly and independently by the primary beam. In a SSIMS experiment where <0.1 per cent of the monolayer is removed and statistically no bombardment site is hit twice this is a reasonable approximation. In dynamic SIMS, as has been shown in Chapter 5, this assumption is no longer valid, and mixing occurs due to the advancing collision cascades, which results in interface broadening in depth profiles.

It is possible, however, under SSIMS conditions to damage the surface. Delicate organic structures or inorganic lattices may be disrupted by the incoming primary species to such an extent as to lose molecular structure. Different materials are damaged at different rates. As many of these materials are insulating, surface potential may be a problem and recent work has shown that under identical flux densities of primary ions and atoms, a spectrum from ion bombardment of polystyrene loses molecular character at a rate approximately four times that for atom bombardment.[16] This is a major problem for SIMS imaging which uses relatively high *ion* beam flux densities.

Other factors to be borne in mind are the relationship between beam current and spatial resolution and between sensitivity and damage.

For a 10 keV gallium ion probe the spot size varies with beam current as shown in Fig. 4.6.[17] For best lateral resolution (and minimum damage), the ion source must be operated with the lowest beam current. Thus the secondary ion signals observed will be correspondingly lower. However, the amount of material present in a monolayer at different magnifications varies, as shown in Table 8.2.

Clearly, a 1 p.p.m. contaminant at 50 000 magnification (analysis area 30

TABLE 8.2

The number of atoms per monolayer in the area analysed as a function of magnification

Nominal magnification	Image width (μm)	Atoms (monolayer)
100	12,000	1.4×10^{13}
200	6000	3.6×10^{12}
500	3000	5.8×10^{11}
1000	1200	1.4×10^{11}
2000	600	3.6×10^{10}
5000	300	5.8×10^{9}
10,000	120	1.4×10^{9}
20,000	60	3.6×10^{8}
50,000	30	5.8×10^{7}

$\times$ 30 μm^2) would only correspond to 580 atoms. If the ionization probability is 10^{-2} (as is not unreasonable) then even with a ToF analyser, having a transmission of 10 per cent, it would be necessary to consume two monolayers to detect a single ion. A quadrupole analyser, with a typical transmission of 0.1 per cent, will obviously be several orders of magnitude worse.

Thus when it comes to gathering meaningful data sets the extent of the limitation can be seen. A typical spectrum of m/z = 0–300 acquired with a m/z step of 0.2 contains 6000 data points. To obtain reasonable statistics at each data channel for such a spectrum would require many monolayers to be consumed. For imaging the situation is even worse. A 256 $\times$ 256 pixel image contains 65 536 data points. To obtain statistically meaningful information for each pixel would require a considerable depth to be probed, as at 50 000 magnification each pixel (assuming they are discrete areas with no overlap) would contain $<$ 900 atoms.

Clearly, the parallel mass detection advantage of the ToF analyser becomes important at this stage. Although the p.p.m.–p.p.b. sensitivity of SIMS is impossible for very small analysis areas, this is still an extremely sensitive technique and the excellent signal to noise ratio and high transmission ToF analysers coupled with this parallel detection mean that for major surface components *SSIMS* images can be acquired with sub-micron lateral resolution.

8.4. Contrast mechanisms

8.4.1. Introduction

The various effects that lead to contrast in secondary electron images have been extensively studied. Much of this work was recently reviewed and restudied by Lai *et al.*[13] By comparison, there are very few studies on the contrast mechanisms involved in secondary ion images although these are just as critical in an understanding of the meaning behind the images. Correct interpretation of a SIMS image can often depend on an accurate appreciation of the factors involved in producing contrast in the image. Similarly, the reverse may often be true, in that the reasons an image shows a lack of contrast may be equally significant.

As is the case with secondary electron imaging, there are several mechanisms which can produce contrast and quite often an image will contain contrast due to more than one of these sources. Much of the work on secondary electron contrast divides these mechanisms up as follows:

(1) topographic contrast,

(2) material contrast,

(3) crystallographic contrast,

(4) voltage contrast,

(5) magnetic contrast.

In describing mechanisms for secondary ion contrast, the same broad divisions can still be used, although it becomes apparent that the relative importance of the mechanisms varies for secondary ion and secondary electron images.

In addition to the above contrast mechanisms, there are other mechanisms to consider in an investigation of secondary ion image contrast. Isotopic contrast and chemical contrast are special cases of material contrast which depend in the first case on the isotopic separation possible with a mass spectrometer and in the second case on the effect of surface chemistry on the secondary ion emission probabilities.[18]

The nature of the primary beam can also have an effect on the secondary ion contrast observed, particularly if a comparison is made between images obtained with an inert (e.g. Ga^+ or Ar^+) or a reactive beam (e.g. O_2^+) due to induced chemical contrast.

Due to the mass difference between electrons and ions, any effects due to magnetic interference will be three to five orders of magnitude less apparent for ions than for electrons. Magnetic contrast then is not a serious cause of contrast in secondary ion images.

8.4.2. Topographic contrast

One of the major factors in producing contrast in secondary electron images is sample topography. The different physical orientation of different regions of an uneven sample will lead to differing probabilities for secondary ion emission. This will lead to a different intensity in the final image, i.e. *contrast*.

This assumes that once a secondary species has been emitted it will be detected. Indeed, the collection efficiency for secondary electrons of a scintillator–photomultiplier detector with a positive bias voltage on the front end is often incorrectly quoted as 100 per cent.

The collection efficiency of a secondary ion optical system is certainly less than this. Furthermore, the ion optics are usually arranged so that the most likely collection is on the axis of the first apertures and is distributed conically about this axis. It is therefore possible that a secondary ion emitted from a given point on the sample will be more likely or less likely to reach the detector than an ion emitted from a different location on the sample. This is then the basis for another, independent source of topographic contrast—instrument geometry.

Topographic contrast can therefore have two components:

(1) instrumental geometry, and

(2) specimen topography.

If an image is obtained from a sample containing no specimen topography, such as a section of silicon wafer, then the contrast due to the instrumental geometry can be defined. The example shown in Fig. 8.1 shows a Si^+ SIMS image acquired on a scanning SIMS system. Produced originally to be used in subsequent growth of semiconductor devices, the sample has an extremely well-defined, highly polished, flat surface. Because it is a chemically pure, uniform, monocrystalline specimen it is also free from material, crystalline, and voltage contrast.

The prime source of contrast in the SIMS image is therefore the instrumental geometry. It can be clearly seen that the image shows a considerable decrease in intensity towards its edge. This is because the image area is greater than the uniform acceptance area of the secondary ion optics (in this case, a parallel plate capacitor after the design of Wittmaack *et al.*[19]). The image is, in effect, a projection of the acceptance cone geometry onto the flat sample. It should be noted, however, that if a line scan across this image is

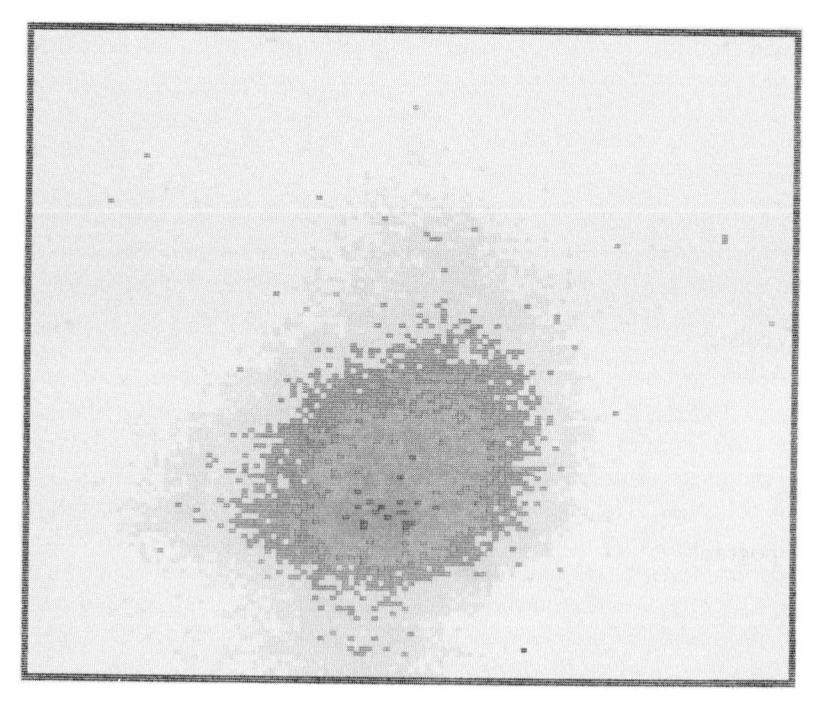

FIG 8.1. Si^+ SIMS image from silicon wafer at low magnification, illustrating decrease in signal at edges of image field due to instrumental geometry.

plotted, the full width at half height is 3 mm and as most scanning SIMS is performed over much smaller analysis areas the field of view falls well within the uniform acceptance cone of the analyser.

One cause of specimen topography is shadowing. This can be either **primary shadowing**, where an area is obscured from the primary beam, or **secondary shadowing**, where the area is obscured from the secondary ion optics. This can be minimized if the angle at the specimen between the primary ion source and the secondary ion optical path is reduced. Obviously, this is only possible at the instrument design stage.

A less drastic form of specimen topography is produced when adjacent areas of the same sample are at slightly different angles with respect to the secondary ion path. It is well known that the secondary ion signal varies with primary beam impact angle (Chapter 2) and this will become important in curved or angled samples where the signal will fall as the angle between the surface and the primary beam falls.

This topographic contrast, being a function of the sample, is unavoidable. Distinguishing the two may not always be easy. In many cases the SEM generated by the primary ion beam is a useful guide to the physical geometry of the sample surface. This is not always possible because if the sample is non-conducting the secondary electrons will be trapped at the surface once the potential has risen more than a few volts positive. Flooding the surface with low-energy primary electrons can stabilize the surface potential to allow secondary ion emission, but the secondary electron detector would be swamped by the primary electrons and would be unable to extract useful secondary electron information. In this situation, the only way to identify topographic contrast is to examine the effect on a range of secondary ion species. If specimen topography is the prime contrast mechanism operating, then all ion species will be affected to a similar degree. It is then possible to greatly reduce the topographic contrast by mathematical means. If a suitable data system is used to acquire an image for a different ion species or a total ion image from the same area, then dividing the required image by this second image should produce a first-order correction of the topography, leaving a more uniform image.[20] Gross contrast effects in the ratioed image may then be due to other contrast mechanisms.

8.4.3. Material contrast

Material contrast is produced by variations in the yield of secondary species due to chemical differences between different parts of the sample. While material contrast is of great importance in secondary electron imaging it is of fundamental importance in SIMS imaging where chemical differences are being characterized.

In addition to the material contrast produced due to differing concentrations of the species being imaged at different locations on the surface there can also be variations due to the chemical environment of that species. This is then a particular form of material contrast due to the chemical state of the surface and it is therefore referred to as **chemical contrast**. For a common example of, chemical contrast refer to the applications example (Plate 8.1). In this example the yield of Cr^+ is much greater from the (Fe, $Cr)_3O_4$ area than from the chromium metal although the latter obviously contains a higher percentage of chromium atoms. Thus the enhanced secondary ion yield in the presence of oxides[4] which is well noted in static and dynamic SIMS is here manifested in scanning SIMS as a contrast difference.

Another variation on material contrast is **isotopic contrast**. This is a form of material contrast that is specific to secondary ion imaging due to the mass spectrometric nature of the technique. In acquiring a SIMS image for multi-isotopic elements, the imaged species is usually of the most abundant isotope for the obvious reason that this gives the best signal to noise ratio. Only if there was a molecular interference or the signal was in danger of saturating the detector would there be a need to use another isotope. This is quite acceptable as there are no naturally occurring contrast differences between the different isotopes. However, in isotopic tracer studies, such differences can be induced, and can then be of crucial importance in understanding chemical mechanisms. The sensitivity of SIMS imaging to isotopic contrast can be used to good effect. The previously mentioned example of chemical contrast also contains isotopic contrast.

8.4.4. Crystallographic contrast

Crystallographic contrast has been shown to be of considerable importance in secondary electron imaging. Secondary ion signals from a single crystal depend on the orientation between the sample and the primary beam[21] and it might therefore be expected that crystallographic contrast is a major feature of SIMS images. In fact this is not the case. The variations of secondary ion yields from different crystal faces are in extreme cases only two to three times and this is usually buried beneath other effects. Only on very clean poly-crystalline samples where other contrast effects are absent will crystallo-graphic contrast be observed.

8.4.5. Voltage contrast

Secondary ions are emitted from the sample with relatively low energies of a few tens of electron volts. In the ion microscope these ions are accelerated by a high (4.5 kV) target potential into the analyser. However, in the ion micro-probe the sample bias is usually close to 0 V and if the sample is biased a few tens of volts then the ions will be accelerated or retarded according to the charges involved. Due to the narrow energy acceptance window of a quadrupole these ions of altered energy may not be detected. If the surface potential of the sample is non-uniform then secondary ion detection probabilities from different areas may be different and contrast can appear in images. This variation of surface potential can arise from several sources:

(1) sample areas of differing conductivity;

(2) uneven surface topography;

(3) non-uniform sample thickness; or

(4) any combination of the above.

There are obviously links between voltage contrast and both material and topographic contrast.

Voltage contrast in secondary ion images is less commonly observed in secondary ion imaging than in SEM as it is more often obscured by greater material contrast. However, semiconductor device structures which are chemically heterogeneous in a regular pattern often show voltage contrast. An example of secondary ion image contrast for such a structure is shown in Fig. 8.2. This shows a pair of $SiOH^+$ images of a memory chip. As the target bias is changed from -30 to -45 V the observed secondary ion contrast is reversed. Returning the voltage to -30 V produced the original image, confirming that the change is due to the voltage and is not an irreversible change due to sample damage.[22]

Problems of sample charging can also cause surface potential contrast effects. These are at their most significant when the sample has adjacent areas of widely differing conductivity. Materials such as semiconductor devices and composites are classic examples of systems which often show wide and discrete conductivity variations across a small area.

In order to obtain meaningful SIMS images from both high and low conductivity areas simultaneously the surface potential must be balanced with great precision. The electron flux must be kept at the minimum necessary to ensure neutralization of the charging in the non-conducting regions with the beam scanning at the normal acquisition rate. The acquisition must then be performed as quickly as possible (computer control) to ensure complete data capture before either the surface potential drifts away from the established optimum or the specimen degrades under the combined ion–electron bombardment.

The problems associated with sample charging in the imaging of insulators or electrically heterogeneous structures is a major one and the following section (Section 8.5) will consider these in more detail and will set down a protocol for acquisition in these circumstances.

(a)

(b)

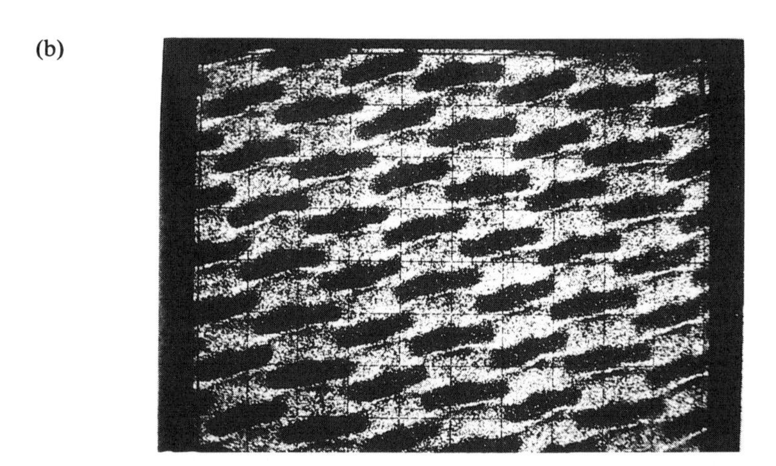

FIG 8.2. $SiOH^+$ image from semiconductor device, illustrating change in viewed contrast with adjustment of target voltage from (a) −30 V to (b) −45V.

8.4.6. Primary beam contrast

As well as contrast effects within a single image due to the mechanisms just described there can also be contrast effects due to the primary beam. As the identity of the primary particle has an effect on the secondary ion yield it can have an effect on the image obtained.

It is well known that the mass and energy of the primary particles affect the secondary ion yields. Hearn and Briggs[23] have compared argon, gallium, and

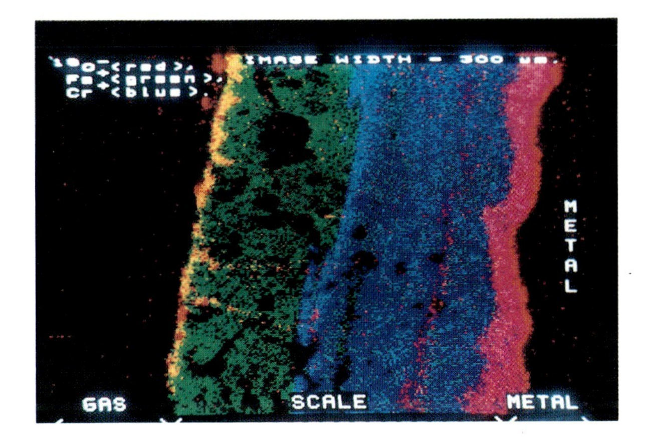

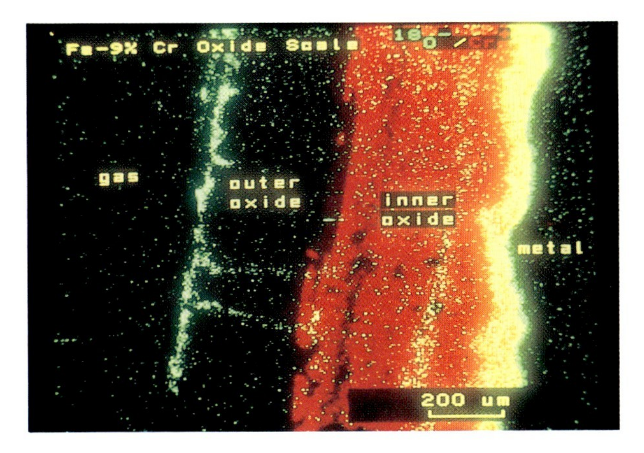

PLATE 8.1. Colour overlay of SIMS images from oxide scale cross-section.

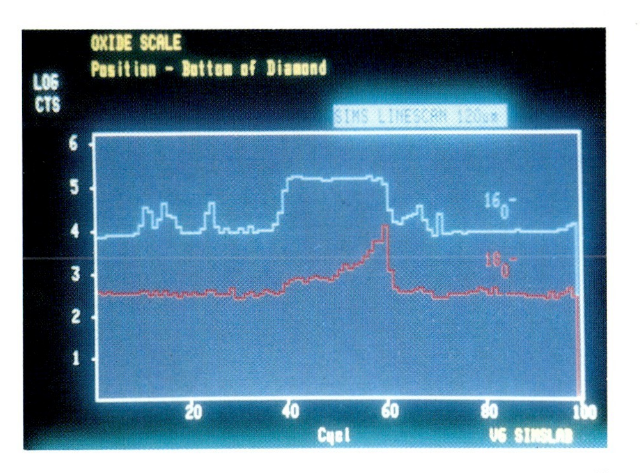

PLATE 8.2. Linescan across oxide scale showing diffusion profile for $^{18}O^-$ through ^{16}oxide. 100 measurement *cycles* = a linescan of 120 μm.

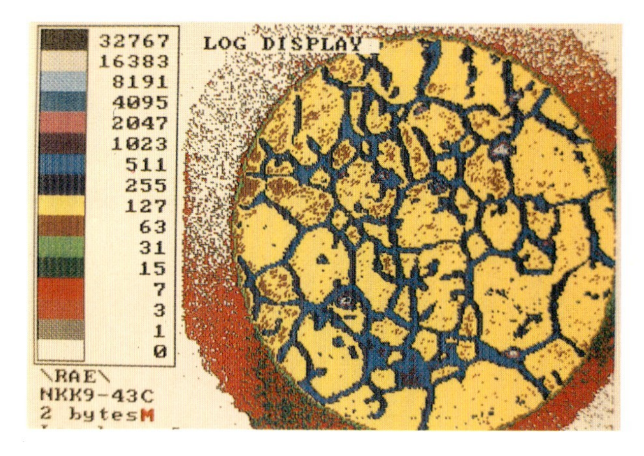

PLATE 8.3. Computer-processed false-colour image of boron segregation in stainless steel acquired on an ion microscope.

(a)

(b)

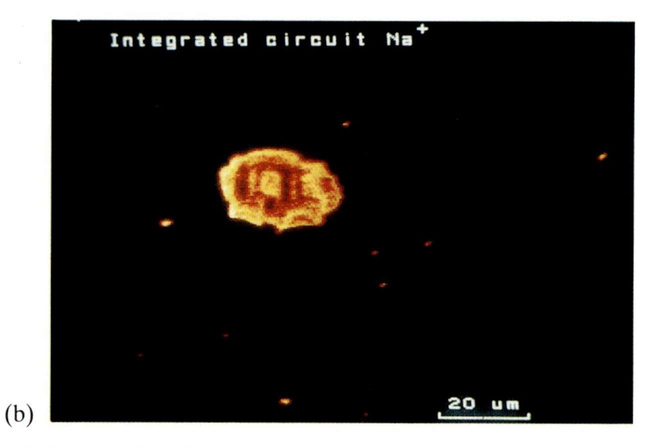

PLATE 8.4. Scanning SIMS images of a semiconductor power transistor array showing (a) aluminium tracks ($A1^+$) and (b) adventitious sodium contamination (Na^+).

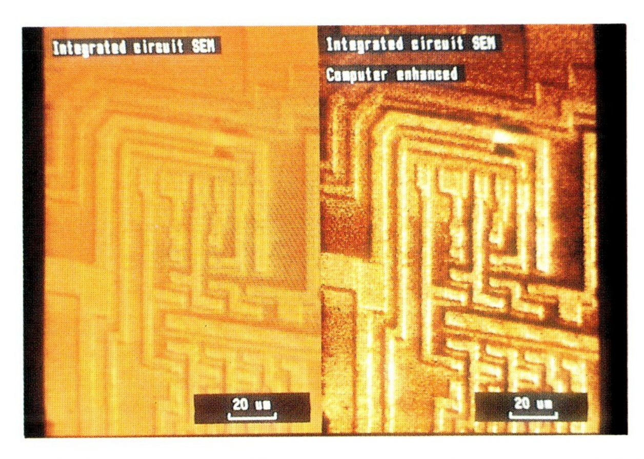

PLATE 8.5. SEM from sample in Plate 8.4 produced by scanning SIMS instrument shown as received (left) and after computer contrast enhancement (right).

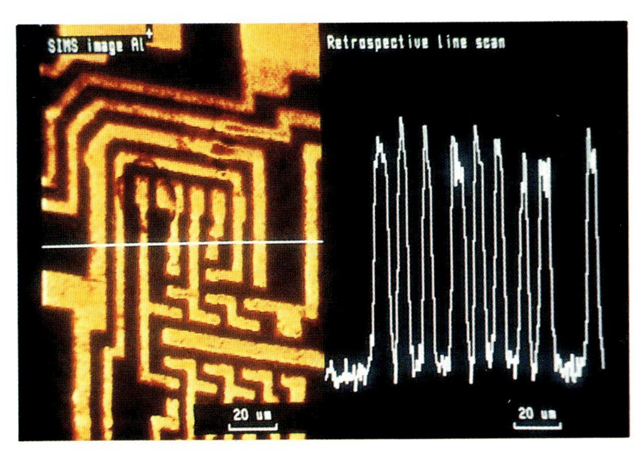

PLATE 8.6. A1^{+} SIMS image from Plate 8.4 (a) (left) with linescan along indicated horizontal produced after acquisition by computer processing.

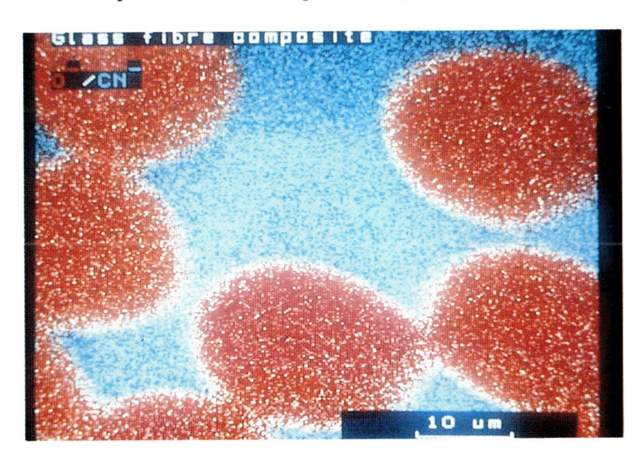

PLATE 8.7. Two-colour overlay of cross-section through glass fibre composite showing fibres indicated by O^- image and resin by CN^-.

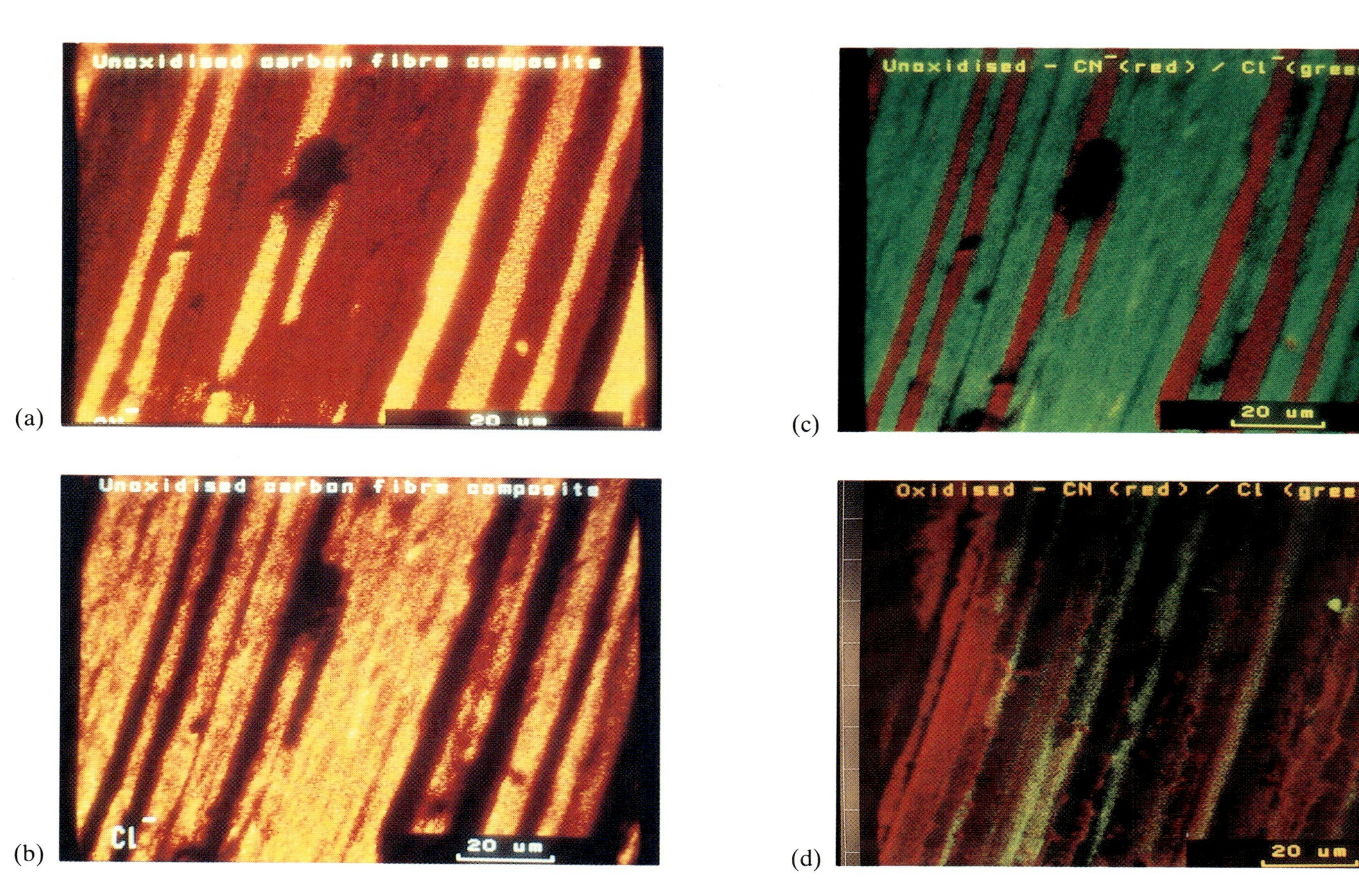

PLATE 8.8. SIMS images from carbon fibre composite showing (a) CN^- distribution (from fibres), (b) Cl^- distribution (from resin), (c) colour overlay of (a) (red) on (b) (green), and (d) colour overlay from oxidized fibre composite showing resin overlayer remaining on fibres, indicating better adhesion.

xenon bombardment at a range of energies (see Chapter 7). These yield differences will be present in images as well but would only become apparent in a direct comparison of the same image obtained under conditions that were identical apart from the identity of the primary beam. This would, however, not affect the contrast within the images, only the relative intensity between the two acquisition conditions.

The primary beam could, however, induce chemical contrast differences if the results from an inert beam (e.g. Ga^+ or Ar^+) are compared with those from a reactive beam (e.g. O_2^+). The reactive beam would produce yield enhancements similar to those produced by the presence of a surface oxide. Certain areas of the sample may be more prone to enhancement due to the surface chemical state and so would generate new contrast differences.

8.5. Specimen charging and image acquisition

8.5.1. Introduction

The problems due to specimen charging have been discussed in some detail in Chapter 7. The difficulties can be quite pronounced in SIMS imaging, and in particular in scanning SIMS, where one is usually constrained to using an ion beam which is often directed into a very small analysis area. The localized charge build-up can then be very intense. Problems may also be experienced with electrically heterogeneous samples due to differential charging between neighbouring regions within an image.

In order to obviate these difficulties and to provide a rapid and reproducible method for acquiring SIMS images from such specimens a protocol for establishing the optimum acquisition parameters has been devised.[18]

8.5.2. Acquisition protocol

The following protocol has been devised for the acquisition of SIMS images from non-conducting or electrically heterogeneous samples. This procedure, as outlined in the following sections, refers specifically to acquisition on a quadrupole-based scanning SIMS instrument. However, the general form is applicable to a wide range of instrumentation.

The setting-up procedure has, for convenience, been divided into several sections. The first section outlines the initial procedure necessary to obtain a secondary ion signal. Once a signal is obtained it is then optimized using the methods described in Section (*2*) below. That section also contains checks to avoid false acquisition data such as might be due to ESIE. With a stable, optimized signal the final acquisition parameters can then be set. As the parameters are adjusted step-wise towards the final requirements the signal is re-optimized at each stage by repeating the procedure in (2). The final stage is then to acquire the image.

(1) Signal search

(a) The sample is moved into the expected correct position for optimum signal based on experience with conducting samples.

(b) The primary beam is scanned across a large area of the sample. This area may be considerably larger than the finally required analysis area but should not be larger than the acceptance field of the analyser.

(c) The secondary ion optics should be set to the normal operating conditions. If signal is obtained at this stage then charging is minimal and signal optimization (see(*2*)) can proceed.

(d) If no signal is obtained then the analyser should be switched to total ion mode. This is achieved on a quadrupole by applying only r.f. to the rods, and this allows the analyser to act as a broad-band filter. This will often be sufficient to produce a workable signal to be optimized.

(e) If no signal is found in total ion mode then the target bias should be increased with the same sign as the required ions. This then aids emission by repelling the secondary ions towards the detector.

(f) Lack of signal with the target bias maximized is indicative of considerable sample charging. The electron beam should then be used with the flux increased until signal just appears. Care must be taken that any signal generated is not due to ESIE. This can be confirmed by blanking the ion beam at which stage the signal, if due to SIMS-EN, will disappear.

(g) A complete failure to obtain any signal after all the preceding steps will usually be due to a sample position which is suitable for SIMS but is unfavourable for the electron beam. The position should then be changed and the stages (a)–(f) repeated.

(h) If no signal is found after all these stages it will often indicate an instrument fault.

(2) Signal optimization Once a signal has been obtained it is then a matter of optimization. This is an iterative process which involves the gradual adjustment of several variables:

(a) Sample position—the x, y, z, and ϕ can all be adjusted slightly to obtain optimum signal. It should, however, be borne in mind that adjustment of these will tend to change the analysis area, which may be critical in scanning SIMS.

(b) The electron beam can also be adjusted in terms of flux, focus, and position. Care must be taken after any alteration of the electron beam to check for ESIE.

(c) When an optimized signal is achieved the stability over a period of 10–20 s should be observed. A tendency for the signal to decay is usually due to imprecise neutralization or alteration of surface due to sputter removal or damage.

(3) Setting of acquisition parameters Once the signal is stable and optimized the final acquisition parameters can be approached. As each step towards these is made the optimization routine should be repeated to ensure full optimization at each stage. Parameters to be finalized may include:

(a) Resolution—if total ion mode was used in the setting up then this will need to be increased to normal levels.

(b) The mass to be analysed should be set. Often it is best to tune-up on generally more intense low mass signals and move gradually up the mass range, optimizing at several successive points until the desired ion is reached.

(c) The analysis area should be reduced to that finally desired. The system will need re-optimization as the area changes.

(d) If possible, the target bias should also be reduced, compensating with the electron beam where necessary, as this often leads to improved resolution.

(e) The scan rate should be gradually reduced to the rate the computer will use for the final acquisition. Often the slowing down of scan rate will create a need for re-optimization.

(f) Area to be analysed—If possible, it is often better to move the sample to a fresh area for the acquisition, especially if the setting-up procedure has been lengthy.

(4) Acquire image Once the final checks are complete, the image can be acquired.

This procedure has been devised over a number of years to produce the best results from a wide variety of samples in the minimum time. It may appear at first sight to be cumbersome but an experienced operator can often complete the whole protocol in 1–2 min. Experience will allow some short-cuts in the setting-up but the basic checking procedure should always be followed. The success of this protocol has been proven by its successful application to a very large range of samples and its usefulness in the training of instrument operators new to the technique.

8.6. Three-dimensional imaging

The fact that the primary beam is continually removing the surface is usually a problem for the acquisition of SIMS images. However, just as dynamic SIMS makes use of this fact to gain secondary ion depth distributions, so SIMS imaging can use this erosion to provide multidimensional elemental analysis.

The method of three-dimensional SIMS acquisition is as follows (Fig. 8.3). An image is acquired into an appropriate computer system and the beam then erodes the surface for a defined time to etch away several layers. After the etch period a second image is acquired. These alternate etch–acquire cycles are repeated until the computer has a sequence of images representing the lateral distributions of a particular secondary ion at a range of depths. Appropriate software can then arrange these images into a vertical stack representative of a volume of sample. Any of the horizontal slices can be recalled and manipulated as described in the following section. The computer can also reconstruct a vertical slice down a given line to produce a vertical cross-section image. This vertical image can then be processed in exactly the same way as a horizontal image.

A further feature of three-dimensional SIMS is the ability to produce from the information stack a retrospective depth profile. By defining a given area on the surface (perhaps a surface feature), the sum of the counts within this area at each depth can be calculated and the results plotted as secondary ion intensity vs. depth, i.e. a depth profile.

It must be noted, however, that this assumes uniform erosion across the image field; as images may be obtained from areas which are often chemically

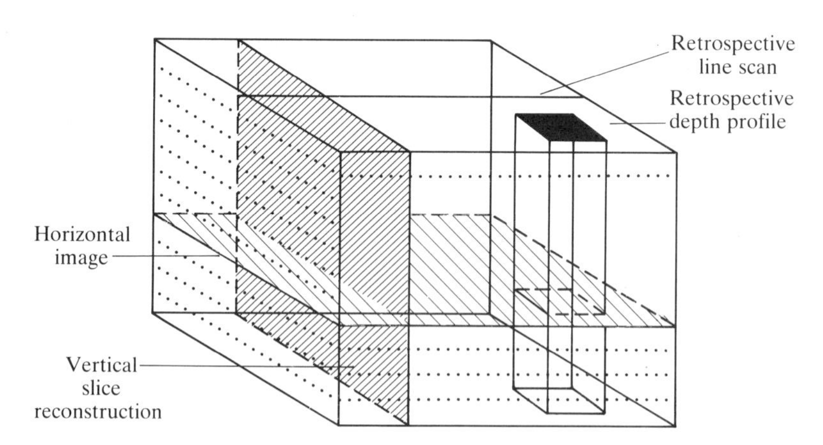

FIG 8.3. Schematic diagram of multidimensional SIMS, illustrating types of information available.

heterogeneous, this may well not be the case and distortions will occur in the image stack. As three-dimensional imaging is a depth-profiling technique, all the physical parameters described previously (Chapter 5) as being important in dynamic SIMS will be important here too. Also, it should be borne in mind that the technique is not suitable for materials that damage easily, such as organics, unless one is interested solely in elemental information.

8.7. Applications of SIMS imaging

8.7.1. Metallurgical samples

There are many applications of SIMS imaging to a wide variety of sample types. This section aims to give a few representative examples.

Metallurgical samples are ideal for showing the abilities of the SIMS imaging technique. The high conductivity means that charging and associated problems are largely eliminated. From a clean surface, high-quality secondary electron images can be generated, which means that finding a particular area on the sample for analysis is very straightforward.

The example shown in Plate 8.1 illustrates the use of induced isotopic contrast by the use of ^{18}O to study the mechanism of oxide growth.

The specimen is a cross-section through an oxide scale formed on the surface of a 90:10 iron–chromium alloy. The alloy is used in the construction of advanced gas-cooled reactors. The oxide scale, once formed, protects the metal from further corrosion and so the growth mechanisms involved in scale formation are very important.

Oxide growth in the presence of CO_2 has been studied. The oxide was grown primarily in $C^{16}O_2$ but in the latter stages this was replaced by $C^{18}O_2$. Oxide has grown initially with ^{16}O but the last oxide to form contains ^{18}O. The oxide scale–metal interface is sectioned following growth and by imaging the distributions of the two isotopes the oxide growth points can be identified. The ^{18}oxide has formed at the ^{16}oxide–metal interface by diffusing through the ^{16}oxide. The image also shows that the presence of small vertical cracks through the oxide can allow the ^{18}O to penetrate easily and form a secondary oxidation zone. This may well be a failure site in the protective oxide scale. The presence of Cr^+ in the inner oxide is indicative of a cation transfer in the opposite direction to the oxide transport. The inner oxide thus formed is an iron–chromium spinel.[24] Thus the oxide formation mechanisms can be deduced, or in some cases confirmed, from the SIMS images. In this case the diffusion of CO_2 through the oxide is an important step in the oxidation mechanism. By scanning the primary beam across a single oxide band on a similar sample while the computer follows the $^{16}O^-$ and $^{18}O^-$, a profile of the distribution of both oxides is obtained. In the example shown (Plate 8.2) the oxide band is clearly delineated by the square rise in the $^{16}O^-$ signal. The $^{18}O^-$ signal illustrates the diffusion profile of the labelled oxygen into the oxide.

This method of cross-sectioning samples and using scanning SIMS to linescan ion distributions is a useful technique which is largely complementary to dynamic SIMS. Due to the limits of resolution (~ 50 nm) it is not suitable for distances < 1 μm unless the section is finely tapered, but this range is adequately covered by dynamic SIMS. Time considerations mean that depth profiling to > 10 μm is impractical, whereas by line-scanning sectioned samples 'depths' of 10–1000 μm can be investigated in seconds.

Oxide growth studies are only one example of the type of metallurgical problems being investigated by SIMS imaging. Corrosion studies, weld analyses, stress–failure cracking and grain boundary segregation (see below) are all areas where the analytical power of SIMS imaging can be used to good effect in the elucidation of the various mechanisms involved.

Whilst ion microscopes are perhaps more commonly used primarily for dynamic SIMS studies, they are capable of producing very elegant results when operated in imaging mode. The example shown in Plate 8.3 is a sample of stainless steel examined in a Cameca ion microscope.[14] The image is of the boron distribution and in this false-colour presentation the segregation of the boron to the intergranular boundaries is clearly visible. This example was acquired, processed, and displayed using the Charles Evans & Associates data system described in Section 8.2.3.

8.7.2. *Semiconductor devices*

Semiconductors are another area where SIMS imaging is finding major application. The technique of SIMS in its many forms is ideally suited to applications in the world of semiconductor materials and device characterization.

Dynamic SIMS has long been used to very good effect in this area. Very often the substrates are doped with various materials or are grown as discrete layers of different composition. It is then that dynamic SIMS has been applied to obtain profiles through the dopant distribution or layer structure (Chapter 5). Static SIMS–FABMS has also been applied in studies of surface cleaning of wafers prior to implantation or device growth (Chapter 7).[25]

Imaging SIMS is important in dynamic SIMS assessment as the depth profile assumes that the materials found are evenly distributed in the lateral directions. If this is not the case then the profile may be misleading.

It is in device characterization, however, that SIMS imaging becomes a very powerful technique. The trend to Very Large Scale Integration (VLSI) means that increasing numbers of devices are being fitted into a wafer of given size. Concurrent with this increasingly sophisticated technology for device fabrication there must be a similarly increasing sophistication of the technology for device charcterization. Scanning SIMS in part provides this.

It might at first sight appear that semiconductor devices are ideal samples for scanning SIMS, having well-defined, regular structures. This is often not the case as the structures in question are usually conducting features on an insulating background, and the inhomogeneous surface potential that this can cause due to localized charge build-ups on bombardment by an ion beam can seriously disrupt the SIMS images obtained.

Additionally, the features of interest are often present in very small structures that are extremely thin. The small size necessitates a high magnification, and the sputter rate, as mentioned previously, then becomes critical, especially given the delicate nature of many of these structures.

With careful charge neutralization such as has been described previously (Section 8.5) it is usually possible to obtain well-defined images from semiconductor devices. Once a stable signal has been established over the area of interest then a series of images can be obtained of the species of interest.[26] Plate 8.4(a,b) shows a pair of images obtained from the same area of a semiconductor device. The tracks are clearly seen in the aluminium image and there is some indication of unevenness on one track junction to the left of centre. A positive secondary ion spectrum acquired from this image area showed a very high sodium signal. Subsequent imaging of this signal gave the image in Plate 8.4(b) which indicates that the adventitious sodium contamination is highly localized. Using the power of computer processing the two images can be overlain with different colour codings. This allows the precise registry of contaminant location with reference to surface features. This may well be of considerable importance in isolating the cause of the contamination. In this case the sodium-containing contaminant is clearly the cause of the unevenness in the aluminium image.

Additional processing can produce further information. The secondary electron image of the above area is of poor quality. By adjusting the contrast and brightness settings of this image a considerable improvement in clarity can be achieved. Plate 8.5 shows the SEM before (left) and after (right) contrast enhancement.

Further processing can zoom in on features of interest or produce a retrospective linescan across an image (Plate 8.6).

The second example from the semiconductor field illustrates the power of the ion microprobe for small area analysis. The image in Fig. 8.4 shows clearly defined regions of different O^- intensity on a section of memory array. The bright tracks are gate oxide, which gives a high yield of O^-. The darker regions are poly-silicon, which gives a much lower oxygen signal and hence appears dark in the image. Having acquired the image it is then possible to direct the primary beam to precise points on the surface and acquire

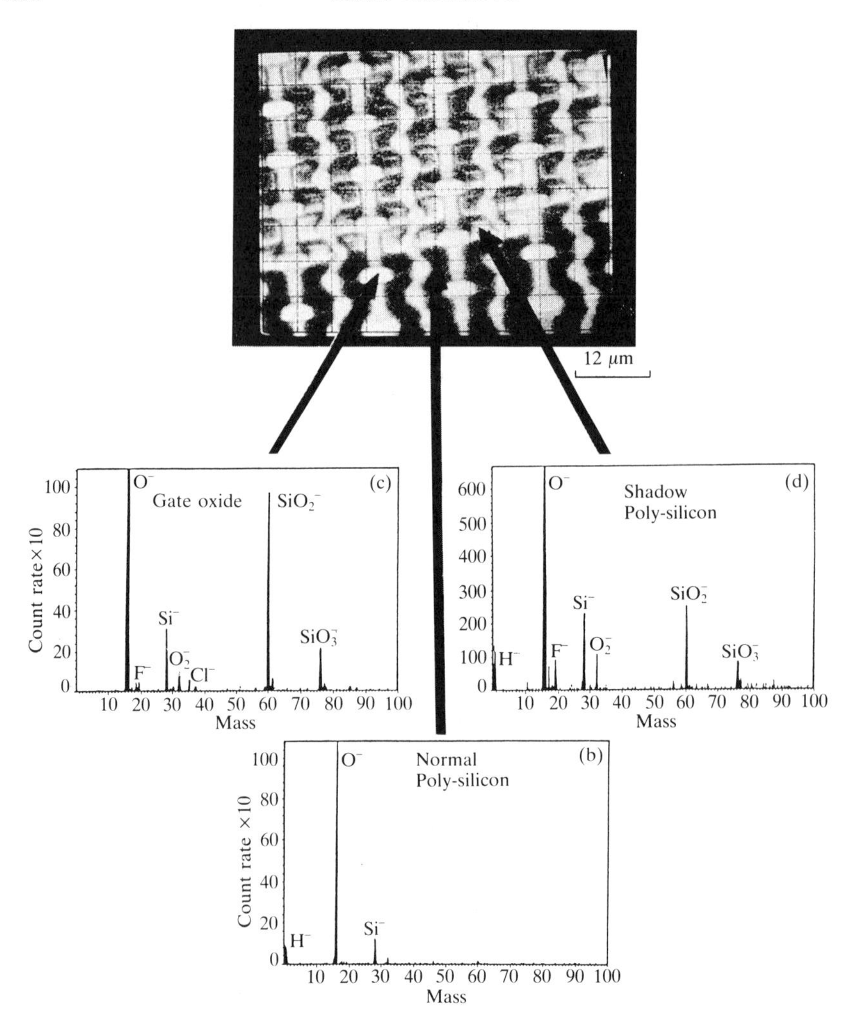

FIG 8.4. O^- SIMS image from a semiconductor device with negative ion point spectra acquired from indicated positions. Investigation of 'shadow' zones on 64 *K* RAM devices. $^{16}O^-$ SIMS image (a) reveals low O^- on normal poly-silicon tracks (b), uniform O^-distribution on 'gate' oxide (c), but non-uniform, intense patches of O^- on poly-silicon tracks within 'shadow' zone (d). Mass spectral analysis of each of these areas confirms presence of silicon dioxide on poly-silicon in 'shadow' region. Magnification ×2000.

spectra from these points. The two negative ion spectra from the gate oxide and poly-silicon each show peaks that would be expected from such regions, in accordance with the chemical structure. The gate oxide spectrum shows, in addition, contamination due to F^- and Cl^-. There are, however, unknown regions in the image which should be regions of poly-silicon but in fact appear bright in the SIMS image. These regions, known as 'shadow zones' from their appearance in a SEM, have been investigated in the third spectrum. As can be clearly seen in Fig. 8.4 (d), the shadow poly-silicon shows clear differences in surface chemistry from the normal poly-silicon. In fact, the presence of SiO_2^- and SiO_3^- peaks characteristic of silicon oxide indicate that the shadow silicon has been oxidized.

8.7.3. *Non-conducting samples*

Polymer production is a major segment of the chemical industry. The surface state of polymers is critical to their behaviour. The same can be said of glass-based products. These two areas provide a wealth of samples of an insulating nature. There is a growing need for these samples to be characterized in terms of lateral variations in surface chemical structure. XPS is the 'traditional' technique for polymer analysis although, as was discussed in Chapter 7, SIMS–FABMS is now becoming recognized as a powerful alternative or complementary technique.

In terms of lateral resolution, however, XPS has no great potential; 10 μm has recently been demonstrated[27] but this is obviously very limited compared with results from SIMS. Scanning Auger microprobe is an alternative technique which has much the same lateral resolution as SIMS, but although it has been used with considerable success on conducting samples it is of lower sensitivity and produces much greater sample damage with organics than SIMS. This leaves scanning SIMS as the method of choice for lateral chemical characterization of surfaces.

Imaging of insulating samples provides the ultimate test of the SIMS imaging technique. The problems of sample charging already discussed in general in Chapter 4, and with specific regard to the imaging situation earlier in this chapter, are at their most challenging when restricted to highly localized areas. It was these very problems which led to the development of the microfocused scanned atom source already described.[28,29] This has already produced high-quality SIMS images of insulating samples with minimum sample charging. The limiting spatial resolution of this source, however, means that for samples where lateral resolution of better than 5 μm is needed a microfocused ion beam has to be used. Although this then presents great problems in achieving a balanced surface potential, it is by no means impossible. With careful charge neutralization it is possible to obtain SIMS images of even quite irregularly structured insulating samples.

An example of a SIMS image from an insulating sample of extremely complex structure, both chemically and topographically, is shown in Fig. 8.5. The sample was a common house-fly and the area imaged shows clearly the C_2^- distribution surrounding the large compound eye. This sample was successfully imaged using a gallium ion beam with no need for coating or pretreatment.

An area where much success has recently been demonstrated is the imaging of composite materials and especially fracture surfaces in such materials. Composite materials are, like semiconductors, an area of significant interest in materials technology. Also, in common with semiconductor devices, they often have irregular topography and adjacent areas of differing chemical structure and electrical conductivity. The two examples shown are both fibre composites. The first example is a fracture surface in a unidirectional carbon fibre composite. The image of the carbon fibres has been obtained using the CN^- ion whilst the epoxy resin, which incorporates epichlorhydrin, is identified by the Cl^- ion. In the first three images (Plate 8.8 (a)–(c)) the composite has been formed from unoxidized fibres. The first two images show the fibres ((a)—CN^-) and the resin ((b)—$^{35}Cl^-$). The two images appear to be complementary. This is confirmed when they are overlain (c). The individual fibres (red) and areas of resin (green) are clearly defined, indicating that the fracture has occurred at the resin–fibre interface, with the resin fracturing away from the fibre. This indicates poor bonding between the two components in the composite. This is expected from unoxidized

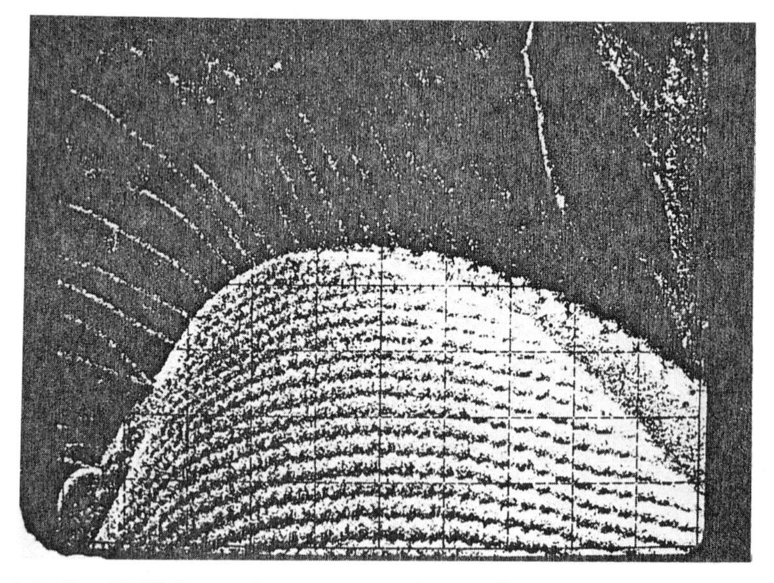

FIG 8.5. C_2^- SIMS image from compound eye of house-fly obtained using a mircofocused ion beam and no sample coating.

fibres. In the fourth image (Plate 8.8 (d)), of a sample manufactured under identical conditions but using oxidized fibres, the demarcation is less distinct and in many areas the fibres show a thin resin overlayer. This overlayer is ~10 nm thick and was undetectable by conventional electron microscopic means. Its presence indicates a fracture failure within the resin, which is a feature of a stronger composite, and, indeed, interlaminar sheer strength tests confirm these observations.[30]

The second composite example is a cross-section through a glass fibre–epoxy resin composite (see Plate 8.7). As in the case of the carbon fibre composite, the individual fibres have diameters of ~7–10 μm. This specimen is a polished transverse section. The circular fibre sections are clearly delineated by the O^- image (yellow) from the silica glass. In this case it is the resin, which contains an amino group, that is identified by the CN^- image (blue).

The ability to probe the all-important interface regions of these materials to define the surface chemistry is critical in developing an understanding of the overall composite performance.

8.8. Future developments

The examples given in the previous section are intended to be illustrative rather than exhaustive. The number of applications areas is ever growing as more and more materials are being examined by SIMS.

The advent of the imaging ToF-SIMS instrument has opened the way for high-sensitivity, sub-micron molecular imaging.[31] In the past, much of the imaging work published has been confined to elemental ions due to a lack of sensitivity for the generally less intense cluster ions. The ToF analyser with its high, mass-independent transmission will allow molecularly specific imaging.

The last few years have also seen a decrease in spot sizes for ion-probes from >1 μm to 20 nm. It is hoped that improvements in technology will allow the useful spot size to be decreased yet further until it would reach a theoretical limit at the diameter of the SIMS collision cascade.[32]

Another area where technological improvements are necessary to reduce spot sizes is in the development of the scanning FABMS sources already mentioned. Currently, the smallest useful spot size is ~5 μm.[29] It is hoped that this will be reduced as instrumental developments progress.

The potential of multidimensional SIMS is an area that has only been touched on in this chapter. The technique is just becoming widely applied to industrially relevant samples and this is a development that will become increasingly important in the near future.

One final area in which the technique of SIMS imaging may well be enhanced in the near future is that of quantification. The use of lasers or electron gases to post-ionize secondary neutrals for quantification in SIMS will be discussed in Chapter 9. There would be technical difficulties in adapting these methods to a scanning mode, but these difficulties are by no means insoluble and we may well see quantitative chemical imaging within the next few years.

References

1. Werner, H. (1975). *Surf. Sci.*, **47**, 301.
2. Drummond, L.W. and Long, J.V.P. (1967). *Nature*, **215**, 5277.
3. Liebl, H. and Herzog, R.F.K. (1963). *J. Appl. Phys.*, **34**, 2893.
4. Castaing, R. and Slodzian, G. (1962). *J. Microsc.*, **1**, 395.
5. Levi-Setti, R. Crow, G., and Wang, Y.L. (1986). *Secondary Ion Mass Spectrometry, SIMS V*, Springer Series in Chemical Physics, Vol 44 (ed. A. Benninghoven, R.J. Colton, D.S. Simons, and H.W. Werner), p. 132. Springer Verlag, Berlin.
6. Bernius, M.T., Ling, Y–C., and Morrison, G.H. (1986). *Secondary Ion Mass Spectrometry, SIMS V*, Springer Series in Chemical Physics, Vol 44 (ed. A. Benninghoven, R.J. Colton, D.S. Simons, and H.W. Werner), p. 245. Springer Verlag, Berlin.
7. Prewett, P.D. and Jefferies, D.K. (1980). *Inst. Phys. Conf. Ser. No. 54*, Chapter 7, p. 316. The Institute of Physics, London.
8. Brown, A. and Vickerman, J.C. (1984). *Analyst*, **109**, 851.
9. Rudenauer, F.G. (1982). *Secondary Ion Mass Spectrometry, SIMS III*, Springer Series in Chemical Physics, Vol. 19 (ed. A. Benninghoven, J. Giber, J. Laszlo, M. Riedel, and H.W. Werner), p. 12. Springer, Berlin.
10. Waugh, A.R., Bayley, A.R., and Anderson, K. (1984). *Vacuum*, **34**, 103.
11. Lam, K., Fox, T.R., and Levi-Setti, R. (1981). *Proc. 28th Int. Field Emission Symp*, Beaverton, Oregon.
12. Anazawa, N., Aihara, R., Okunuki, M., and Shimizu, R. (1982). *Scanning Electron Microscopy*, 1443.
13. Lai, S.Y., Briggs, D., Brown, A., and Vickerman, J.C. (1986). *Surf. Interface Anal.*, **8**, 93.
14. Evans, C.A. (1988). *Secondary Ion Mass Spectrometry, SIMS VI*, Versailles, September 1987 (ed. A. Benninghoven, A.M. Huber, and H.W. Werner), p. 13. John Wiley, Chichester.
15. Fassett, J.D. and Morrison, G.H. (1978). *Anal. Chem.*, **50**, 1861.
16. Brown, A., van den Berg, J.A., and Vickerman, J.C. (1985). *Spectrochimica Acta*, **40B**, 871.
17. Bayly, A.R., Waugh, A.R., and Vohralik, P. (1984). *Spectrochimica Acta*, **40B**, 717.
18. Humphrey, P. (1988). Ph.D thesis, UMIST.
19. Wittmaack, K., Maul, J., and Schulz, F. (1973). *Int. J. Mass Spectrom. Ion Phys.*, **11**, 23.

20. Bayly, A.R., Fathers, D.J., Vohralik, P., Walls, J.M., Waugh, A.R., and Wolstenholme, J. (1986). *Secondary Ion Mass Spectrometry, SIMS V*, Springer Series in Chemical Physics, Vol 44 (ed. A. Benninghoven, R.J. Colton, D.S. Simons, and H.W. Werner), p. 245. Springer Verlag, Berlin.
21. Benninghoven, A. and Müller, A. (1973). *Surf. Sci.*, **39**, 416.
22. Brown, A., Eccles A.J., and Vickerman, J.C. (1986). *Secondary Ion Mass Spectrometry, SIMS V*, Springer Series in Chemical Physics, Vol 44 (ed. A. Benninghoven, R.J. Colton, D.S. Simons, and H.W. Werner), p. 245. Springer Verlag, Berlin.
23. Hearn, M.J. and Briggs, D. (1986). *Surf. Interface Anal.*, **9**, 411.
24. Lees, D.G., Rowlands, P.C., and Brown, A. (1988). *Inst. Phys. Conf. Ser.* **90** 57.
25. Brown, A., Humphrey, P., Hunt, N., Patterson, A.M., Vickerman, J.C., and Williams, J.O. (1986). *Chemtronics*, **1**, 11.
26. Brown, A., Humphrey, P., and Vickerman, J.C. (1988). *Secondary Ion Mass Spectrometry, SIMS VI*, Versailles, September 1987 (ed. A. Benninghoven, A.M. Huber, and H.W. Werner), p. 393. John Wiley, Chichester.
27. V.G. Scientific, East Grinstead, Sussex, UK.
28. Eccles, A.J. (1986). PhD thesis, UMIST.
29. Eccles, A.J., van den Berg. J.A., Brown, A., and Vickerman, J.C. (1986). *J. Vac. Sci. Technol.*, **A4(4)**, 1888.
30. Dennison, P., Jones, F.R., Brown, A., Humphrey, P., and Watts, J.F. (1988). *6th Int. Conf. on Composite Materials (ICCOM-VI)*, London, July 1987, *J. Mat. Sci.*, **23**, 2153.
31. Eccles, A.J. and Vickerman, J.C. (1988). *Secondary Ion Mass Spectrometry, SIMS VI*, Versailles, September 1987 (ed. A. Benninghoven, A.M. Hober, and H.W. Werner), p. 239. John Wiley, Chichester.
32. Slodzian, G. (1988). *Secondary Ion Mass Spectrometry, SIMS VI*, Versailles, September 1987 (ed. A. Benninghoven A.M. Huber, and H.W. Werner), p. 3. John Wiley, Chichester.

9

SIMS - RELATED TECHNIQUES

Other particle desorption techniques have been referred to in the course of our discussion of SIMS. They have similarities in that secondary ion data are produced, although the mechanism by which the secondary ions are generated is rather different in each case.

The difficulties associated with quantification in SIMS have stimulated the investigation of post-ionization of the sputtered neutrals to decouple the ionization and emission processes. Sputtered (or secondary) neutral mass spectrometry, SNMS, is an emerging technique which, with development, is sure to become an important variant of SIMS, possibly even superseding it in dynamic SIMS.

In many respects, SSIMS of organic compounds and ^{232}Cf fission fragment plasma desorption mass spectrometry (PDMS) have developed in parallel. They contrast sharply in the character of the primary bombarding particle. On the one hand we have keV inert gas atoms, whilst on the other MeV multi-charged high-mass fragments are used. Despite this, the spectral data observed from the two techniques are similar. PDMS seems to be preferable when very high molecular weight ions ($>10\ 000$) are sought.

Laser microprobe has also been developing in parallel with SSIMS. The desorption process seems at first sight to be rather different from the collisional mechanism of SIMS since pulses of UV radiation are used which appear to thermally desorb and ionize elements and clusters. The data generated, however, again frequently appear to be very similar, and the technique has the merit of being *very* fast (70 μs for a spectrum). It is, however, very destructive because, in its usual mode of operation, material is removed to a depth of $\sim 1\ \mu m$ and an area of $\sim 1\ \mu m^2$.

9.1. Sputtered neutral mass spectrometry

Although SIMS is an extremely sensitive technique, the fact that it relies on secondary ions generated during the sputtering process has two fundamental consequences for quantification. First, as we have already seen (§S 2.2), the relationship between elemental composition and secondary ion intensity is very complicated and dependent on surface electronic state (ionization efficiency varies from 10^{-1} to 10^{-5}). Second, as most of the emitted particles are neutral, almost all of the potentially useful information is not used.

The elemental sputter yields vary only by a factor of three to five across the Periodic Table (see Fig. 2.1.5); thus, if an efficient method of post-ionization could be found, SNMS might be very sensitive and rather more easily quantifiable. The yield equation of a SNMS signal $I(X^0)$ for a neutral sputtered species X is given by

$$I(X^0) = I_p Y_X \alpha_X^0 \eta_X (1 - \alpha_X^+ - \alpha_X^-) \tag{9.1.1}$$

where Y_X is the partial sputter yield of the species X; I_p is the bombarding primary beam current; α_X^0 is the post-ionization coefficient for X and includes a contribution for the ionization potential and the cross-section for ionization using the particular ionization method; η_X is the transmission factor for X for the particular apparatus; and α_X^+ and α_X^- are the secondary ion probability factors which are usually much less than unity. When $\alpha_X^\pm \ll 1$, which is true for most elements, although not for alkali metals, then

$$I(X^0) = I_p Y_X D_X^0. \tag{9.1.2}$$

If most of the sputtered species are atoms and there is negligible yield of molecular species, the partial sputter yield of X is given by

$$Y_X = C_X Y_{tot} \tag{9.1.3}$$

and under steady-state conditions,

$$Y_{tot} = \Sigma\, Y_X. \tag{9.1.4}$$

Thus $$C_i/C_j = [I(X_i^0).\, D_j^0]/[I(X_j^0).\, D_i^0]. \tag{9.1.5}$$

Knowing that $\Sigma\, C_X = 1$, if the relative detection constants D_i^0/D_j^0 of the instrument are known, the concentrations of elements in the sample can be determined.

A number of post-ionization methods are being explored.[1] The approaches used and their potential will be briefly described.

9.1.1. *Electron bombardment post-ionization*

Simple electron bombardment was tried many years ago by Hönig. It is compatible with UHV and involves a relatively simple modification to a SIMS system. An electron beam ionizer is incorporated in the secondary ion optics (see Fig. 9.1). This can be similar in form to the type of ion source used for residual gas analysis, and is basically a beam of electrons emitted from a hot filament, accelerated to 30 – 100 eV across the path of the secondary neutrals. Ionization occurs within the electron cloud by the same electron bombardment phenomenon as discussed in Section 4.3.3.

The region of ion production is small and this means that relatively few neutrals are sputtered in the correct direction to drift into this region.

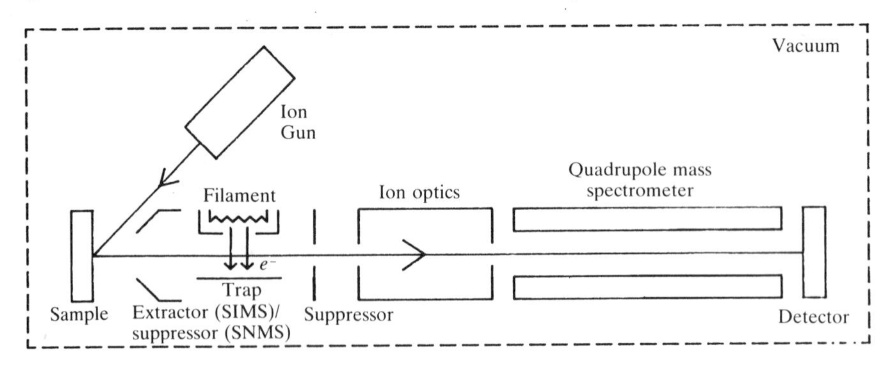

FIG 9.1. Schematic diagram of electron bombardment SNMS.

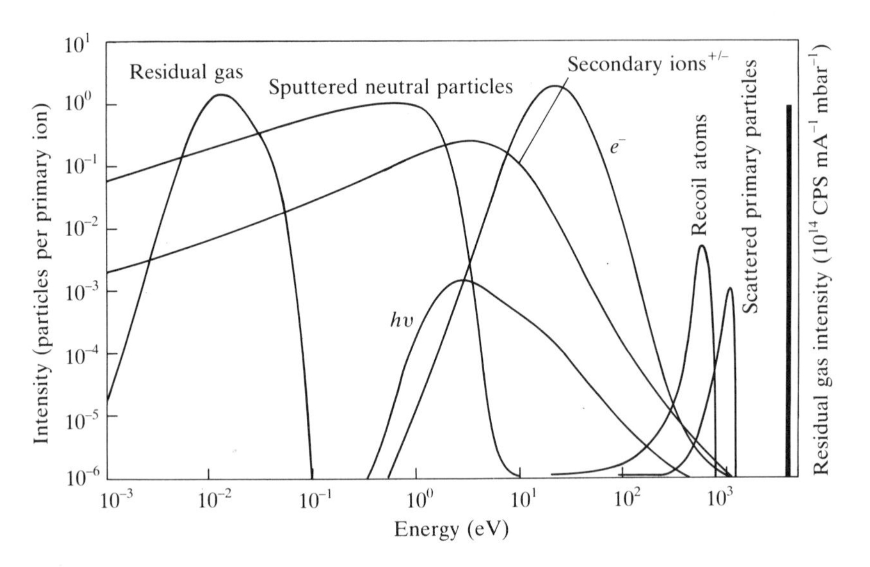

FIG 9.2. Schematic diagram of the energy distributions of residual gas ions, secondary ions, sputtered neutrals, electrons, and other particles. Reproduced with permission from Ref.2.

Additionally, the probability of ionization is rather low ($\simeq 10^{-4}$), so that for most elements sensitivity is significantly lower than SIMS. Some improvement can be obtained using magnetic confinement of the electrons. The ionization yield, however, is always low, $\sim 10^{-4}$–10^{-3}.

Interfering signals arise from a variety of sources. Secondary ion emission from the target can be suppressed by biasing the target and the extraction field. This suppression needs to be very effective (so that the residual secondary ion signal is no more than 10^{-6} of its intensity in SIMS), and in some cases can be quite difficult because of the wide energy spread of elemental ions. Residual gas ions are a significant source of interference. Fortunately, these ions have only thermal energies (<0.1 eV) when formed. Figure 9.2 shows how their energies are widely separated from those of the secondary ions and sputtered neutrals ($\simeq 10$ eV). An appropriate retarding potential just after the ionizer can suppress these ions. A final background signal often arises from the ionization of neutrals by grazing angle collisions with aperture edges and grids. This can be reduced by geometry changes and retarding potentials.

Figure 9.3 shows an ionizer design incorporated in a parallel-plate energy analyser. Figure 9.4(a) shows the comparative spectra of GaAs obtained

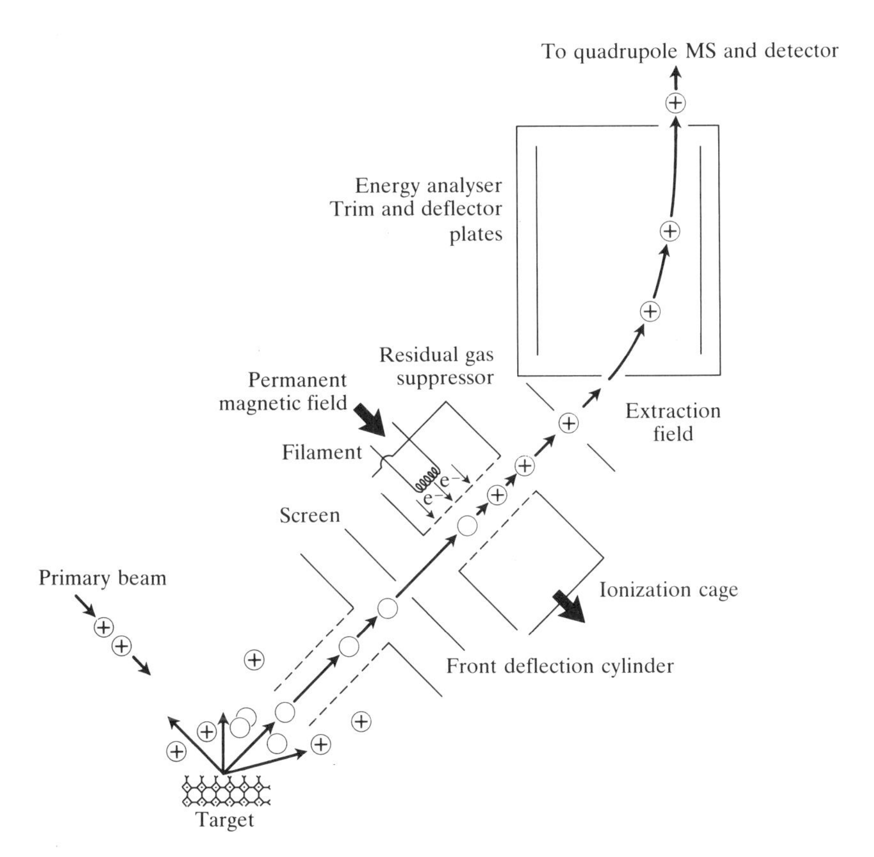

FIG 9.3. Schematic diagram of an electron bombardment ionizer combined with a parallel-plate energy analyser.

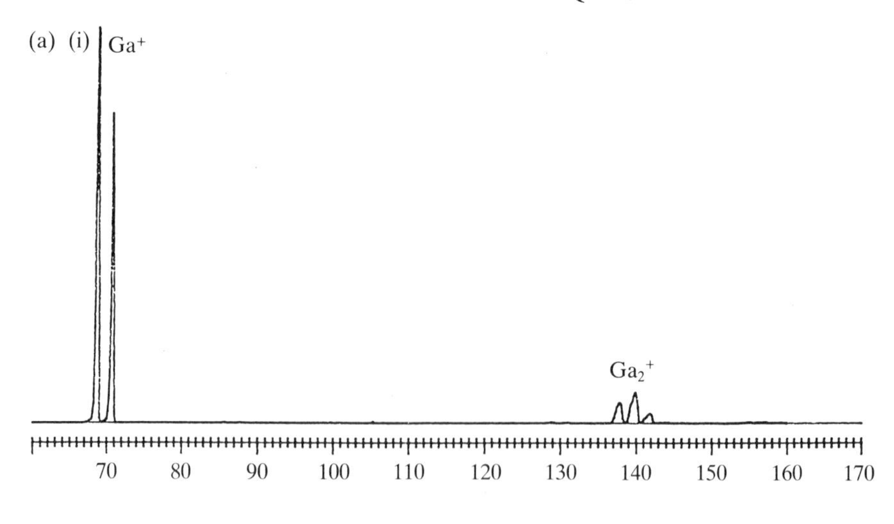

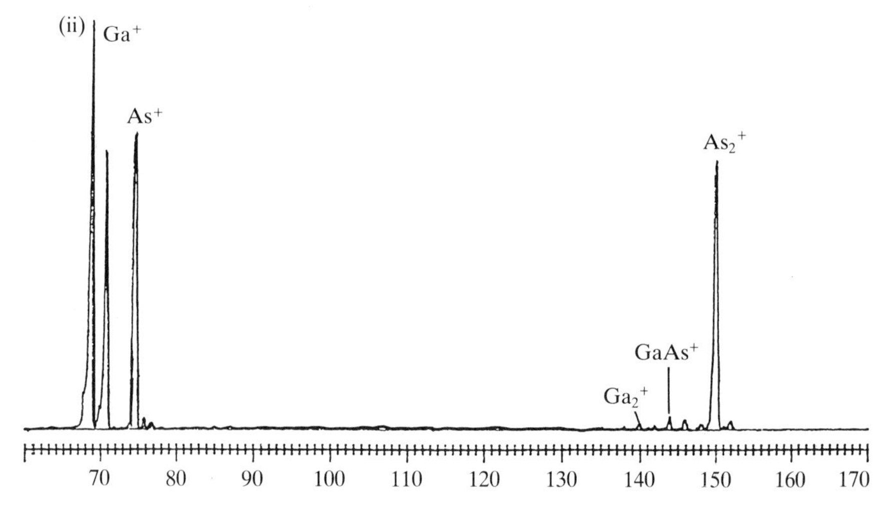

FIG 9.4. (a) (i) SIMS spectrum of a GaAs surface, $I_p = 1$ nA, 5 keV Ar^+; f.s.d.3×10^4 cps. (ii) SNMS spectrum of a GaAs surface, $I_p = 10\ \mu A$, 5 keV Ar^+; f.s.d 1.3×10^4 cps.

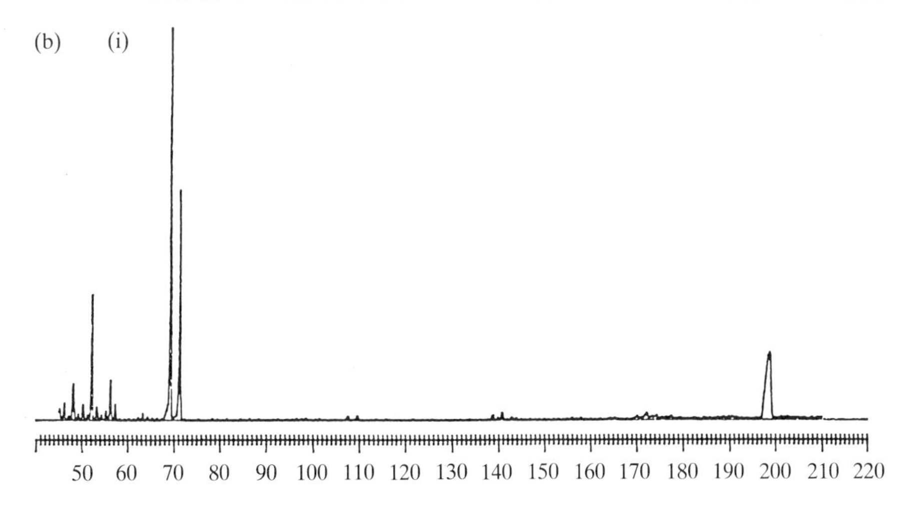

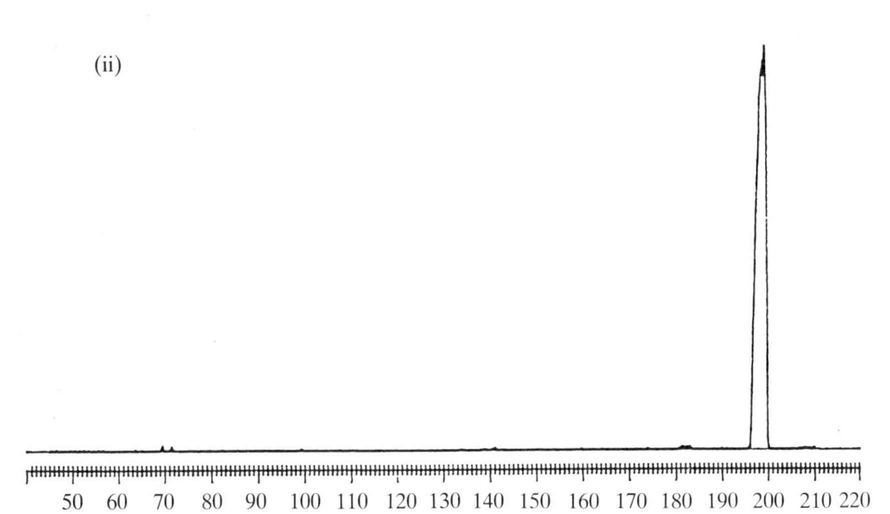

(b) (i) SIMS spectrum of a gold surface, $I_p = 1\ \mu A$, 5 keV Ar^+; f.s.d. 1.5×10^4 cps. (ii) SNMS spectrum of a gold surface, $I_p = 1 \mu A$, 5 keV Ar^+; f.s.d. 0.9×10^4 cps; anode – 10 V.

using this arrangement in the SIMS and SNMS modes. It can be seen that, whereas the As^+ signal is negligible in the SIMS mode due to its very low ionization probability, in SNMS its intensity is comparable with Ga^+, as it should be if the intensities reflect the true surface composition. It is interesting to note the cluster ions generated, e.g. $GaAs_x^+$. It might be expected that electron bombardment would dissociate these ions; however, there is no evidence that increasing the electron bombardment energy causes them to decrease significantly. As yet, little is known about their relationship to composition.

Figure 9.1.4(b) show similar comparative spectra of gold. The yield of Au^+ in SNMS is some 10–30 times higher than in SIMS.

The relative sensitivity factors obtained by comparing a SNMS analysis of an iron sample with a wet chemistry analysis are displayed in Table 9.1.[2]

TABLE 9.1

The relative sensitivity factors of elements contained in a steel sample calibrated according to wet chemical analysis

Element	Atom (per cent)	Intensity (cps $\mu A^{-1} mA^{-}$)	Sensitivity factor (D_{Fe}/D_X)
C	1.17	34	4
N	0.0092	Mostly Si_{dc}	—
O	0.0223†	5	0.58†
Al	0.224	69	0.38
Si	1.39	458	0.36
P	0.0461	6	0.86
S	0.0583	6	1.14
Ti	0.31	48	0.75
V	0.518	81	0.75
Cr	0.81	100	0.95
Fe	92.41	1.1×10^4	1
Mn	0.51	<97	‡
Ni	0.863	116	0.87
Co	0.317	35	1.08
Cu	0.6	41	1.75
As	0.0572	8	0.85
Mo	0.476	57	0.98
Nb	0.112	11	1.15
Sn	0.0315	2	1.92

†Quoted concentration is not trusted, because in all other oxygen-containing metals $D_{me}/D_O > 2$ was observed dc(subscript) = double charged.
‡ Not mass resolved at same sensitivity.

Relative to Fe the sensitivity factors (D_{Fe}/D_X) vary by at most a factor if only four. Elemental detection limits in the region of 30 p.p.m. have been reported. The ionization efficiency and hence the sensitivity will always be relatively low in this approach, yet it is very effective as a calibration facility in a dynamic SIMS system.

9.1.2. *Electron plasma post-ionization*

The low ionization probability can be improved by a factor of $\sim 10^2$ by increasing the electron density. Oechsner *et al.* have used a low-pressure (10^{-4} mbar) argon plasma, or hot electron gas, to increase the electron density and hence to increase ionization efficiency to $\sim 10^{-2}$.[3] Plasma excitation is by Electron Cyclotron Wave Resonance (ECWR), which is established when a weak magnetic field (10 G) is imposed on a plasma generated by a high-frequency (27.12 MHz) r.f. generator. Figure 9.5 shows the arrangement schematically. The plasma chamber contains no excitation electrodes and simply consists of a flat 5–10 cm long cylinder having a diameter of $\sim$ 10–20 cm. Using 10^{-4} mbar Ar as the working gas, electron temperatures, T_e, of $\sim$ 10–15 eV are attained. The sputtered neutrals traverse the plasma cylinder along its axis. The ionization volume is large and hence the ionization probability is greatly increased. The path length in the plasma at the working pressures is, however, an order of magnitude *less* than the mean free path for heavy particle (i.e. ion or atom) collisions. Thus scattering of ions, redeposition of emitted particles and the formation of new species due to ion molecule reactions should be at a minimum.

The great advantage of this method is that the electron space charge is compensated by the background of positive ions. High electron densities (10^{10} cm^{-3}) are thus generated. The post-ionization factor α_X^0 is dependent on the plasma parameters n_e, the density of electrons, T_e; the electron impact

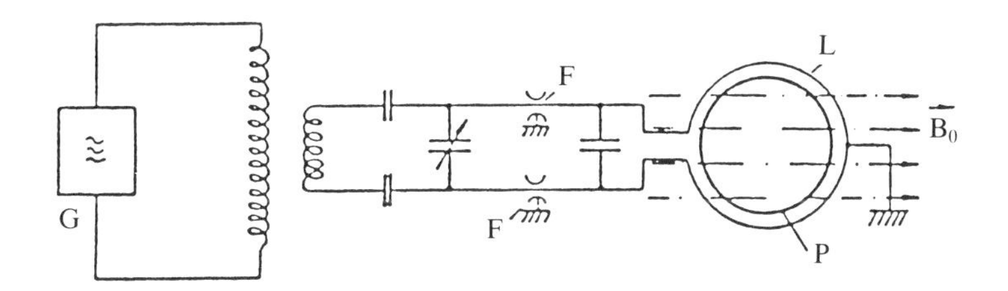

FIG 9.5. Schematic diagram for the plasma excitation by Electron Cyclotron Wave Resonance (ECWR). G, High-frequency generator (27.12 MHz); F, electrical UHV feed-throughs; L, single-turn excitation coil; P, wall of plasma chamber; B_0, magnetic field. Reproduced with permission from Ref. 3.

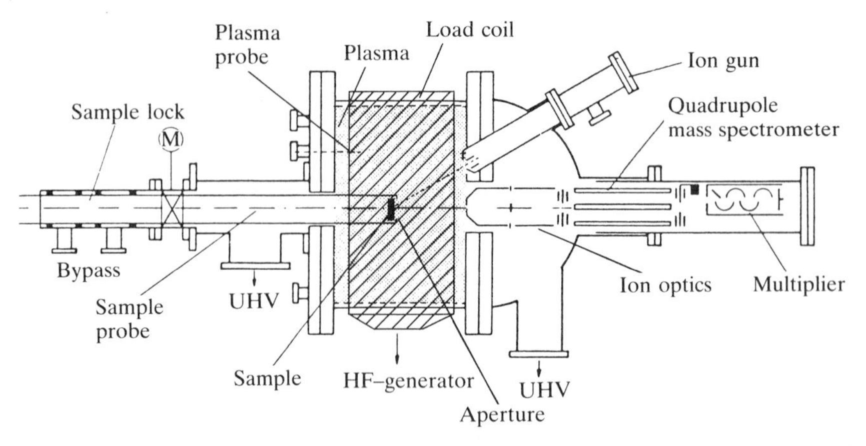

FIG 9.6. Schematic diagram of a commercial combined SNMS–SIMS instrument, from Leybold. With the plasma gas removed, SIMS is possible using the ion gun.

ionization function of X; and the path length of X through the ionizer (i.e. on the mass of X and on the energy distribution of X^0).[4]

This method has the advantage that the plasma may also be used as a low-energy primary beam source. By placing the sample in the plasma region and biasing the sample to attract positive ions the surface of the sample may be sputtered. The density of ions is very high and the biasing can be quite low so that primary beam energies of just a few hundred volts can be used. The technique has the potential for *very* low atomic mixing in depth profiling and hence high depth resolution. This mode of operation is termed the direct bombardment mode (DBM). A schematic diagram of a commercial version of this instrument is shown in Fig 9.6. In Fig. 9.7(a) the quantification possibilities are demonstrated; the certified concentrations in an NBS standard steel are plotted against the measured SNMS signals. Figure 9.1.7(b) shows that detectability is in the low p.p.m. region for magnesium. This is rather impressive when it is realized that a quadrupole analyser was used and the collection efficiency cannot be very great because of the long drift path to the entrance to the mass filter.

The depth profiling capability is shown in Fig. 9.8 for the Ta_2O_5–Ta standard. The measured profile sharpness is close to the physical limit. Unfortunately, it is not possible to gate the analysed area, which is specified solely by an aperture. In consequence, the collection of material from the crater edges is inevitable, and is indeed possible from the material of the aperture. Thus high dynamic range ($> 10^3$) is very difficult to obtain.

High pressure (from 0.1 to a few mbar) glow discharge mass spectrometry (GDMS) can also be used, and again the discharge may be used as a sputter source. Usually this source has been used with a magnetic sector instrument

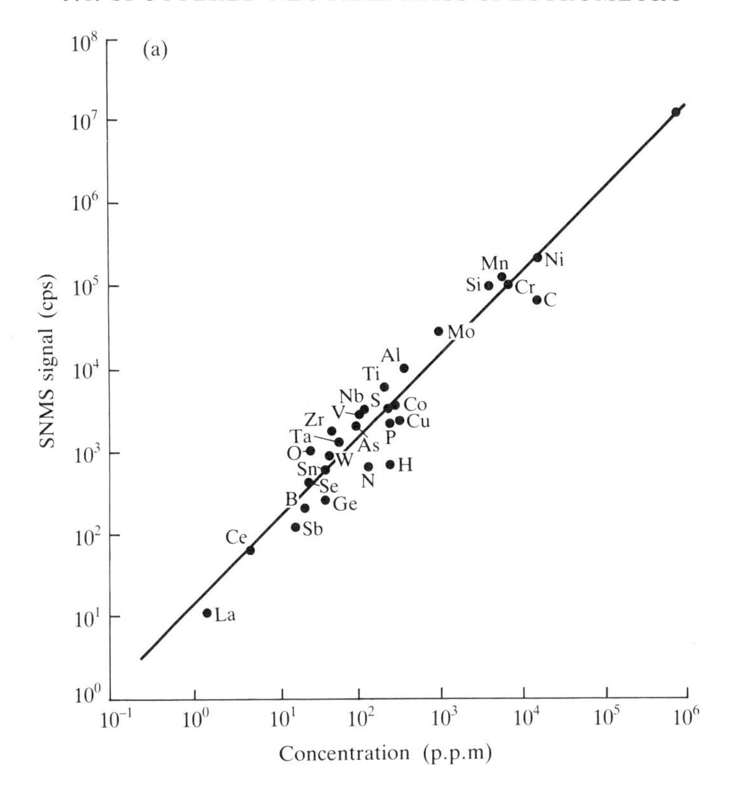

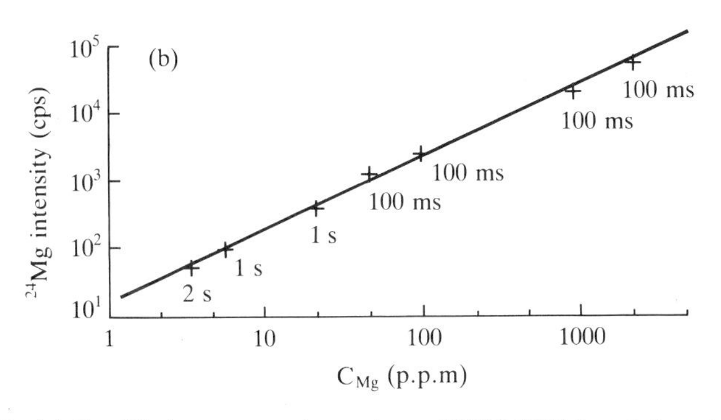

FIG 9.7. (a) Certified concentration values of NBS 1261A stainless steel constituents vs. measured uncorrected SNMS signal using an ECWR instrument. (b) Measured SNMS signals vs. specified concentrations of Mg in aluminium. The required collection times are given for each point. Reproduced with permission from Leybold.

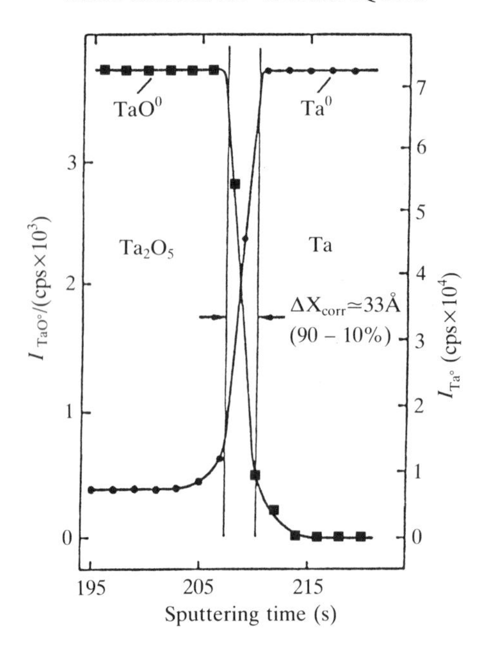

FIG 9.8. Low-energy SNMS depth profiling of the interface between an anodic Ta_2O_5 layer on Ta by normally incident Ar^+ ions of 202 eV in an ECWR plasma instrument. Reproduced with permission from Ref.3.

and, whilst the ionization efficiency is no greater than the hot electron gas, the effective useful yield is significantly higher than with a quadrupole analyser (see Chapter 4), yielding detection limits in the p.p.b. range.

9.1.3. Laser-induced post-ionization

The most elegant method of post-ionization is to use a pulsed high-energy laser to ionize the neutrals as they leave the surface. This means that the SIMS system must use a ToFMS (see Fig. 9.9). Ionization requires that the energy of the highest energy electron be elevated to the vacuum level. This may be accomplished by a progressive 'climb' up the energy levels (see, for example, Fig. 9.10), by carefully tuning the laser energy to the energy differences between the energy levels. Sometimes only one photon will be required. Unfortunately, for some of the most interesting elements, several photons of different energy will be required to achieve ionization (Si and Sb, four photons; P, As, and Se, five photons),[5] which necessitates an array of several different lasers tuned to specified frequencies. This is known as multi-

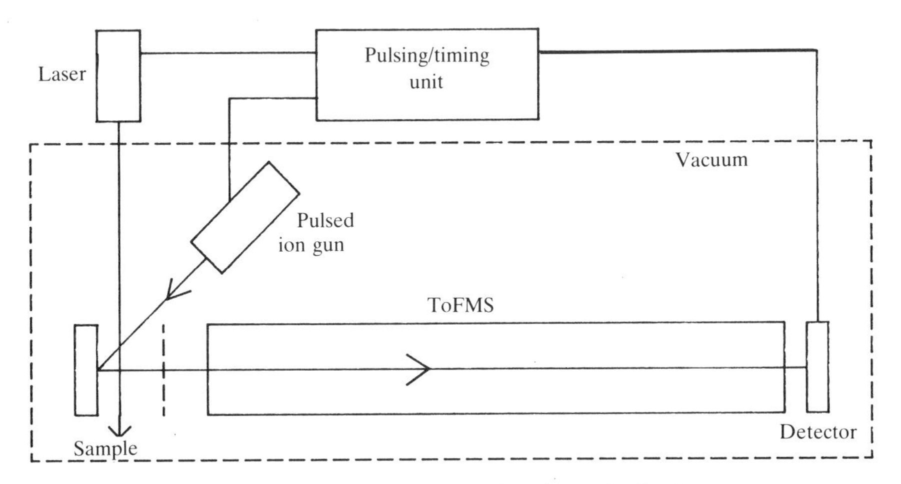

FIG 9.9. Schematic diagram of a laser-induced post-ionization apparatus.

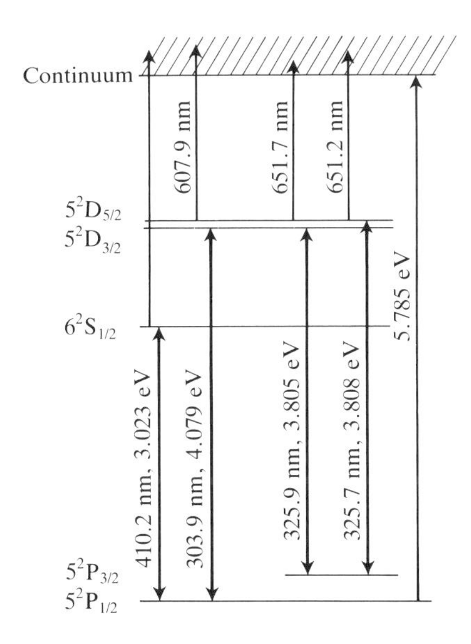

FIG 9.10. Partial electronic structure of indium. Ground-state indium can be multiphoton ionized in either a one-colour experiment (two photons of 410.2 nm) or a two-colour experiment (one photon of 303.9 nm followed by a second photon of 607.9 nm). The first excited state can be investigated by either of the two-colour MPRI schemes shown. Reproduced with permission from Ref. 7.

photon *resonance* ionization (MPRI). Clearly, as all the elements have unique energy level arrangements, the process will be element specific. It can, however, be extremely efficient, close to 100 per cent.

The laser beam is directed parallel to and a few mm above the sample surface, and is pulsed $\leq 10^{-6}$ s after the moment of secondary ion emission. The pulsed nature of the operation requires a ToFMS and in practice the primary beam source is also pulsed to minimize sample consumption. Pulse timing is crucial. Figure 9.11 is a schematic diagram of how the pulsing of the ion beam, laser beam, and secondary ions are controlled. The flash lamp trigger in the ionizing laser acts as an initiator. Some 200 μs after (D1) the argon ion pulse starts its journery from the ion gun. After a primary ion flight

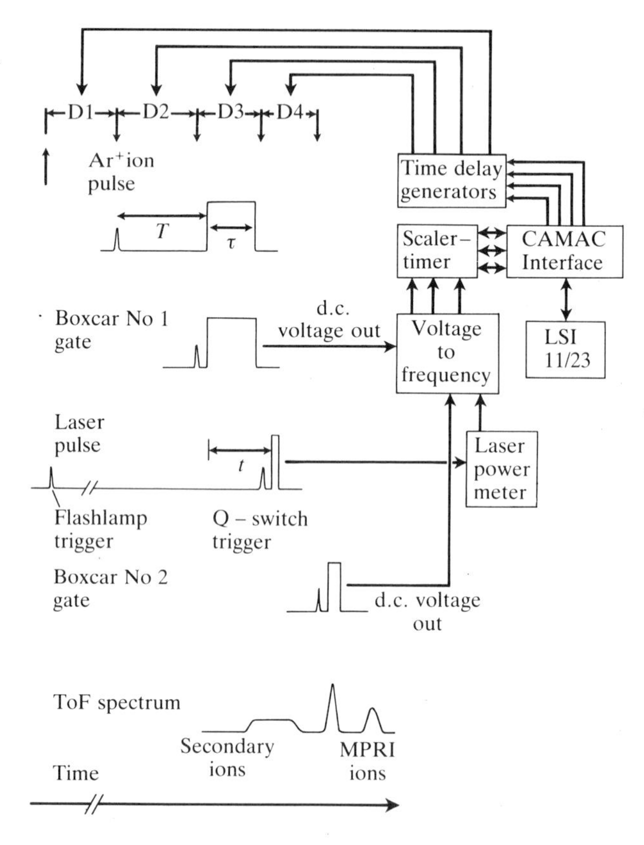

FIG 9.11. Schematic diagram of the pulse timing and data collection in the MPRI experiment. Typical delay times are D1 $=200\ \mu$s, D2 $=5\ \mu$s, D3 $=0.5\ \mu$s and D4 $=4\ \mu$s. T is the primary ion flight time to the sample, τ is the primary ion pulse width and t is the laser firing time. The laser pulse width is ~ 6 ns. Reproduced with permission from Ref. 7.

time, T, it arrives at the sample for a period $\tau \leq 10\ \mu s$. The Boxcar integrator measures the total primary beam flux. Time-delay generator D3 ensures that the laser pulse is triggered not more than 500 ns after the ion pulse is complete so that the laser pulse (width $t \simeq 6$ ns) intercepts as large a proportion as possible of the emitted neutral plume. Finally, after $\sim 4\ \mu s$, D4 triggers the detector to collect the MPRI-generated ions as distinct from the secondary ions generated at the surface.

The high cross-section for the resonant ionization process means that the power density requirements are only moderate (10^5–10^7 W cm^{-2}), so that long laser pulses (100 ns) or defocused beams ($\simeq 1$ cm) can be used. A high proportion ($\geq$ 20 per cent) of the sputtered neutrals can then be intercepted by the ionizing radiation and can be used for analysis. Although long pulse times and large analysis volumes will degrade ToF mass resolution this is not critical because the mass analyser is not absolutely necessary as the ionizing process is element specific.

The basic equation for the ionized neutral current of element X is given by

$$I_X = DI_p Y_X C_X U_X \eta_X \tag{9.1.6}$$

where D is the duty cycle of the laser, which will be $\simeq 3 \times 10^{-4}$, assuming a typical repetition rate for a UV laser of 10 Hz and a primary ion pulse length of 10 μs. I_p is the primary ion beam current, which for trace analysis must be 1–10 μA in the pulse or 10 mA continuous. Y_X is the sputter yield; C_X is the concentration of element X.

U_X is the fraction of X atoms in the energy state investigated by the MPRI process used. The process can only study one energy state at a time; any atoms in excited states as a consequence of the sputtering process will not be included. The yield of metastable excited states increases with primary ion energy. However, if several hundred ns elapse between ion pulse and laser pulse most neutrals should have returned to the ground state.

Detection efficiency, η_X, depends on three basic parameters:

1. *The overlap between the spatial distribution of sputtered atoms and the photon field.* This will be a function of the pulse width, energy and energy distribution of the emitted neutrals, the photon beam diameter, timing of the laser pulse, and the closeness of the laser beam to the surface.

2. *The ionization of the atoms in the photon field.* To achieve 100 per cent ionization requires ~ 100 mJ cm^{-2} per pulse. The power available increases with focusing but that then reduces the number of particles in the beam. Clearly, maximum ionization requires a knowledge of the correct ionization scheme. The most advantageous is to have a two-step process which requires a low-power UV photon for resonance, and a high-power visible laser for ionization. Where many steps are involved high powers will be necessary. Problems arise where energy steps are not known.

3. *The extraction and detection of generated ions*. If detector efficiency is not high, with the low laser duty cycle, the advantage of high-efficiency ionization may be lost and ion yields may be no higher than in SIMS or electron bombardment SNMS. Thus the high-transmission ToF mass spectrometers are necessary.

Winograd *et al.*[7] have shown that an easily ionized element such as Ga could be detected at the 1 p.p.b. level yielding $I_{Ga} \sim 150$ cs^{-1}, whereas As, which is rather difficult to ionize and would require a highly focused beam to generate the necessary photon power, would only yield $I_{As} \sim 1$ cs^{-1}.

The element-specific nature of the process can be exploited in trace analysis. For example, the analysis of traces of Fe in silicon is difficult because of the mass overlap between Si_2^+ and Fe^+, both nominally $m/z = 56$. Using MPRI tuned to Fe ionization it has been possible to obtain a depth profile of a ^{56}Fe implant in silicon down to the 2 p.p.b. level (see Fig. 9.12). Experimentally, it was necessary to perform the crater profiling with a continuous beam and switch to the pulsed mode of operation for analysis at regular intervals.

MPRI is very element specific; however, the process can be accomplished by the use of a high-energy, high-power UV laser in an element-non-specific manner. The aim is to deposit so much energy into the neutral cloud as it leaves the surface that all elements will be ionized. This is known as multi-

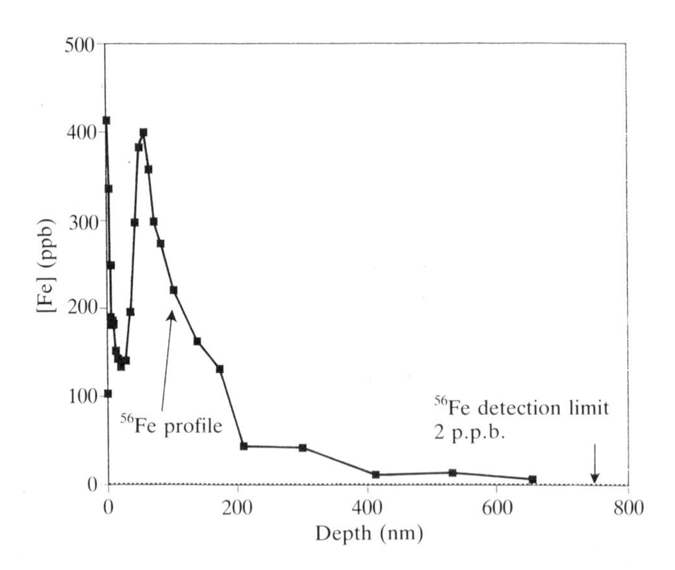

FIG 9.12. Depth profile of ^{56}Fe-implanted silicon (60 keV Fe^+ dose; 10^{11} atoms cm^{-2}); analysis performed by MPRI of sputtered Fe atoms. Reproduced with permission from Ref.6.

photon *non-resonant* ionization,[8] and would appear to be more useful for applied analysis. Again, the process is potentially very efficient, ionization efficiencies close to 1 being reported *in* the photon beam. The same parameters influence the experimental ion yield as for MPRI.

Because the process is non-resonant, cross-sections for ionization are significantly lower than for MPRI and consequently higher power densities are required to achieve saturation of ionization.[9] Thus the photon beam has to be more tightly focused to attain the required power densities of 10^9–10^{10} W cm^{-2}. The sensitivities obtainable are therefore potentially slightly lower than for MPRI.

At present, one of the problems for the successful application of the method is that high-power pulsed UV lasers are rather slow, 10–100 Hz, which increases analysis time. New lasers are appearing which have higher repetition rates and promise to reduce this problem.

The use of post-ionized *cluster* ions in quantitative analysis is beginning to be explored. The use of carefully tuned ionization to ionize molecular and cluster neutrals to provide chemical state quantification has been shown to hold great promise. This is demonstrated by studies of the ionization of molecular adsorbates desorbed from glass surfaces using a CO_2 laser and then ionized by MPI using UV radiation from a frequency quadrupled Nd–YAG laser.[10] Here it was shown that surface coverages of molecules such as adenine and β-oestradiol could be quantified at the femtomole level (see Fig. 9.13).

There is, however, evidence that use of very high fluences of photons in the post-ionization step may be counter-productive. Becker has observed that attempting to post-ionize neutrals sputtered by argon ions from poly (methyl-methacrylate) (PMMA) films, using 248-nm photons from a KrF eximer laser, results in complete destruction of the clusters and production of carbon ions (see Fig. 9.14(a)).[11] If the power level is reduced there is no evidence of ion production. This observation has been attributed to the excitation of the emitted clusters by a number of photons whose energy is not quite sufficient to ionize the cluster, but which increase the vibrational energy so that the molecule falls apart. It was therefore suggested that a photon beam should be used which had sufficient energy to ionize the cluster, but was of sufficiently low flux that the absorption of a second photon which might fragment the cluster ion was unlikely. To do this, 355 nm photons from a Nd–YAG laser were frequency tripled in xenon phase-matched with argon to yield 118 nm photons (see Fig. 9.14(b)). The efficiency of the process is low ($\simeq 10^{-4}$) but, with a 20 mJ input, 10 ns pulses of 1.3×10^{12} photons per pulse were obtained. These photons have energy of 9.5 eV, just above the ionization potential of most organic clusters. Using this radiation source, a spectrum was obtained from sputtered neutral clusters which was meaningful in terms of the PMMA structure (see Fig. 9.14(c)). The spectrum is similar to the

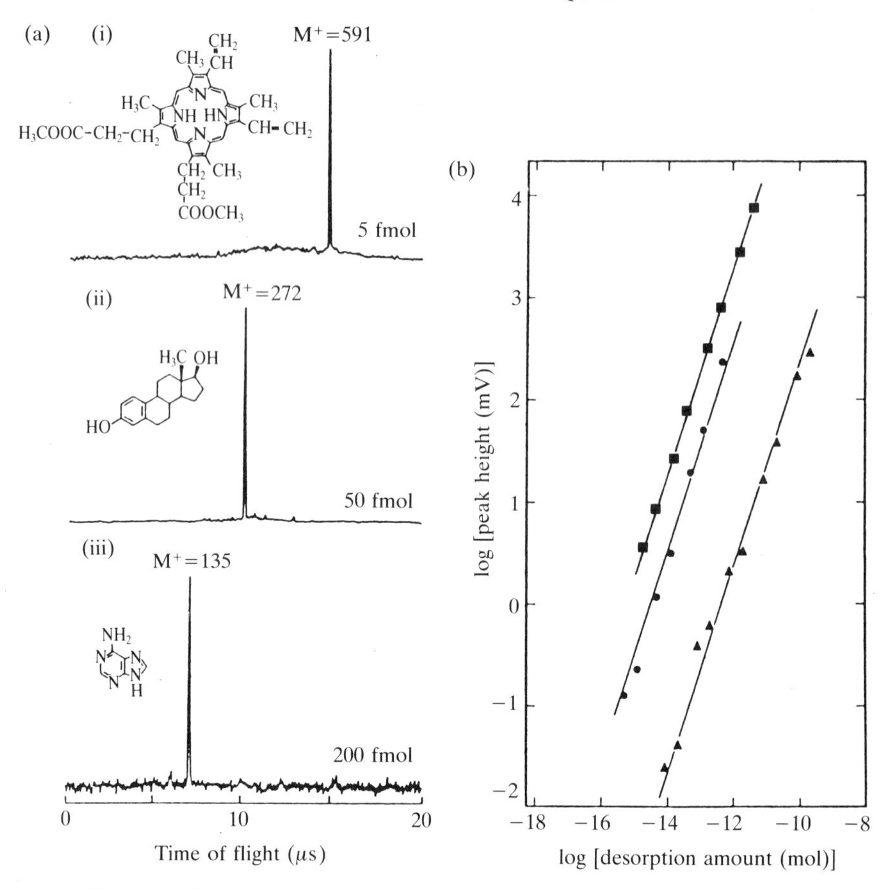

FIG 9.13. Laser desorption–multiphoton ionization spectra of a (i) protoporphyrin IX dimethylester, (ii) β-oestradiol, and (iii) adenine. Operating conditions: CO_2 laser fluence ~400 mJ cm^{-2}, Nd–YAG laser fluence ~3 mJ cm^{-2}, duty cycle 10 Hz, and signal averaging time 20 s. (b) Parent ion signal for (●) protoporphyrin ☒ dimethylester, (■) β-oestradiol, (▲) adenine vs. amount desorbed per laser pulse. Reproduced with permission from Ref. 10.

FIG 9.14. (a) Laser post-ionization spectrum of poly(methyl-methacrylate) (PMMA) obtained with pulsed Ar^+ sputtering with multi-photon ionization (258 nm at 1×10^7 W cm^{-2}). Spectra were obtained using 1000 pulses. (b) Schematic diagram of the photon-induced post-ionization arrangement using VUV (118 nm) radiation for ionization. (c) Spectrum of PMMA obtained with pulsed Ar^+ sputtering with single photon post-ionization (118 nm at 3×10^3 W cm^{-2}). Spectra were obtained using 1000 pulses. Reproduced with permission from Ref. 11.

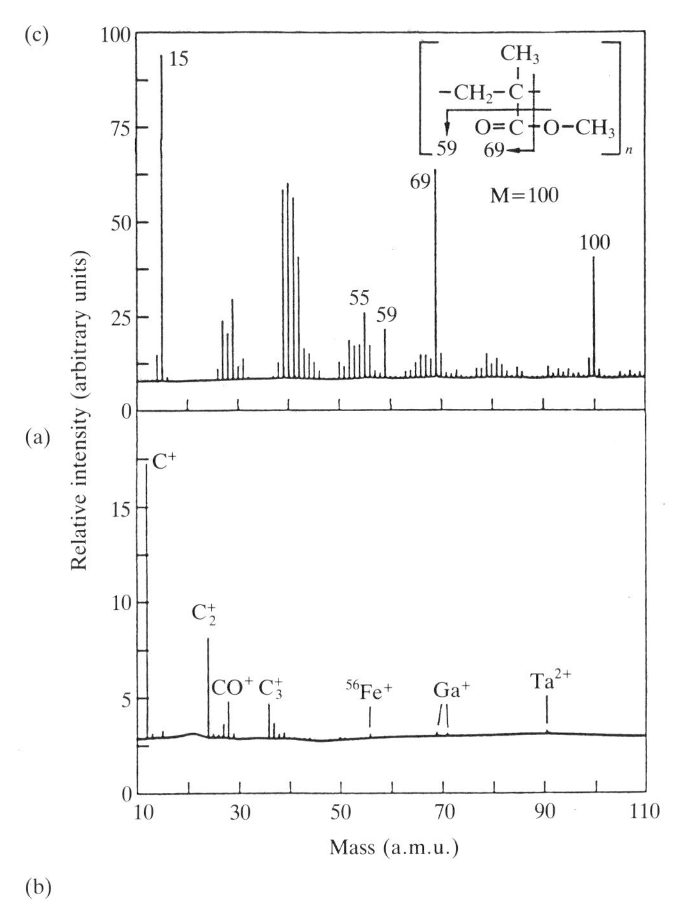
(c)
(a)
Relative intensity (arbitrary units)
CH3
-CH2-C-
O=C-O-CH3
59 69
n
M=100
15
55
59
69
100
C+
C2+
CO+ C3+
56Fe+
Ga+
Ta2+
Mass (a.m.u.)
100
75
50
25
0
15
10
5
0
10
30
50
70
90
110

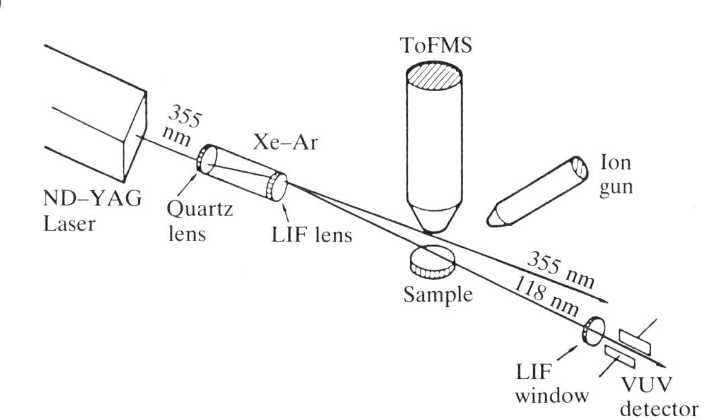
(b)
ToFMS
355 nm
Xe–Ar
Ion gun
ND–YAG Laser
Quartz lens
LIF lens
Sample
355 nm
118 nm
LIF window
VUV detector

corresponding SSIMS spectrum (see Fig. 7.15); however, the relative peak intensities are rather different. In particular, the monomer ion at $m/z = 100$ is significantly more intense and is a major feature of the spectrum.

This approach has real potential for quantification of chemical state, although the efficiency of ionization of the emitted neutrals is probably rather low. An estimate of 1 per cent of neutrals in the laser beam has been made by Becker *et al.* Thus, for some materials, overall sensitivity may not be greater than by SIMS. However, the example above shows that there may well be cases where cluster ions are difficult to generate by the SSIMS process but can be formed by gentle post-ionization.

The method also offers the possibility to study the identity, stability, and spectroscopy of emitted clusters to further our understanding of the sputtering process.

9.1.4. Summary

To date, the main interest in SNMS has been in elemental quantification, and it does seem that matrix effects are greatly reduced. They are not eliminated, however, because sputter yield is sensitive to the angular distribution of emission, which may vary from matrix to matrix, but probably only by a factor of two to three. Accurate quantification can, however, be disturbed by two other parameters. First, if secondary ion emission is very significant, as it can be for the alkali metals, the neutral yield will be significantly reduced and will be affected by the variations in secondary ion yield. Second, a significant yield of atom *clusters* may also distort the elemental yield.

From the standpoint of quantitative surface chemistry, post-ionization of neutral clusters holds considerable potential.

Table 9.2 summarizes the important parameters in the main methods of SNMS. Each technique has its advantages. As yet, none has reached the level of maturity attained by SIMS but rapid progress is being made.

9.2. ^{252}Cf plasma desorption mass spectrometry

This approach uses the fact that 3 per cent of the ^{252}Cf decay involves a fission branch. When spontaneous fission occurs, two fission fragments are emitted at 180° to each other. The Cf source is mounted behind a foil on which the material to be analysed has been deposited (see Fig. 9.15).[12] When a fission fragment passes *through* the sample it deposits energy into the surface region, which causes the emission of atomic and molecular ions. These are accelerated through a grid and the ToF of the desorbed ions is measured relative to the detection time of the complementary fission fragment emitted

TABLE 9.2

SNMS comparison

Technique	Ionization	Ionization efficiency	Neutrals in ionization region	Mass spectrometry transmission	Relative sensitivity
1. e^- bombardment	General	10^{-4}–10^{-3}	10^{-2}	10^{-4}–10^{-3}	10^{-9}
2. e^- discharge	General	10^{-3}–10^{-2}	≤ 1	10^{-6}–10^{-5}	10^{-8}
3. Resonant multiphoton	Specific	$\simeq 1$	$< 10^{-1}$	< 1	10^{-2}
4. Non-resonant UV multiphoton	General	< 1	10^{-2}	< 1	10^{-3}
5. Non-resonant VUV single photon	General	$< 10^{-1}$	$< 10^{-2}$	< 1	10^{-4}

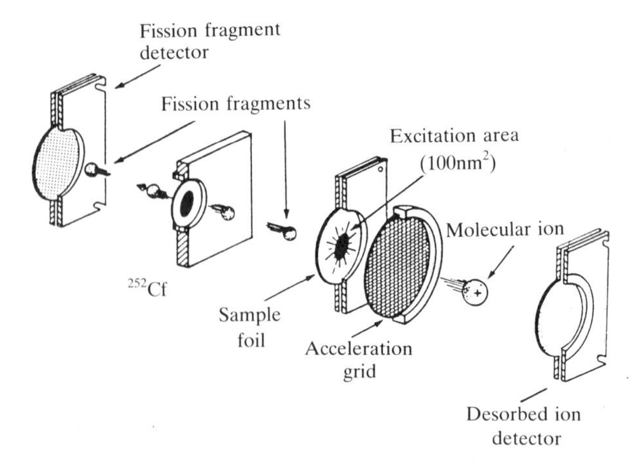

FIG 9.15. Schematic diagram of the operation of the ^{252}Cf PDMS systems. Reproduced with permission from Ref. 12.

at 180°. A variant on the technique uses MeV multi-charge ions, e.g. 90 MeV $^{127}I^{14+}$, generated by tandem accelerators; the resulting spectra are identical.[13]

Whilst the ion interaction with the bombarded matrix in SIMS is mainly a nuclear stopping phenomenon, the interaction of a fission fragment when it passes through a solid is *electronic* and is due to a combination of the effect of the charge of the ion (typically +20) and its velocity. The high-energy ions produce a linear damage track ~15 μm in length. It can be quite clearly visualized with electron microscopy. The profile of a heavy ion track contains an inner core of radius 0.1 nm that includes all the atoms and molecules which were excited by the primary ion. This is surrounded by a cylinder of energetic electrons (delta rays), extending to ~1 nm, which were ejected from the inner core in the initial excitation process. The lifetime of the track structure is $\sim 10^{-17}$ s. The linear energy transfer (LET) to the matrix is a measure of the density of energy deposition, which is a function of the charge and velocity of the primary ion. It can be shown that only energy deposited within the top six to ten monolayers of, for example, a valine matrix, is contributing to the emission of secondary molecular ions. Macfarlane has pointed out that this means that for a 100 MeV fission fragment only 50 keV are used in the emission of secondary ions, giving a LET of 10 keV nm^{-1}. Of course, this is concentrated in a very small volume and in a very short time interval, giving an effective power density of 10^{15} W cm^{-2}.

Although the power input is considerably greater than for SIMS or FAB, the spectra produced by the three methods are essentially very similar. It is clear that the particular power of PDMS is that it can lift off higher mass fragments and whereas, with FAB, molecular ion yields fall several orders of magnitude per m/z = 1000, this is not observed with PDMS. The average number of secondary ions produced per incident ion is a factor of four higher for MeV primaries, and *multiplicities* as high as 36 have been observed for some fission tracks. Sundqvist has shown that high-energy densities are required to desorb large molecules. The yield of insulin molecular ions varies as the LET to the sixth power, while for small molecules it varies as LET to power one or two. The LET of a typical SIMS primary ion is a factor of 10 lower than for a ^{252}Cf fission fragment. This means that the yield of insulin molecular ions will be six orders higher in ^{252}Cf PDMS compared with SIMS; hence the success of the former in desorbing and analysing very high mass molecules to m/z = 40 000 and beyond.

Although PDMS seems to be even more violent than SIMS, it is gentle enough to be able to monitor the behaviour of an adsorbed surface layer on a polymer substrate. This is illustrated by a study of the adsorption of Rhodamine 6-G hydrochloride on PET.[14] The spectra generated are very similar to those from SIMS.

The Rhodamine 6-G, which exists as a cation, was added to the polymer surface from ethanol, and by varying the solution concentration from 10^{-5} to 10^{-2} M and following the molecular ion (m/z = 443) intensity, a Langmuir adsorption isotherm is generated. Monolayer coverage was attained at $\sim 4 \times 10^{-4}$ M (see Fig. 9.16(a)). Optical measurements were used to confirm the number density of molecules. From the linear portion of the curve it is possible to infer that the area of excitation for ion emission is <20 nm.

It is interesting to note that there is no evidence of the Cl^- counter-ion until beyond monolayer coverage. The adsorption sites on the PET surface are clearly negative and the cation is preferentially adsorbed. Only in the multi-layer region where molecular adsorption occurs is the Cl^- present and evident in the spectrum. At low coverages (see Fig. 9.16(b, i)) there is no evidence of a peak in the dimer (M_2^+) ion region; at intermediate coverages (ii) metastable ions appear probably due to randomly oriented contact pairs at the surface. Finally, at high multi-layer concentrations (iii) stable dimers are emitted. The successive stages of adsorption are characterized by ion emission reflecting the binding state of the molecule at the surface.

The mutual interaction of molecules at the surface has been shown to affect their fragmentation pattern in many areas of SIMS. Rhodamine-B is related to Rhodamine 6-G; however, it has a carboxylic group which can be lost to yield a distinctive $(M-45)^+$ fragment ion. In a similar experiment, as adsorption progresses it is observed that at low coverage the M^+ and $(M-45)^+$ ions are of similar intensity, whereas at high coverage the $(M-45)^+$ ion is significantly lower. This suggests that when the molecules are in close proximity the excitation energy can be shared and internal excitation is reduced, resulting in lower fragmentation.

It is clear that although the excitation processes are rather different in

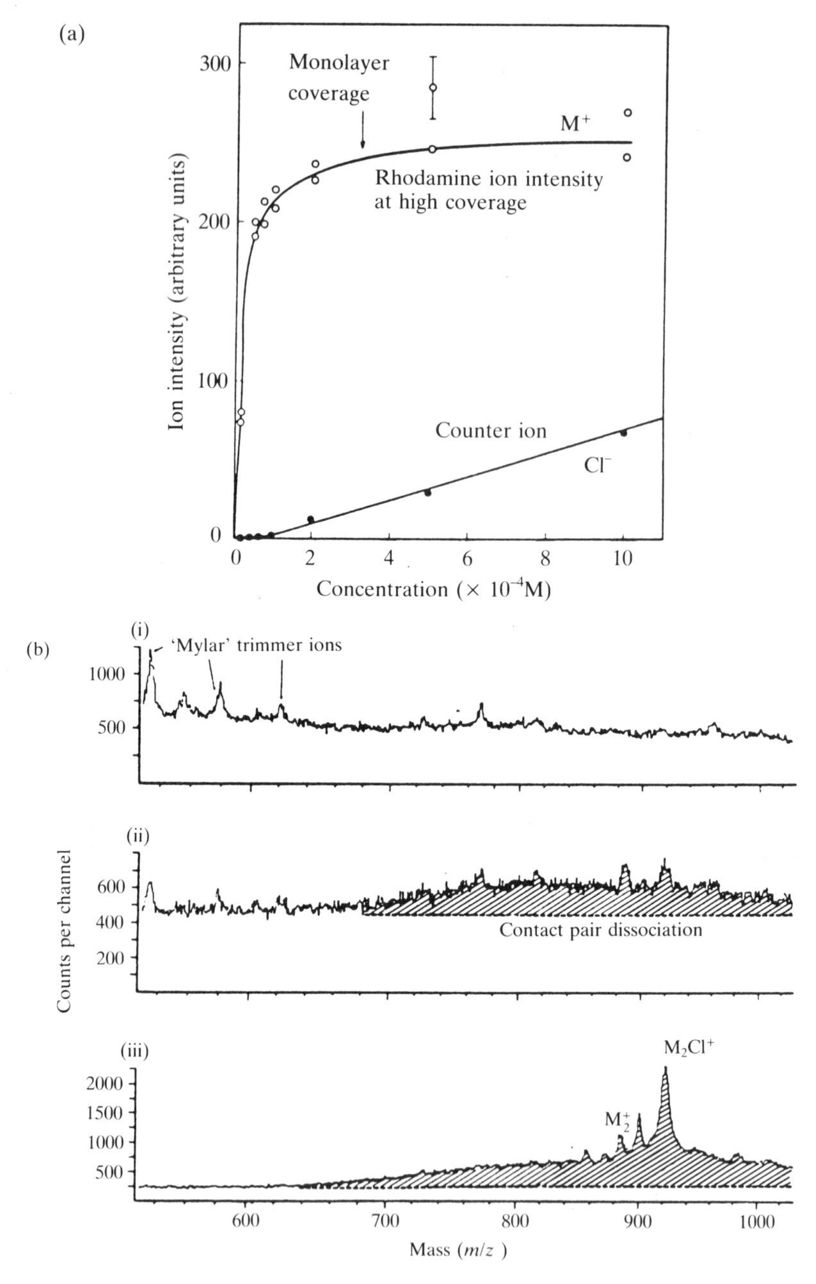

FIG 9.16. (a) Dependence of the molecular ion and Cl^- intensity of Rhodamine 6-G on ethanol solution concentration. (b) ^{252}Cf PDMS spectra of Rhodamine 6-G as a function of concentration, showing evidence of dimer adsorption at high surface concentrations. (i) 1×10^{-5} M; (ii) 1×10^{-4} M; (iii) high coverage; 1×10^{-3} M. Reproduced with permission from Ref. 14.

SIMS, FABMS, and PDMS, the effects at the molecular level during ion emission are very similar.

9.3. Laser desorption mass spectrometry

This approach to ion generation from solid surfaces generally uses a high-energy pulsed UV or visible laser to both desorb and ionize the evaporated species. As the ions are produced in pulses, ToFMS is used for analysis and detection. An energy-compensated reflectron analyser is normally used because of the kinetic energy distribution of the emitted ions.

Initially, the technique was mainly exploited as a microprobe with very high laser irradiances for elemental analysis in geology, the life sciences, etc.[15] The use of lower laser powers has permitted the analysis of non-volatile organic materials and the investigation of surface layers.[16]

Figure 9.17 shows the generally used experimental arrangement. The laser usually used is Nd–YAG, Q-switched and frequency tripled or quadrupled, giving pulses of ~10 ns focused to around 1 μm. The laser may be directed normal to the surface either from the rear or from the analyser side. In the former case, the sample has to be thin enough to be perforated by the laser beam. Clearly, this mode of operation provides bulk analysis. More recently, to enable greater surface specificity, instruments have been developed with 30° incidence on the front surface. The ions generated are mass analysed using the ToF analyser, collected by an electron multiplier, and recorded using a transient recorder or by single ion counting. Spectral acquisition is very fast, ~70 μs being required for a spectrum. Rapid (though crude) depth profiles can be obtained by rapid pulsing on one point. The laser power levels can be varied, typically between 0.01 and 1000 W μm^{-2}. The power per pulse is, however, somewhat variable and some spectral irreproducibility is often encountered from pulse to pulse in the absolute yields of particular peaks and in the relative yields between spectral features. Quantitative measurements require the average of many pulses to be determined.

Hillenkamp has distinguished four ion formation processes possible under laser irradiation:

(1) thermal evaporation of ions from the solid;

(2) thermal evaporation of neutral molecules from the solid followed by ionization in the gas phase;

(3) laser desorption; and

(4) ion formation in a laser-generated plasma.

The first two processes are essentially caused by a heating effect of the laser. The evaporation of cations from quaternary ammonium salts is well documented. The second process relies on ionization by ion–molecule reactions in the gas phase. Thus alkali metal ions would need to be

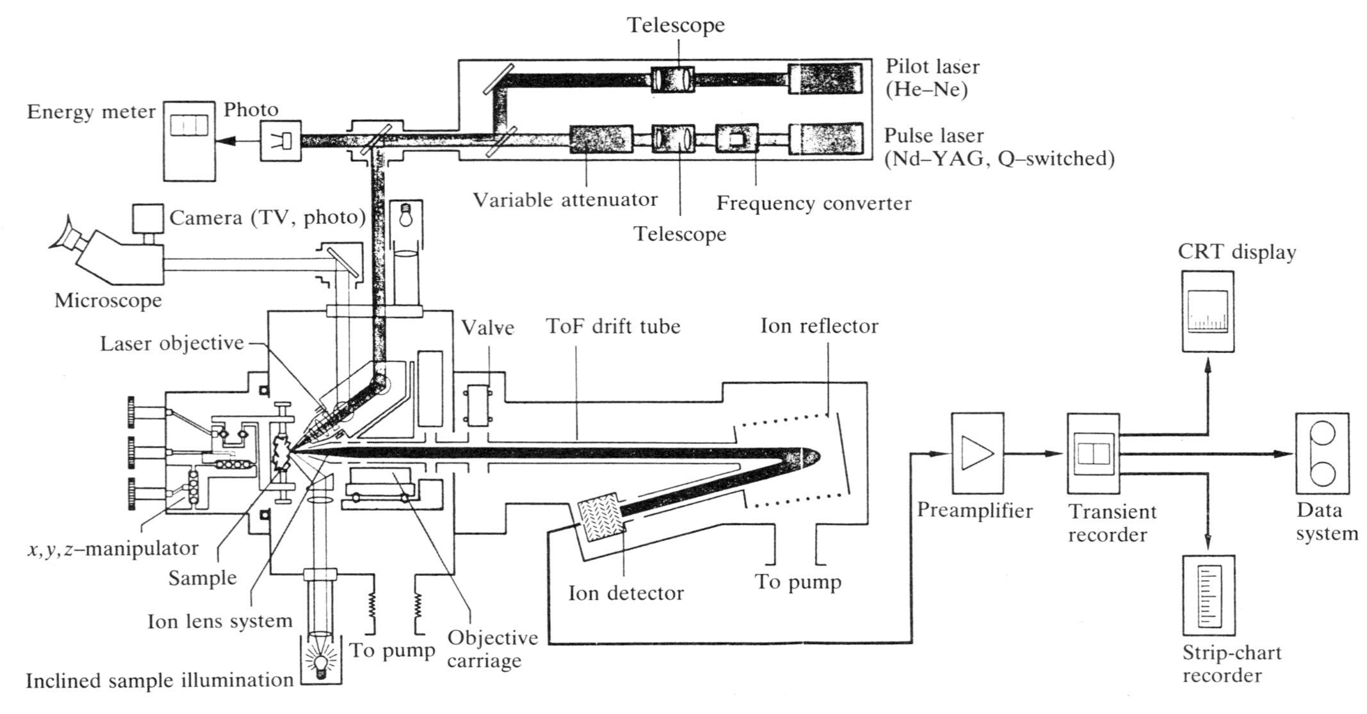

FIG 9.17. Schematic diagram of a laser ionization mass analyser.

evaporated together with the neutral molecules. These ion formation mechanisms have been studied in the main using continuous-wave CO_2 lasers at power levels of $\sim 10^6$ W cm^{-2}.

The third process is akin to the sputtering phenomenon observed in SIMS or PDMS. This is a non-equilibrium collective process. The UV laser pulse delivers a large amount of energy into a very small area. Calculation shows that the affected volume of about ~ 0.1 cm^3 expands at speeds of $\sim 10^5$ cm s^{-1}. This is very similar to the velocity which a $m = 200$ molecule would have if it was sputtered in a SIMS experiment with a kinetic energy of 2 eV. Thus it is to be expected that the generated spectra might be rather similar to those found in SSIMS or FABMS. This is found to be the case when UV pulses with power levels of 10^7–10^{10} W cm^{-2} are used.

Most of the observations outlined in earlier chapters for molecular SIMS have been reported for the generation of molecular spectra using this mode of laser desorption. Cluster ions reflect the chemistry of the material being analysed; their stability reflects the bond energies involved; cationization frequently occurs via metal ions which are present. Ionic organic salts generate even-electron ions. However, non-ionic compounds produce *both* even- and odd-electron parent ions. The latter will tend to undergo further fragmentation until an even-electron state is produced.[17]

Situations have been reported, however, where some organic materials will not generate a meaningful spectrum. At low powers no ions are produced, and when the energy is increased only a hydrocarbon spectrum is produced. In other cases, for example, for some peptides, low power produces the molecular ion whereas increasing the power to generate fragment ions destroys the material and produces a hydrocarbon spectrum. Since the process is usually initiated by a UV laser it may be that the photon energy is not always appropriate to the gentle desorption and ionization of all molecules. Very little work has been done over a range of laser wavelengths to investigate the effect of photon energy on the generation of spectra. The reader will recall that a very similar observation has been mentioned with regard to laser post-ionization in Section 9.1.3. It is possible that a similar explanation holds, and clusters are either not ionized because the photon energy is inappropriate, or are destroyed by absorption of several photons. Since in the present technique a single laser desorbs *and* ionizes, little control can be exercised over the two processes separately.

It is often suggested that laser desorption does not result in surface charging. It is noticeable, however, that the generation of negative ion spectra is often reported to be difficult even when negative ions are expected. Photon bombardment will result in electron emission, with the likelihood of consequential development of positive surface charge. This would effectively inhibit the emission of negative ions, as was found in SIMS and FABMS.

When laser power levels above 10^{10} W cm^{-2} are used, a dense plasma results, and all molecules are broken down into their constituent atoms. This mode of operation has great advantages for inorganic analysis where elemental concentrations are required and the data sought are rather similar to those provided by dynamic SIMS. There is evidence that ion yields become almost uniform throughout the Periodic Table, although, as in SIMS, the only reliable method of quantitative analysis is to use standards of a similar matrix to that being analysed.[15]

References

1. Reuter, W. (1986). In *Secondary Ion Mass Spectrometry, SIMS V*, Springer Series in Chemical Physics, Vol 44 (ed. A. Benninghoven, R.J. Colton, D.S. Simons, and H.W. Werner), p. 94. Springer Verlag, Berlin.
2. Lipisky, D., Jede, R., Ganshow, O., and Benninghoven, A. (1985). *J. Vac. Sci. Technol.*, **A3**, 2007.
3. Oechsner, H., Rühe, W., and Stumpe, E. (1979). *Surf. Sci.*, **85**, 289.
4. Gerhard, W. and Oechsner, H. (1975). *Z. Physik*, **B22**, 41.
5. Hurst, G.S., Payne, M.G., Kramer, S.D., and Young, J.P. (1979). *Rev. Mod. Phys.*, **51**, 767.
6. Young, C.E., Pellin, M.J., Calaway, W.F., Jorgensen, B., Schweitzer, E.L., and Gruen, D.M. (1987). *Nucl. Instrum. Methods Phys. Res.*, **B27**, 119.
7. Kimock, F.M., Baxter, J.P., Pappas, D.L., Kobrin, P.H., and Winograd, N. (1984). *Anal. Chem.*, **56**, 2782.
8. Becker, C.H. and Gillen, K.T. (1984). *Appl. Phys Lett.*, **45**, 1063.
9. Morellec, J., Normand, D., and Petite, G. (1982). *Adv. Atomic Mol. Phys.*, **18**, 97.
10. Hahn, J.H., Zenobi, R., and Zare, R.N. (1987) *J. Am. Chem. Soc.*, **109**, 2842.
11. Pallix, J.B., Schühle, U., Becker, C.H., and Huestis, D.L. (1989). *Anal. Chem.*, in press.
12. Macfarlane, R.D. (1983). *Anal. Chem.*, **55**, 1247A.
13. Sunqvist, B., Hedin, A., Hokansson, P., Jonsson, G., Salepour, M., Säve, G., and Roepstorff P. (1986). In *Ion Formation from Organic Solids (IFOS III)*, Springer Series in Physics, Vol 9 (ed. A. Benninghoven), p. 6. Springer Verlag, Berlin.
14. Macfarlane, R.D. (1985). In *Desorption Mass Spectrometry, Am. Chem. Soc. Symp. Ser.*, **291**, 56.
15. Verbueken, A.H., Bruynseels, F.J., and Van Grieken, R.E. (1985). *Biomed. Mass Spectrom.*, **12**, 438.
16. Hillenkamp, F. (1983). In *Ion Formation from Organic Solids (IFOS II)*, Springer Series in Chemical Physics, Vol 25 (ed. A. Benninghoven), p. 190. Springer Verlag, Berlin.
17. van Vaeck, L., Claereboudt, J., De Waele, J., Esmans, E., and Gijbels, R. (1985). *Anal. Chem.*, **57**, 2944.

APPENDIX 1

Physical constants and useful relations

General data

Speed of light	c	2.99792458×10^{8} m s^{-1}
Charge of proton (charge on the electron is $-e$)	e	1.60218×10^{-19} C
Faraday constant	$F = eL$	9.64853×10^{-19} C mol^{-1}
Boltzmann constant	k	1.38066×10^{-23} J K^{-1}
Gas constant	$R = kL$	8.31451 J K^{-1} mol^{-1}
		1.98717 cal K^{-1} mol^{-1}
		8.20575×10^{-2} dm^3 atm K^{-1} mol^{-1}
Planck constant	h	6.62608×10^{-34} J s
	$\hbar = h/2\pi$	1.05457×10^{-34} J s
Avogadro constant	L	6.02214×10^{23} mol^{-1}
Atomic mass unit	$u = 10^{-3}$ kg/(L mol)	1.66054×10^{-27} kg
Mass of electron	m_e	9.10939×10^{-31} kg
proton	m_p	1.67262×10^{-27} kg
neutron	m_n	1.67493×10^{-27} kg
nuclide	$m = M_r u$	$1.66054 \times 10^{-27} \times M_r$ kg
Vacuum permitivity	ϵ_0	8.854188×10^{-12} J^{-1} C^{-2} m^{-1}
	$4\pi\epsilon_0$	1.112650×10^{-10} J^{-1} C^2 m^{-1}
Vacuum permeability (Note that $\epsilon_0\mu_0 = 1/c^2$)	μ_0	$4\pi \times 10^{-7}$ J s^2 C^{-2} m^{-1}
Bohr magneton	$\mu_B = e\hbar/2m_e$	9.27402×10^{-24} J T^{-1}
Nuclear magneton	$\mu_N = e\hbar/2m_p$	5.05078×10^{-27} J T^{-1}
Bohr radius	$a_0 = 4\pi\epsilon_0\hbar^2/m_e e^2$	5.29177×10^{-11} m
Rydberg constant	$R_\infty = m_e e^4/8h^2\epsilon_0^2$	2.179908×10^{-23} J
	R_∞/hc	1.097373×10^{5} cm^{-1}
Gravitational constant	G	6.6726×10^{-11} N m^2 kg^{-2}

Useful relations

$1\ \text{N} = 1\ \text{kg m s}^{-2}$
$\phantom{1\ \text{N}} = 1\ \text{J m}^{-1} \triangleq 10^5\ \text{dynes}$

$1\ \text{J} \triangleq 10^7\ \text{erg}$
$1\ \text{J} = 1\ \text{A V s}$

$1\ \text{A} = 1\ \text{C s}^{-1}$
$1\ \text{A} \simeq 6.2 \times 10^{18}\ \text{particles s}^{-1}$

$1\ \text{W} = 1\ \text{J s}^{-1}$

$1\ \text{eV} \triangleq 1.602 \times 10^{-19}\ \text{J}$
$96.485\ \text{kJ mol}^{-1}$
$8065.5\ \text{cm}^{-1}$

At 298.15 K, $RT = 2.4789\ \text{kJ mol}^{-1}$
$kT/hc = 207.223\ \text{cm}^{-1}$

$1\ \text{atm} = 101.325\ \text{Pa}$
$= 1.01325 \times 10^5\ \text{N m}^{-2}$

$1\ \text{mm Hg} \triangleq 133.322\ \text{N m}^{-2}$

APPENDIX 2

Atomic weights of the elements based on the ^{12}C standard

(due to Vacuum Generators)

Symbol	Mass	Abundance	Symbol	Mass	Abundance
^{1}H	1.007825037	99.985	^{40}Ca	39.9625907	96.941
^{2}H	2.014101787	0.015	^{42}Ca	41.9586218	0.647
^{3}He	3.016029297	0.00014	^{43}Ca	42.9587704	0.135
^{4}He	4.00260325	99.99986	^{44}Ca	43.9554848	2.086
^{6}Li	6.0151232	7.5	^{46}Ca	45.963689	0.004
^{7}Li	7.0160045	92.5	^{48}Ca	47.952532	0.187
^{9}Be	9.0121825	100.00	^{45}Sc	44.9559136	100.00
^{10}B	10.0129380	19.9	^{45}Ti	45.9526327	8.0
^{11}B	11.0093053	80.1	^{47}Ti	46.9517649	7.3
^{12}C	12.00000000	98.90	^{48}Ti	47.9479467	73.8
^{13}C	13.003354839	1.10	^{49}Ti	48.9478705	5.5
^{14}N	14.003074008	99.634	^{50}Ti	49.9447858	5.4
^{15}N	15.000108978	0.366	^{50}V	49.9471613	0.250
^{16}O	15.99491464	99.762	^{51}V	50.9439625	99.750
^{17}O	16.9991306	0.038	^{50}Cr	49.9460463	4.35
^{18}O	17.99915939	0.200	^{52}Cr	51.9405097	83.79
^{19}F	18.99840325	100.00	^{53}Cr	52.9406510	9.50
^{20}Ne	19.9924391	90.51	^{54}Cr	53.9388822	2.36
^{21}Ne	20.9938453	0.27	^{55}Mn	54.9380463	100.00
^{22}Ne	21.9913837	9.22	^{54}Fe	53.9396121	5.8
^{23}Na	22.9897697	100.00	^{56}Fe	55.9349393	91.72
^{24}Mg	23.9850450	78.99	^{57}Fe	56.9353957	2.2
^{25}Mg	24.9858392	10.00	^{58}Fe	57.9332778	0.28
^{26}Mg	25.9825954	11.01	^{59}Co	58.9331978	100.00
^{27}Al	26.9815413	100.00	^{58}Ni	57.9353471	68.27
^{28}Si	27.9769284	92.23	^{60}Ni	59.9307890	26.10
^{29}Si	28.9764964	4.67	^{61}Ni	60.9310586	1.13
^{30}Si	29.9737717	3.10	^{62}Ni	61.9283464	3.59
^{31}P	30.9737634	100.00	^{64}Ni	63.9279680	0.91
^{32}S	31.9720718	95.02	^{63}Cu	96.9295992	69.17
^{33}S	32.9714591	0.75	^{65}Cu	64.9277924	30.83
^{34}S	33.96786774	4.21	^{64}Zn	63.9291454	48.6
^{36}S	35.9670790	0.02	^{66}Zn	65.9260352	27.9
^{35}Cl	34.968852729	75.77	^{67}Zn	66.9271289	4.1
^{37}Cl	36.965902624	24.23	^{68}Zn	67.9248458	18.8
^{36}Ar	35.967545605	0.337	^{70}Zn	69.9253249	0.6
^{38}Ar	37.9627322	0.063	^{69}Ga	68.9255809	60.1
^{40}Ar	39.9623831	99.600	^{71}Ga	70.9247006	39.9
^{39}K	38.9637079	93.2581	^{70}Ge	69.9242498	20.5
^{40}K	39.9639988	0.0117	^{72}Ge	71.9220800	27.4
^{41}K	40.9618254	6.7302	^{73}Ge	72.9234639	7.8

Symbol	Mass	Abundance	Symbol	Mass	Abundance
^{74}Ge	73.9211788	36.5	^{110}Cd	109.903007	12.49
^{76}Ge	75.9214027	7.8	^{111}Cd	110.904182	12.80
^{75}As	74.9215955	100.00	^{112}Cd	111.9027614	24.13
^{74}Se	73.9224771	0.9	^{113}Cd	112.9044013	12.22
^{76}Se	75.9192066	9.0	^{114}Cd	113.9033607	28.73
^{77}Se	76.9199077	7.6	^{116}Cd	115.904758	7.49
^{78}Se	77.9173040	23.5	^{113}In	112.904056	4.3
^{80}Se	79.9165205	49.6	^{115}In	114.903875	95.7
^{82}Se	81.916709	9.4	^{112}Sn	111.904823	1.0
^{79}Br	78.9183361	50.69	^{114}Sn	113.902781	0.7
^{81}Br	80.916290	49.31	^{115}Sn	114.9033441	0.4
^{78}Kr	77.920397	0.35	^{116}Sn	115.9017435	14.7
^{80}Kr	79.916375	2.25	^{117}Sn	116.9029536	7.7
^{82}Kr	81.913483	11.6	^{118}Sn	117.9016066	24.3
^{83}Kr	82.914134	11.5	^{119}Sn	118.9033102	8.6
^{84}Kr	83.9115064	57.0	^{120}Sn	119.9021990	32.4
^{86}Kr	85.910614	17.3	^{122}Sn	121.903440	4.6
^{85}Rb	84.9117996	72.165	^{124}Sn	123.905271	5.6
^{87}Rb	86.9091836	27.835	^{121}Sb	120.9038237	57.3
^{84}Sr	83.913428	0.56	^{123}Sb	122.904222	42.7
^{86}Sr	85.9092732	9.86	^{120}Te	119.904021	0.096
^{87}Sr	86.9088902	7.00	^{122}Te	121.903055	2.60
^{88}Sr	87.9056249	82.58	^{123}Te	122.904278	0.908
^{89}Y	88.9058560	100.00	^{124}Te	123.902825	4.816
^{90}Zr	89.9047080	51.45	^{125}Te	124.904435	7.14
^{91}Zr	90.9056442	11.27	^{126}Te	125.903310	18.95
^{92}Zr	91.9050392	17.17	^{128}Te	127.904464	31.69
^{94}Zr	93.9063191	17.33	^{130}Te	129.906229	33.80
^{96}Zr	95.908272	2.78	^{127}I	126.904477	100.00
^{93}Nb	92.9063780	100.00	^{124}Xe	123.90612	0.10
^{92}Mo	91.906809	14.84	^{126}Xe	125.904281	0.09
^{94}Mo	93.9050862	9.25	^{128}Xe	127.9035308	1.91
^{95}Mo	94.9058379	15.92	^{129}Xe	128.9047801	26.4
^{96}Mo	95.9046755	16.68	^{130}Xe	129.9035095	4.1
^{97}Mo	96.9060179	9.55	^{131}Xe	130.905076	21.2
^{98}Mo	97.9054050	24.13	^{132}Xe	131.904148	26.9
^{100}Mo	99.907473	9.83	^{134}Xe	133.905395	10.4
^{96}Ru	95.907596	5.52	^{136}Xe	135.907219	8.9
^{98}Ru	97.905287	1.88	^{133}Cs	132.905433	100.00
^{99}Ru	98.9059371	12.7	^{130}Ba	129.906277	0.106
^{100}Ru	99.9042175	12.6	^{132}Ba	131.905042	0.101
^{101}Ru	100.9055808	17.0	^{134}Ba	133.904490	2.417
^{102}Ru	101.9043475	31.6	^{135}Ba	134.905668	6.592
^{104}Ru	103.905422	18.7	^{136}Ba	135.904556	7.854
^{103}Rh	102.905503	100.00	^{137}Ba	136.905816	11.23
^{102}Pd	101.905609	1.02	^{138}Ba	137.905236	71.70
^{104}Pd	103.904026	11.14	^{138}La	137.907114	0.09
^{105}Pd	104.905075	22.33	^{139}La	138.906355	99.91
^{106}Pd	105.903475	27.33	^{136}Ce	135.90714	0.19
^{108}Pd	107.903894	26.46	^{138}Ce	137.905996	0.25
^{110}Pd	109.905169	11.72	^{140}Ce	139.905442	88.48
^{107}Ag	106.905095	51.839	^{142}Ce	141.909249	11.08
^{109}Ag	108.904754	48.262	^{141}Pr	140.907657	100.00
^{106}Cd	105.906461	1.25	^{142}Nd	141.907731	27.13
^{108}Cd	107.904186	0.89	^{143}Nd	142.909823	12.18

Symbol	Mass	Abundance	Symbol	Mass	Abundance
^{144}Nd	143.910096	23.80	^{177}Hf	176.943233	18.6
^{145}Nd	144.912582	8.30	^{178}Hf	177.943710	27.1
^{146}Nd	145.913126	17.19	^{179}Hf	178.945827	13.74
^{148}Nd	147.916901	5.76	^{180}Hf	179.946561	35.2
^{150}Nd	149.920900	5.64	^{180}Ta	179.947489	0.012
^{144}Sm	143.912009	3.1	^{181}Ta	180.948014	99.988
^{147}Sm	146.914907	15.0	^{180}W	179.946727	0.13
^{148}Sm	147.914832	11.3	^{182}W	181.948225	26.3
^{149}Sm	148.917193	13.8	^{183}W	182.950245	14.3
^{150}Sm	149.917285	7.4	^{184}W	183.950953	30.67
^{152}Sm	151.919741	26.7	^{186}W	185.954377	23.6
^{154}Sm	153.922218	22.7	^{185}Re	184.952977	37.40
^{151}Eu	150.919860	47.8	^{187}Re	186.955765	62.60
^{153}Eu	152.921243	52.5	^{184}Os	183.952514	0.02
^{152}Gd	151.919803	0.20	^{186}Os	185.953852	1.58
^{154}Gd	153.920876	2.18	^{187}Os	186.955762	1.6
^{155}Gd	154.922629	14.80	^{188}Os	187.955850	13.3
^{156}Gd	155.922130	20.47	^{189}Os	188.958156	16.1
^{157}Gd	156.923967	15.65	^{190}Os	189.958455	26.4
^{158}Gd	157.924111	24.84	^{192}Os	191.961487	41.0
^{160}Gd	159.927061	21.86	^{191}Ir	190.960603	37.3
^{159}Tb	158.925350	100.00	^{193}Ir	192.962942	62.7
^{156}Dy	155.924287	0.06	^{190}Pt	189.959937	0.01
^{158}Dy	157.924412	0.10	^{192}Pt	191.961049	0.79
^{160}Dy	159.925203	2.34	^{194}Pt	193.962679	32.9
^{161}Dy	160.926939	18.9	^{195}Pt	194.964785	33.8
^{162}Dy	161.926805	25.5	^{196}Pt	195.964947	25.3
^{163}Dy	162.928737	24.9	^{198}Pt	197.967879	7.2
^{164}Dy	163.929183	28.2	^{197}Au	196.966560	100.00
^{165}Ho	164.930332	100.00	^{196}Hg	195.965812	0.15
^{162}Er	161.928787	0.14	^{198}Hg	197.966760	10.1
^{164}Er	163.929211	1.61	^{199}Hg	198.968269	17.0
^{166}Er	165.930305	33.6	^{200}Hg	199.968316	23.1
^{167}Er	166.932061	22.95	^{201}Hg	200.970293	13.2
^{168}Er	167.932383	26.8	^{202}Hg	201.970632	29.65
^{170}Er	169.935476	14.9	^{204}Hg	203.973481	6.8
^{169}Tm	168.934225	100.00	^{203}Tl	202.972336	29.524
^{168}Yb	167.933908	0.13	^{205}Tl	204.974410	70.476
^{170}Yb	169.934774	3.05	^{204}Pb	203.973037	1.4
^{171}Yb	170.936338	14.3	^{206}Pb	205.974455	24.1
^{172}Yb	171.936393	21.9	^{207}Pb	206.975885	22.1
^{173}Yb	172.938222	16.12	^{208}Pb	207.976641	52.4
^{174}Yb	173.938873	31.8	^{209}Bi	208.980388	100.00
^{176}Yb	175.942576	12.7	^{232}Th	232.03805381	100.00
^{175}Lu	174.940785	97.40	^{234}U	234.04094740	0.0055
^{176}Lu	175.942694	2.60	^{235}U	235.04392525	0.7200
^{174}Hf	173.940065	0.16	^{238}U	238.05078578	99.2745
^{176}Hf	175.941420	5.2			

2.1 Calculation of isotope distribution patterns

The matrix set out below demonstrates one method used for calculating isotope patterns. Each isotope (with the associated fractional abundance shown in parentheses) is set out along the matrix as shown, leaving a space for any 'missing' isotope. Each possible combination of masses is then fitted into the matrix; the fractional abundances are calulated by multiplying together the abundances of the masses composing the cluster.

Table (matrix) for Appendix 2

Ru Ru	96 (0.05)	98 (0.02)	99 (0.13)	100 (0.12)	101 (0.17)	102 (0.32)	104 (0.19)
96 (0.05)	192 (0.0025)	194 (0.001)	195 (0.0065)	196 (0.0060)	197 (0.0085)	198 (0.016)	200 (0.0095)
98 (0.02)	194 (0.001)	196 (0.0004)	197 (0.0026)	198 (0.0024)	199 (0.0034)	200 (0.0064)	202 (0.0038)
99 (0.13)	195 (0.0065)	197 (0.0026)	198 (0.0169)	199 (0.0156)	200 (0.0221)	201 (0.0416)	203 (0.0247)
100 (0.12)	196 (0.006)	198 (0.0024)	199 (0.0156)	200 (0.0144)	201 (0.0204)	201 (0.0384)	203 (0.0228)
101 (0.17)	197 (0.0085)	199 (0.0034)	200 (0.0221)	201 (0.0204)	202 (0.0289)	203 (0.0544)	205 (0.0323)
102 (0.32)	198 (0.016)	200 (0.0064)	201 (0.0416)	202 (0.0384)	203 (0.0544)	204 (0.1024)	206 (0.0608)
104 (0.19)	200 (0.0095)	202 (0.0038)	203 (0.0247)	204 (0.0228)	205 (0.0323)	206 (0.0608)	208 (0.0361)

Total abundance of cluster 200 = (0.0904)

By summing the abundances along each diagonal the fractional abundance of that particular mass combination can be calculated and the expected isotopic pattern of a cluster can be determined as shown for Ru_2^+ (See Fig. A2.1).

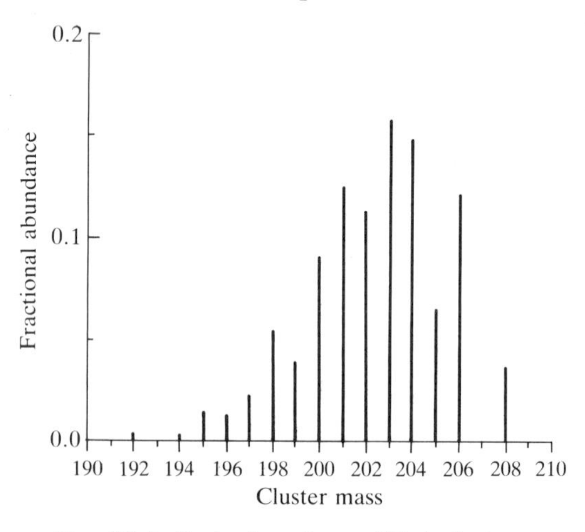

FIG A2.1. Isotopic pattern of Ru_2^+ cluster.

APPENDIX 3

Erosion rates and monolayer lifetimes as a function of sputtering parameters

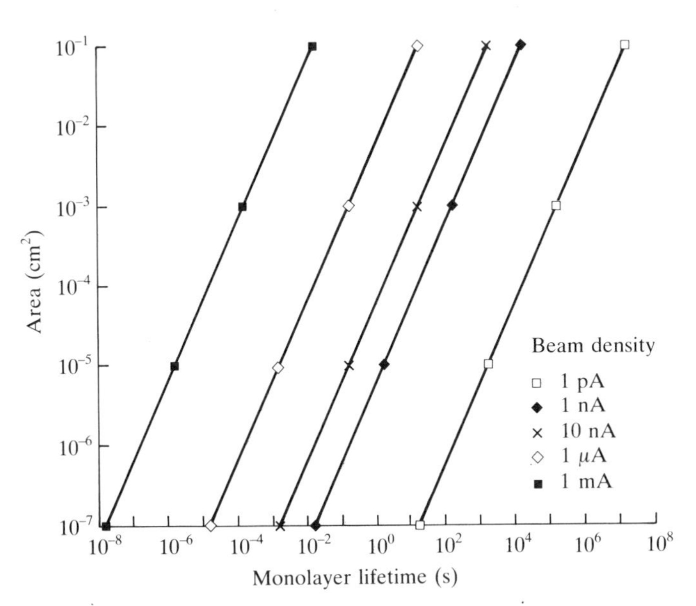

FIG A3.1. Monolayer lifetime vs. area analysed as a function of beam density assuming a sputter yield of one.

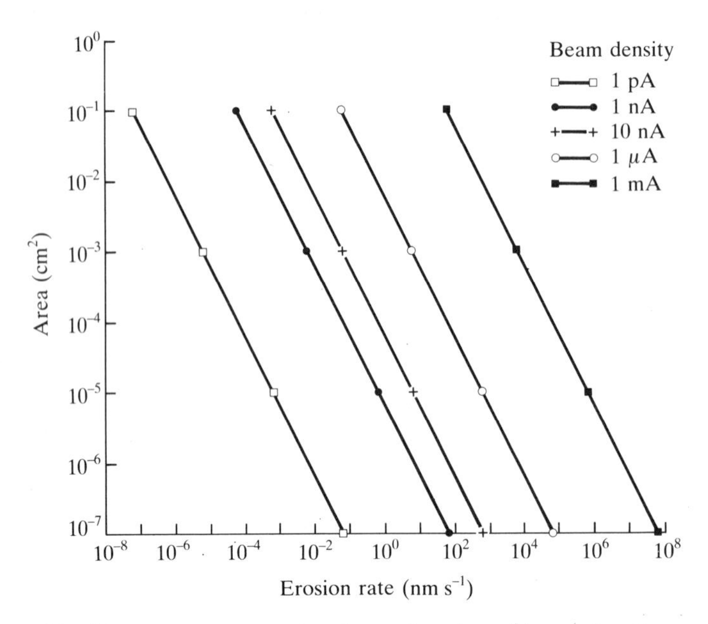

FIG A3.2. Erosion rate vs. area analysed as a function of beam density assuming a sputter yield of one.

APPENDIX 4

Ion velocities

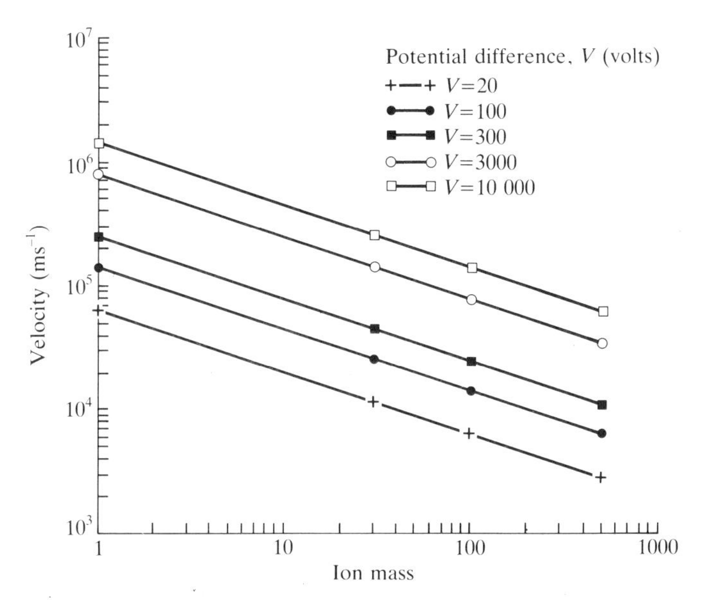

FIGS A4.1 Ion mass vs. velocity of a charged ion as a function of potential difference through which the ion is accelerated.

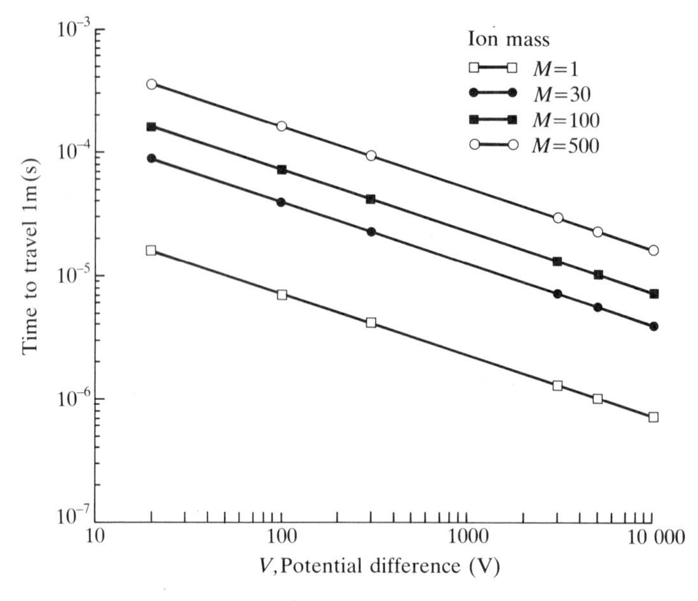

FIG A4.2. Time required to travel 1 m for ions of mass M as a function of potential difference.

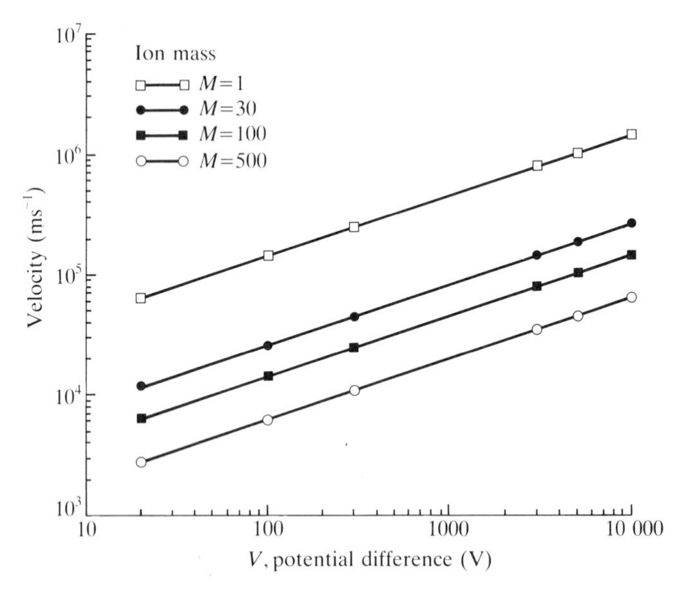

FIG A4.3. Potential difference vs. velocity of a singly charged ion as a function of ion mass accelerated through a field of potential difference V (volts).

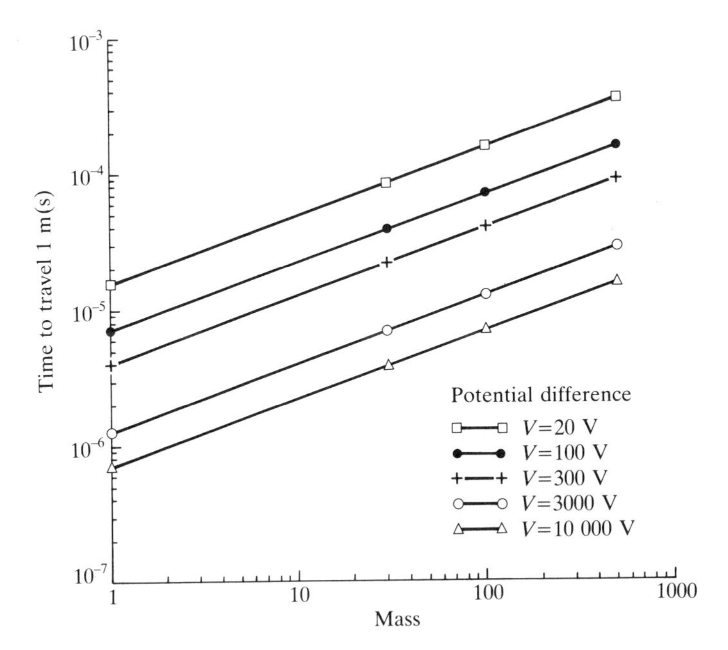

FIG A4.4 Time required to travel 1 m when accelerated through potential difference V (volts), as a function of mass.

APPENDIX 5

Standard static SIMS spectra and their acquisition

Standard SSIMS spectra

These spectra can be used to assess the performance of a SSIMS instrument. They have been obtained under low damage conditions and are characteristic of the range of materials which might be encountered when using SSIMS for surface analysis.

Instrument settings

By far the most widely used analyser in SSIMS is the quadrupole mass filter. The operating parameters which were considered when acquiring the present spectra on a quadrupole-based instrument are set out below. The procedures set out were defined by the UK SIMS User's Forum. Increasingly, ToF instruments are being used but a consistent operating mode has yet to be defined.

The mass spectrometer

1. Mass resolution

All quadrupole mass analysers provide for an electronic resolution control which sets the d.c./r.f. voltage ratios applied to the quadrupole rods. In addition, a control, usually labelled 'ΔM' is often provided which allows accurate zeroing of the d.c. voltage amplifiers at zero mass to reduce the unwanted signal that is transmitted when the r.f. fields are collapsed to zero. This offset affects the resolution as the voltage ratio is now

$$(V_{\text{d.c.}} + V_{\text{offset}})/V_{\text{r.f.}}.$$

V_{offset} is set at a fixed level while $V_{\text{d.c.}}/V_{\text{r.f.}}$ is proportional to mass. Therefore V_{offset} has a stronger effect at lower masses than at high mass. This means that the mass width is generally not quite constant over the low masses. For normal spectral purposes the controls are set to the level giving peak widths of $m/z = 1$ at 5 per cent of peak height.

2. Pole bias

Another d.c. bias is generally applied to the quadrupole rods which, unlike the bipolar 'resolution' d.c. levels, has the same polarity on all four rods and does not directly determine resolution. Its function is to set the energy of ions injected into the quadrupole. This determines the 'quality' of mass resolution attained since the maximum resolution is proportional to (time of flight)^2. At resolutions below maximum, i.e. at low masses, the observed effect of pole bias is in the mass peak shape and peak height.

The precise effect of pole bias will depend on the type of energy filter, quadrupole and possible detector arrangement, as the pole bias affects not only the ion energy but also inevitably the input position and angle of entry of the ion into the quadrupole. Practically, the pole bias is set to a level which gives the best peak shapes and peak heights consistent with the 5 per cent width definition of resolution.

3. *Setting the energy filter and target potential*

Single element peaks from solids exhibit a broad energy spectrum whereas multi-mass peaks, arising from a different sputtering mechansim, have a narrower energy spectrum, both spectra peaking at very low energies (see Fig 2.6). Thus if the mass spectrometer is set to 'see' high-energy ions the low mass (monomer) peaks will be strong whereas the high mass (cluster) peaks will be weak and vice versa. If the pass energy of thee filter is set to maximize a cluster ion, such as $K(KBr)^+$ from KBr or MoO^+ from oxidized molybedenum, this will provide a defined energy acceptance condition which will be suitable to obtain SSIMS cluster ion spectra. If the energy filter also provides the facility to increase the energy window, ΔE, increasing ΔE will increase the relative intensity of monomer peaks and the overall spectral intensity. Care has to be exercised, however, because there may be a degradation of mass resolution if ΔE is increased too far.

For systems where the target (sample) bias is adjustable, control of the ion input energy into the quadrupole is possible since

$$E_i = E_0 + eV_t$$

Where E_0 is the initial energy of the secondary ion and V_t is the target bias. Having set up the optimum mass and energy resolution conditions, variation of V_t allows maximum ion intensity to be obtained.

Clearly, the precise operation will depend on the analyser used but the basic procedure outlined above seeks to ensure that the operator is aware of the important mass analyser parameters.

4. *Setting up the channeltron*

The channeltron is part of the detection circuit and does not count all ions at 100 per cent efficiency for two main reasons: (1) saturation effects and (2) failure to give an output pulse for each ion impact. Saturation of the channeltron is easy to detect and rectify, and occurs when count rates are too high. Halving the primary current should halve the count rate; typically count rates below 50 Kc s^{-1} should be used. It is not so easy to deal with (2), and this cannot be rectified. The yield is never 100 per cent for ions, protons or electrons. it varies with incident energy, ion mass and charge. The gain of the channeltron falls with increasing ion mass at a constant detector voltage. It is recommended that the detector counting response is measured as a function of applied voltage for a high-mass species. The operating voltage across the channeltron is chosen to be in the central plateau response of this curve.

The primary beam

Static conditions are used. Usually an inert gas beam of argon or xenon is chosen, although, on imaging systems, gallium is increasingly used. The beam energies generally used lie in the range 2–10 keV. The area irradiated by the beam at the sample should be the maximum area (or field of view) from which ions are collected by the energy filter–analyser system. This area will generally lie between 1 and 3 mm diameter. In the present case the diameter is 3 mm. The primary beam density is chosen to give a damage-free spectrum within the time-scale required to acquire the spectrum. The minimum primary beam density consistent with reasonable spectral intensity is required and this should normally be below 1 nA cm^{-2}. For many polymer materials this is a total dose of 10^{13} particles cm^{-2}. Thus a beam density of 1 nA into a 3 mm diameter spot allows only 50 s to acquire the spectrum.

Neutralization. The spectra acquired here used a neutral argon beam. If a neutral beam is used charging is not a problem when acquiring positive ion spectra from insulators. In the negative ion mode some electron flooding is required, as outlined in Chapters 7 and 8. The use of primary ion beam does require the use of electron flooding. The most sensitive method of determining the effectiveness of neutralization is in the negative ion mode. On a material such as PMMA, incomplete neutralization results in a negative ion spectrum consisting only of C^- and O^- ions, whereas good conditions result in a detailed fragmentation pattern.

1. Oxidized molybdenum

High-purity molybdenum foil can be prepared to yield a useful standard. Two basic approaches are used:

(1) the foil may be oxidized in air by slowly drawing it through a bunsen flame such that the foil is heated to red heat in the process;
(2) The foil is installed in the SIMS UHV system, argon ion etched to remove surface oxide and then exposed to 10^{-5} Torr of oxygen in the chamber.
(3) It is found that the most intense yields are generated by leaving the sample under the primary beam for 10 min to remove superficial adsorbates.

Analysis of the oxidized foil using a 1 nA Ar^+ or Ar^0 beam of energy 2–4 keV yields a positive ion SIMS spectrum in the Mo^+ and MoO^+ region as shown Fig. A5.1. Summing the intensities of $^{98}Mo^+$ and $^{114}(MoO)^+$ gives a measure of sensitivity performance. For unit mass resolution at the 10 per cent level, performance levels have been found in the following ranges:

	Mass spectrometer	Primary beam	Extraction energy	$^{98}Mo^+ + ^{114}(MoO)^+$ Yield counts (nC)
1.	Quadrupole	Argon 2–4 kV	Low 100 V	$2-5 \times 10^4$
2.	Quadrupole	Argon 10 kV	High > 300 V	$2-5 \times 10^5$
3.	ToF	Gallium 30 kV	High 5 kV	$1-3 \times 10^7$
4.	Magnetic sector	Gallium 30 kV	High 5 kV	1×10^7

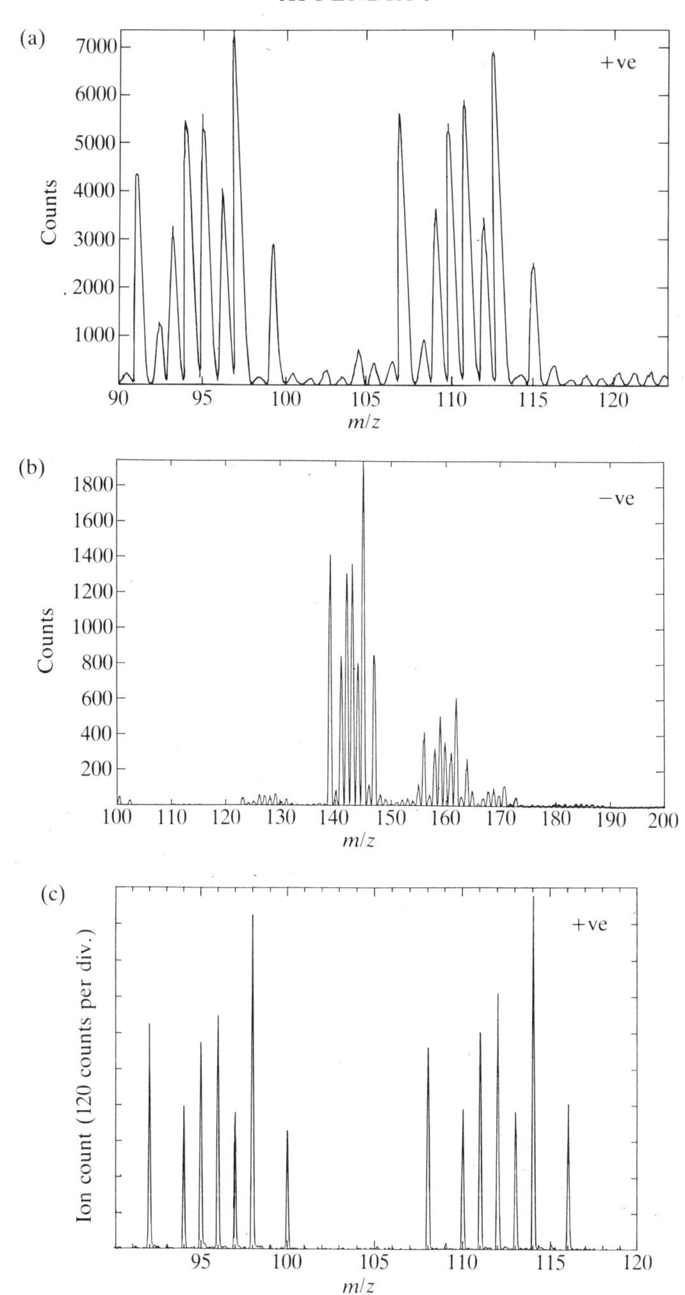

FIG A5.1. Positive (a) and negative (b) ion spectra of oxidized molydebenum. Conditions: 2 keV Ar^0, current equivalent 1 nA cm^{-2}. Analyser: VG 12–12 quadrupole. (c) Positive ion spectrum of oxidized molybdenum. Conditions: 25 keV Ga^+ pulsed beam, dose 10^{10} ions. Analyser: Cambridge Mass Spectrometry reflectron ToF.

2. Potassium bromide

This is another compound which can be used as a semi-standard. Both potassium and bromine have two isotopes. This provides a test of mass resolution with increasing mass. In the positive ion spectra, ions are found at regular intervals corresponding to KBr (*ca.* m/z = 119) $K(KBr)_n{}^+$. In the negative ion spectra, similar distribution is found corresponding to $Br(KBr)_n{}^-$.

The positive ion spectrum obtained with a primary beam of 0.5 nA cm^{-2} Ar^0 is shown in Fig A5.2. In Fig A5.3 is displayed the high-mass spectrum obtained with a 10 nA cm^{-2} Ar^0 beam.

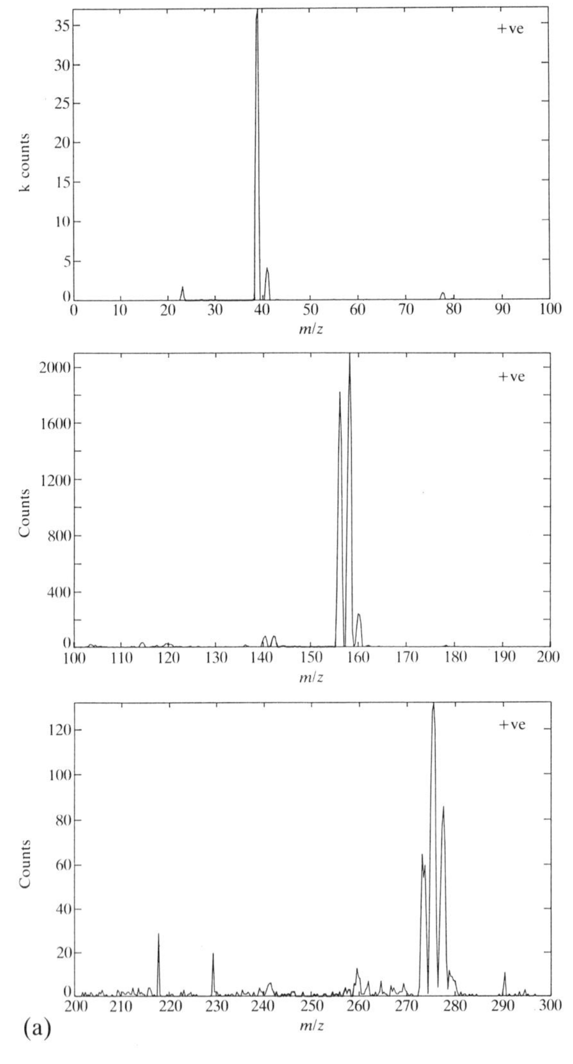

FIG A5.2 Positive (a) and negative (b) ion mass spectra of KBr powder. Conditions: 2 keV Ar^0, current equivalent 0.5 nA cm^{-2}. Analyser: VG 12–12 quadrupole.

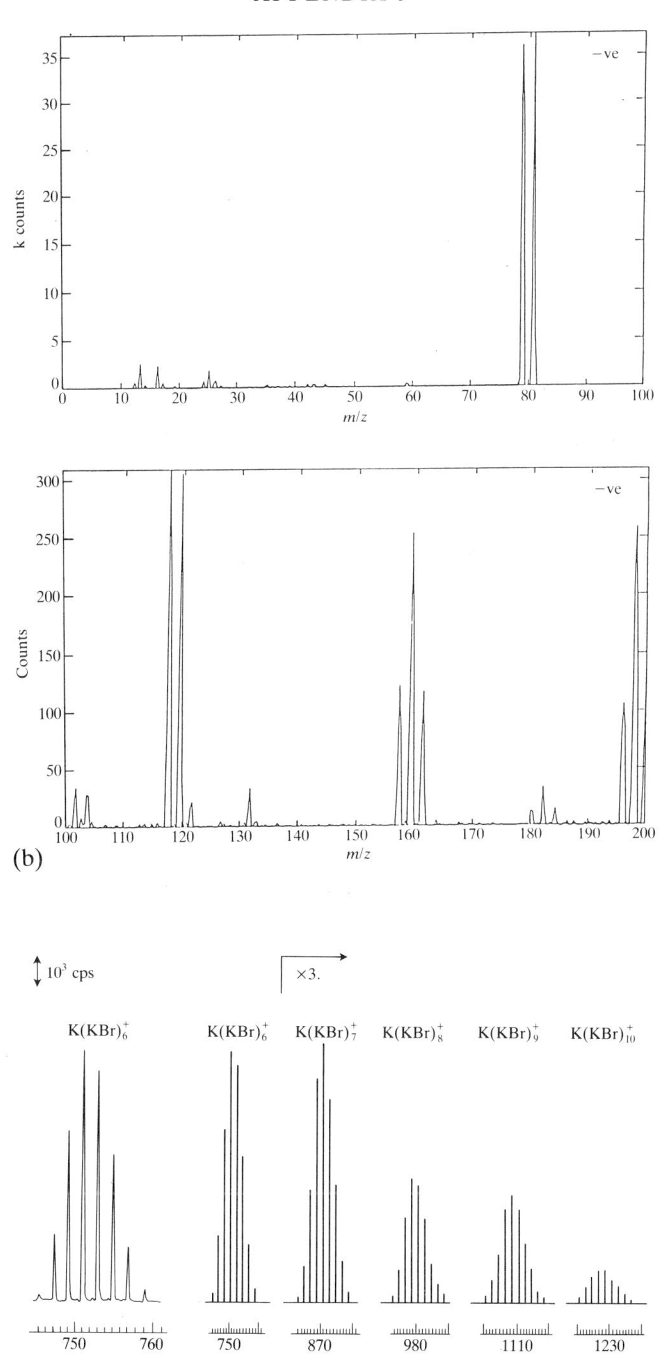

FIG A5.3. Positive ion high-mass spectrum of KBr. Conditions: 2 keV Ar^0, current equivalent 10 nA cm^{-2}. Analyser: VG 12–12 quadrupole.

3. Poly(methylmethacrylate)—PMMA

This polymer has been shown to be a sensitive test material to assess the ability of an instrument–operator combination to obtain SSIMS spectra with minimum damage and with optimum charge neutralization. The polymer can be analysed either in the form of a thin spin cast layer on a metal substrate or as a thick polymer film. The spin casting process can be carried out to yield a non-insulating layer such that spectra can be obtained without charge neutralization. The spin casting procedure is quite difficult in this case. The spectra shown in Fig. A5.4 have been obtained from a thick spin cast layer which is insulating. A 2 keV Ar^0 beam was used with a total beam fluence of 10^{12} particles. Some electron flooding was required to obtain the negative ion spectra. When the charge neutralization is incomplete the negative ion spectra only consist of C^- and O^- peaks.

The development of particle-induced damage can be monitored by the fall in the intensity of the structurally significant ions, for example, the negative ion at $m/z = 185$ and the positive ion at $m/z = 126$. These are accompanied by slow increases in, for example, the positive ions at $m/z = 91$ and 133. These are thought to be cyclic structures formed as a consequence of the loss of side-chains or the generation of unsaturation. As damage accumulates to higher levels, $> 6 \times 10^{13}$ particles cm^{-2}, the intensity of these ions falls relative to simple carbon fragment ions such as $C_xH_y^{+/-}$, where y is quite small.

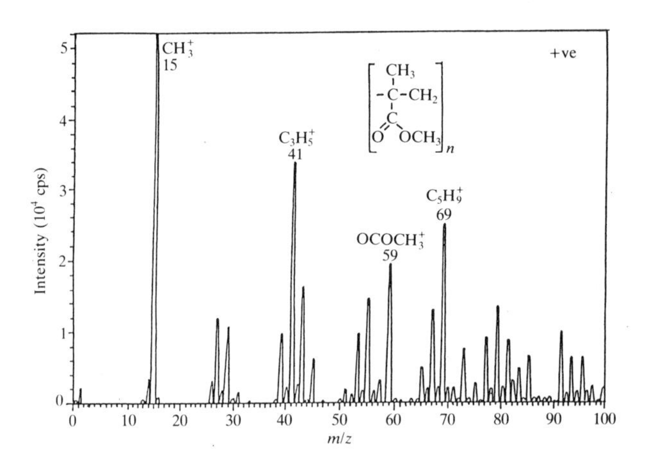

FIG A5.4 Positive (a) and negative (b) ion mass spectra of PMMA. Conditions: 2 keV Ar^0, dose 10^{12} particles cm^{-2}. Analyser: VG 12–12 quadrupole.

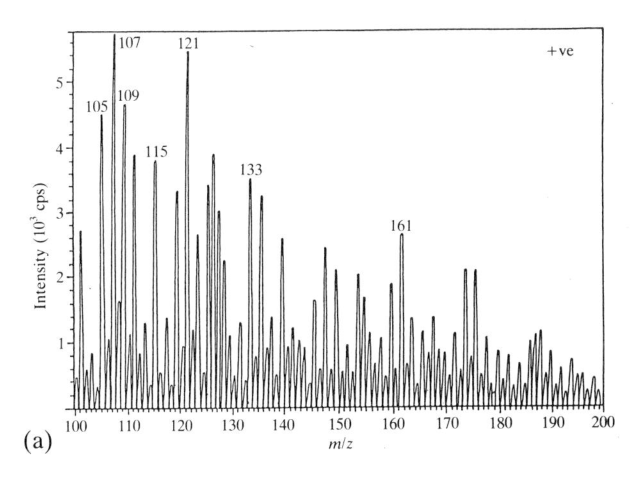
+ve
Intensity (10^3 cps)
107
121
105
109
115
133
161
m/z
(a)

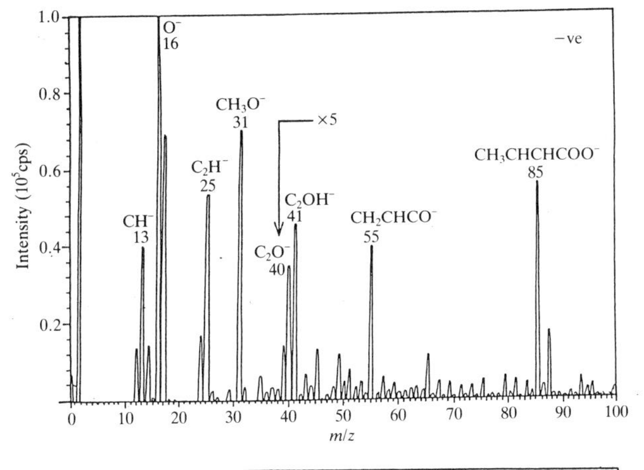
−ve
Intensity (10^5cps)
O^-
16
CH_3O^-
31
×5
C_2H^-
25
$CH_3CHCHCOO^-$
85
C_2OH^-
41
CH^-
13
CH_2CHCO^-
55
C_2O^-
40
m/z

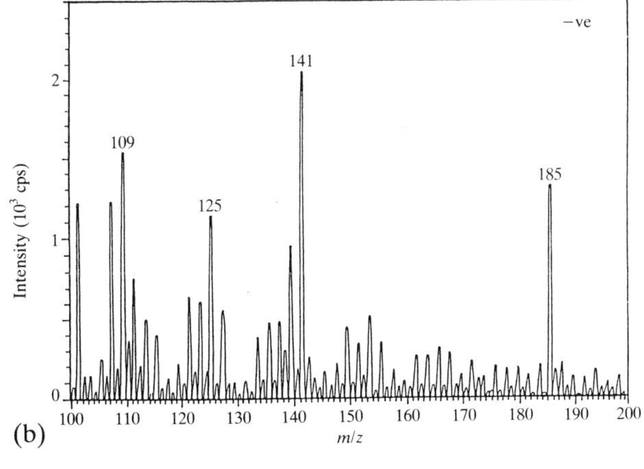
−ve
Intensity (10^3 cps)
141
109
185
125
m/z
(b)

4. Poly(tetrafluoroethylene)—PTFE

This polymer is relatively robust to ion-induced damage. Structurally related cluster ion spectra can be obtained at relatively high primary particle doses up to 10^{14} cm^{-2}. It is also easy to obtain good positive ion spectra with an ion beam with little charge neutralization, although negative ion spectra do require charge neutralization. This is therefore a good polymer to practise on. It gives the operator a feel for the conditions required to obtain good polymer spectra. The relative intensities of the ions in the cluster pattern, which stretches to quite high mass, is a useful test of the settings of the analyser energy filter (see Fig. A5.5).

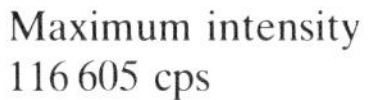

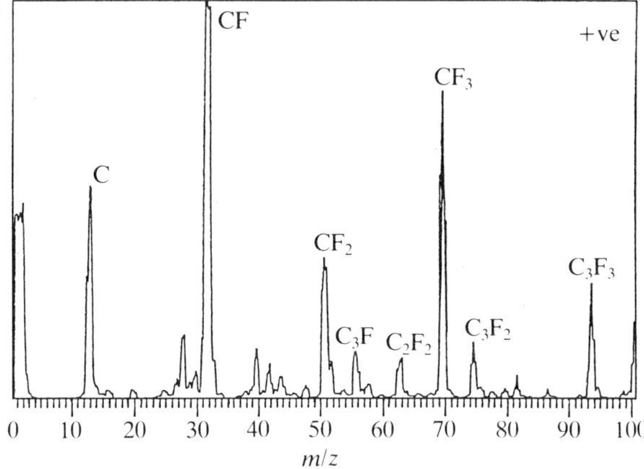

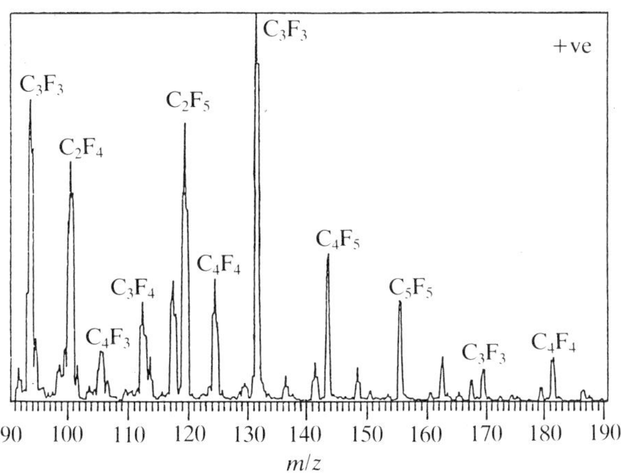

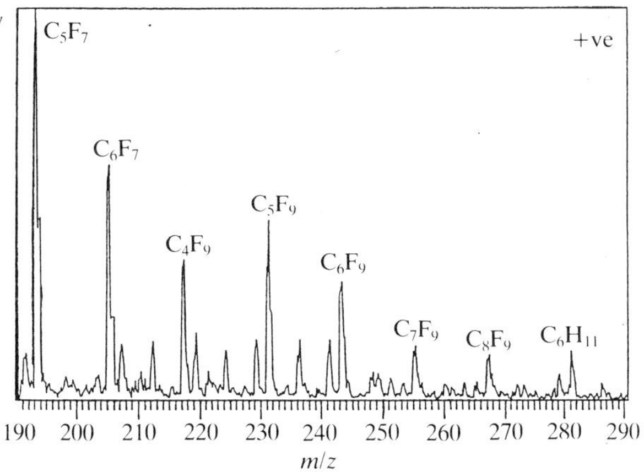

FIG A5.5 Positive ion mass spectrum of PTFE. Condition: 2 keV Ar^0, current equivalent 1 nA cm^{-2}. Analyser: VG 12-12 quadrupole.

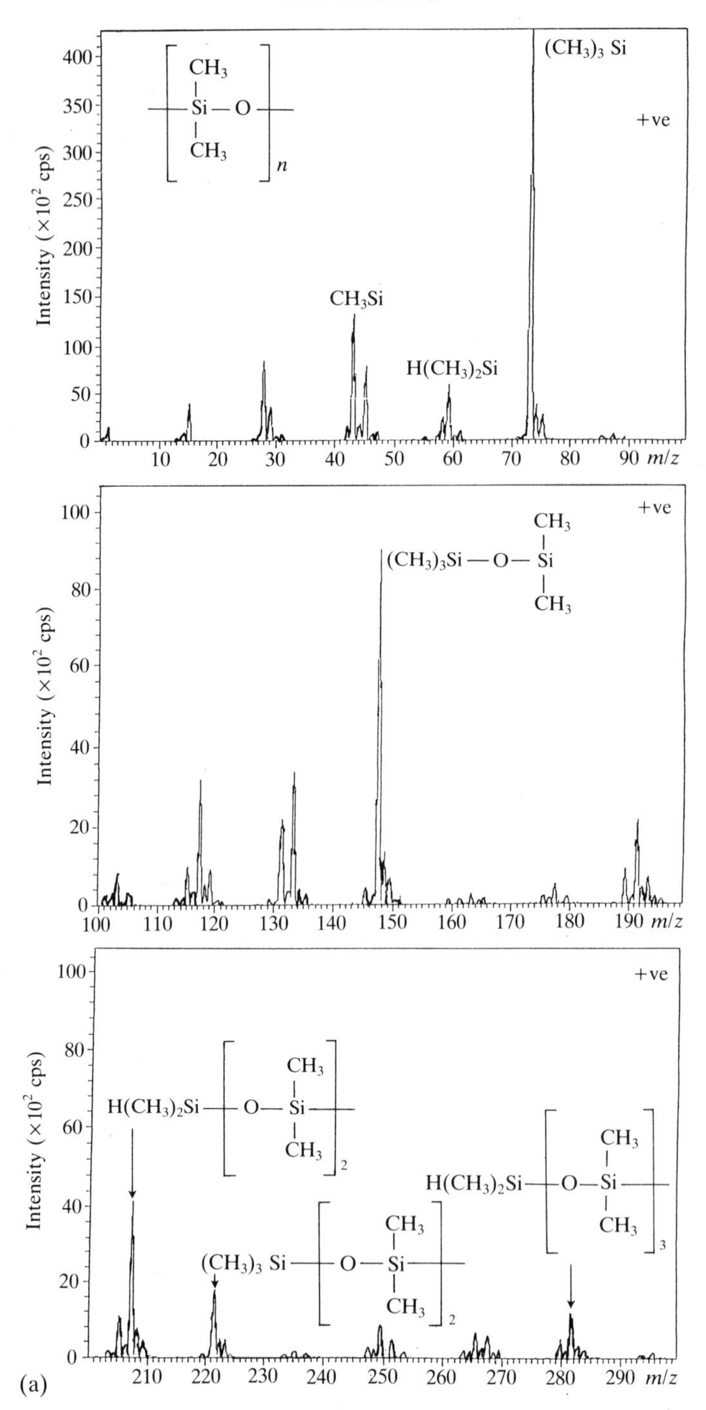
(CH3)3 Si
+ve
CH3Si
H(CH3)2Si
Intensity (×10^2 cps)
m/z
(CH3)3Si — O — Si
H(CH3)2Si — O — Si
(CH3)3 Si — O — Si
CH3
n
2
3
(a)

5. Poly(dimethyl-siloxane)—PDMSO

This is included because it provides the spectrum of the most common organic contaminant found in the course of practical analysis. Silicone oils are used extensively as lubricants and stress cracking agents, and contamination from them is very frequently found on polymer packaging or on samples transported in plastic boxes or envelopes. The characteristic positive ion peaks at m/z = 147 and 73 are frequently diagnostic (see Fig. A5.6.).

Other SSIMS fingerprint spectra can be found in the *Handbook of Static SIMS* by D. Briggs, A. Brown, and J. C. Vickerman, published by John Wiley, Chichester, 1989.

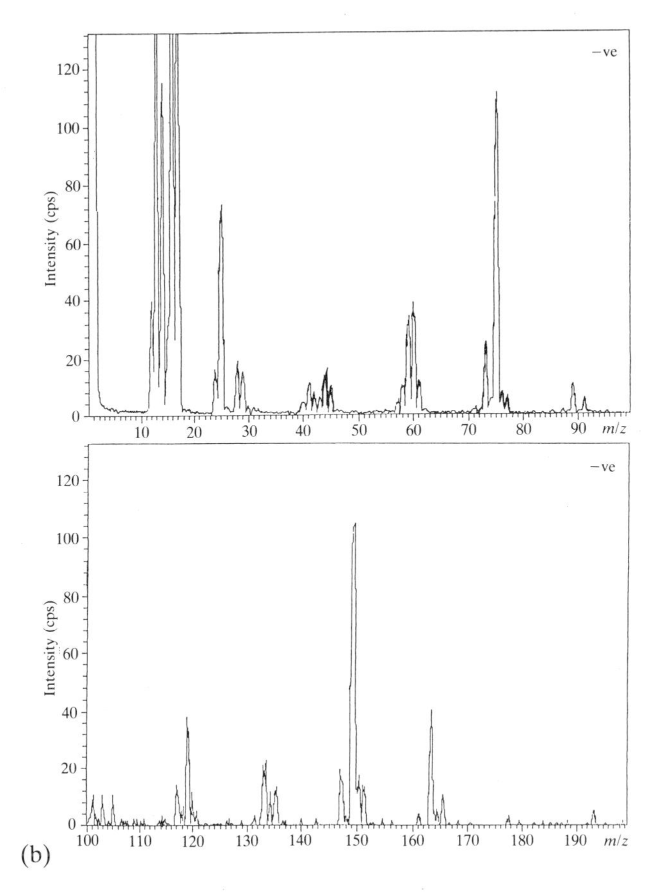

FIG A5.6 Positive (a) and negative (b) ion mass spectra of poly (dimethyl-siloxane). Conditions: 2 keV Ar^0, current equivalent 1 nA cm^{-2}. Analyser: VG 12–12 quadrupole.

APPENDIX 6

Commonly observed fragment ions in SSIMS spectra

This list is only intended to be suggestive rather than exhaustive and represents the most common ions found by analysts working on a wide variety of industrial samples. It is suggested that readers amend the list according to their own experience.

Obviously, specific sample types will give rise to specific ions. In interpreting a spectrum, full use must be made of the chemical fragment information available to avoid mis-assignment of peaks.

m/z	Positive	*m/z*	Negative
1	H^+	1	H^-
2		2	
3		3	
4		4	
5		5	
6		6	
7	Li^+	7	
8		8	
9	Be^+, Al^{3+}	9	
10	$^{10}B^+$	10	
11	B^+	11	
12	C^+	12	C^-
13	CH^+	13	CH^-
14	CH_2^+	14	CH_2^-
15	CH_3^+	15	
16	O^+	16	O^-
17		17	OH^-
18	H_2O^+	18	
19	H_3O^+	19	F^-
20	Ca^+	20	
21		21	
22		22	
23	Na^+	23	
24	Mg^+	24	C_2^-
25		25	C_2H^-
26	$C_2H_2^+$	26	CN^-
27	Al^+, $C_2H_3^+$	27	BO^-
28	Si^+, $CHNH^+$	28	Si^-
29	SiH^+, $C_2H_5^+$, $^{29}Si^+$	29	

m/z Positive	m/z Negative
30 CH_3NH^+	30
31 P^+, CF^+, CH_3O^+	31 P^-, CH_3O^-
32 O_2^+	32 S^-, O_2^-
33	33 SH^-
34	34
35	35 Cl^-
36	36 C_3^-
37	37 $^{37}Cl^+$, C_3H^-
38	38 C_2N^-
39 K^+, $C_3H_3^+$	39
40 Ca^+, Ar^+, CH_2CN^+	40 C_2O^-
41 $C_3H_5^+$	41 C_2OH^-
42	42 CNO^-, SiN^-
43 AlO^+, $C_3H_7^+$, CH_3Si^+	43 AlO^-, BO_2^-, $C_2H_3O^-$
44	44 CO_2^-
45 $SiOH^+$, Sc^+, $C_2H_5O^+$	45 CO_2H^-, $C_2H_5O^-$
46 Na_2^+	46 NO_2^-
47 PO^+, CFO^+	47 PO^-
48 Ti^+	48 C_4^-
49	49 C_4H^-
50 CF_2^+	50 C_3N^-
51 V^+	51
52 Cr^+	52
53 $C_4H_5^+$	53
54 Al_2^+	54
55 Mn^+, $C_4H_7^+$	55 $C_2H_3CO^-$
56 Fe^+	56
57 FeH^+, $C_4H_9^+$, $CaOH^+$	57
58 Ni^+	58 CNS^-
59 Co^+, $C_3H_5O^+$	59 AlO_2^-, $CH_3SiO_2^-$, CH_3COO^-, BO_3^-, $C_3H_7O^-$
60 $^{60}Ni^+$	60 SiO_2^-, C_5^-
61	61 SiO_2H^-, C_5H^-
62	62 NO_3^-
63 Cu^+	63 PO_2^-
64 Zn^+, TiO^+	64 SO_2^-, S_2^-
65 $^{65}Cu^+$	65
66 $^{66}Zn^+$	66 C_3NO^-
67 $C_5H_7^+$	67
68 $^{68}Zn^+$	68
69 Ga^+, CF_3^+, $C_4H_5O^+$, $C_5H_9^+$, $CrOH^+$	69
70	70
71 $^{71}Ga^+$, $C_5H_{11}^+$	71
72 $^{72}Ge^+$,	72 C_6^-

m/z Positive	*m/z* Negative
73 $(CH_3)_3Si^+$, $FeOH^+$	73 C_6H^-, $C_2H_5COO^-$, $C_4H_9O^-$
74 Ge^-	74
75 As^+	75 As^-, $CH_3SiO_2^-$
76	76 SiO_3^-
77 $C_6H_5^+$	77 SiO_3H^-
78 Se^+	78 Se^-
79 $C_6H_7^+$	79 Br^-, PO_3^-
80 $^{80}Se^+$	80 SO_3^-, $^{80}Se^-$
81 $ZnOH^+$, $C_6H_9^+$	81 Br^-, SO_3H^-
82	82
83 $C_6H_{11}^+$	83
84	84 CrO_2^-
85	85 CrO_2H^-, $C_3H_5COO^-$
86	86
87	87 GaO^-
88 Sr^+	88
89 V^+	89
90 Zr^+	90
91 $C_7H_7^+$	91 AsO^+
92	92
93 $C_3F_3^+$, Nb^+, $C_7H_9^+$	93 $C_7H_9^-$, $C_3F_3^-$
94	94
95 $C_7H_{11}^+$	95 PO_4^-
96	96 SO_4^-
97 $C_7H_{13}^+$	97 SO_4H^-
98 Mo^+	98
99	99
100 $C_2F_4^+$	100 CrO_3^-
	127 I^-

Note: For elemental species only the major isotopes have in general been included in this list. Use should be made of the natural isotopic abundances (Appendix 2) to confirm assignments.

Additional ions

Siloxane			
131	$(CH_3)_3SiSi(CH_3)_2^+$	149	$(CH_3)_3SiOSiO_2^-$
147	$(CH_3)_3SiOSi(CH_3)_2^+$		
221	$(CH_3)_3SiOSi(CH_3)_2OSi(CH_3)_2^+$		
295	$(CH_3)_3SiOSi(CH_3)_2OSi(CH_3)_2\ OSi(CH_3)_2^+$		
Phthalate			
104	$C_6H_4CO^+$	120	$C_6H_4COO^-$
105	$C_6H_5CO^+$	121	$C_6H_5COO^-$
Carboxylic acids			
	Caprate	171	$CH_3(CH_2)_8COO^-$
	Laurate	199	$CH_3(CH_2)_{10}COO^-$
	Myristate	227	$CH_3(CH_2)_{12}COO^-$
	Palmitate	255	$CH_3(CH_2)_{14}COO^-$
	Oleate	281	$CH_3(CH_2)_7CH{=}CH(CH_2)_7\ COO^-$
	Stearate	283	$CH_3(CH_2)_{16}COO^-$

APPENDIX 7

Sputter yields under positive ion bombardment

Element	2 keV	4 keV	10 keV
Ar^+ bombardment			
Be	1.2	1.4	
C	1.5	2.0	3.75
Mg	4.5	5.2	6.3
Al	2.8	3.0	3.8
Si	1.1	1.25	1.5
Ca	7.0	8.5	10.0
Sc	3.0	3.5	3.8
Ti	1.2	1.5	2.0
V	2.0	2.5	3.7
Cr	3.0	3.5	4.1
Mn	4.0	4.5	5.0
Fe	2.4	3.0	3.5
Co	2.5	2.8	3.0
Ni	3.2	3.6	4.5
Cu	4.0	5.0	6.0
Zn	10.0	14.0	16.0
Ge	2.0	2.5	3.3
Zr	1.0	1.5	1.7
Nb	1.9	2.0	2.5
Mo	1.3	1.6	2.0
Pd	2.8	3.7	5.0
Ag	6.4	7.8	10.0
In	7.8	8.5	10.0
Sn	8.0	8.2	9.5
Ta	1.25	1.4	1.5
W	1.25	1.9	2.4
Pt	3.4	3.6	4.0
Au	5.5	7.4	9.0
Tl	4.5	6.2	8.0
Bi	0.6	0.7	0.8
U	1.3	1.5	1.6

(continued overleaf)

Element	2 keV	4 keV	10 keV
Xe^+ bombardment			
C	0.9	1.0	1.3
Al	2.4	3.9	6.0
Si	1.5	1.9	2.5
Ti	3.5	4.0	6.0
V	3.5	4.0	6.0
Cr	4.0	5.7	8.0
Fe	3.1	4.0	5.2
Co	4.5	5.5	8.0
Ni	4.5	5.7	7.6
Cu	5.5	7.8	10.0
Ge	5.5	6.5	9.5
Nb	3.1	3.9	5.2
Mo	3.0	3.6	5.9
Pd	4.7	5.0	7.8
Ag	6.8	10.0	15.1
In	2.0	3.0	8.0
Ta	4.0	6.0	8.0
W	2.8	3.7	5.2
O^+ bombardment			
Ti	0.23	0.4	
Fe	0.23	0.28	0.3
Ni	0.55	0.64	
Ag	2.0	3.8	2.7
Sb	2.0	1.8	2.3
W	0.45	0.2	
Au	1.0	1.4	1.75
Cs^+ bombardment			
Al	1.8	2.6	4.0
Ti	0.7	0.9	1.1
Fe	2.8	4.9	7.0
Cu	5.9	7.3	9.8
Nb	2.4	3.7	5.3

APPENDIX 8

Relative secondary ion yields due to Cs^+ and O^-

(Reproduced with permission of the American Chemical Society from H. A. Storms, K. F. Brown and J. D. Stein (1977) *Anal. Chem.*, **49**, 2023.)

	16.5 keV Cs^+		13.5 keV O^-	
Standard	Secondary ion	cps per 10^{-9} A Cs^+	Secondary ion	cps per 10^{-9} A O^-
Be	Be^-	3.4×10^3	Be^+	5.2×10^5
	BeO^-	1.1×10^5	BeO^+	2.4×10^3
BN	B^-	1.6×10^5	B^+	4.6×10^5
	BN^-	3.2×10^4	N^+	20
C	C^-	4.3×10^6	C^+	1.1×10^3
	C_2^-	4.5×10^5		
	H^-	6.1×10^4		
ThO_2	O^-	9.0×10^6	O^+	4.7×10^2
	Th^-	5.3×10^3	Th^+	1.4×10^4
	ThO^-	2.8×10^4	ThO^+	6.6×10^4
	ThO_2^-	4.6×10^4	ThO_2^+	4.4×10^3
$CaWO_4$	O^-	7.9×10^6		
	Ca^-	2.3×10^2	Ca^+	1.4×10^6
	CaO^-	1.2×10^4	CaO^+	3.6×10^4
	W^-	1.8×10^4	W^+	2.9×10^4
	WO^-	1.2×10^5	WO^+	3.0×10^4
$CaPO_4$	O^-	3.1×10^5	Ca^+	4.3×10^5
	P^-	5.4×10^3	P^+	4.7×10^3
			CaO^+	3.8×10^3
MgO	Mg^-	$<2.1 \times 10^4$	Mg^+	5.4×10^5
	O^-	2.0×10^6	MgO^+	1.9×10^3
	MgO^-	1.8×10^3		
Al_2O_3	O^-	2.1×10^5	Al^+	6.3×10^5
	Al^-	3.7×10^3	AlO^+	6.3×10^2
SiO_2	O^-	5.8×10^5	Si^+	1.1×10^6
	Si^-	1.2×10^4	SiO^+	1.0×10^4
$BaTiSi_3O_4$	O^-	2.0×10^6	Si^+	1.6×10^5
	Si^-	6.9×10^4	SiO^+	6.6×10^2
	Ti^-	3.4×10^3	Ti^+	1.9×10^5
			TiO^+	1.0×10^4
			Ba^+	2.8×10^5
			BaO^+	4.5×10^4

Standard	16.5 keV Cs^+ Secondary ion	cps per 10^{-9} A Cs^+	13.5 keV O^- Secondary ion	cps per 10^{-9} A O^-
Al	H^-	1.3×10^4	Al^+	7.0×10^5
	Al^-	2.9×10^4	AlO^+	7.2×10^2
	O^-	6.3×10^5		
	AlO^-	3.0×10^4		
	Al_2^-	2.2×10^4		
	Al_3^-	5.2×10^2		
Si	Si^-	3.8×10^6	Si^+	5.8×10^5
	Si_2^-	1.4×10^5	SiO^+	4.2×10^3
	Si_3^-	6.2×10^2		
	H^-	4.6×10^3		
FeS_2	S^-	1.3×10^7	Fe^+	5.2×10^5
	S_2^-	7.4×10^4	S^+	2.8×10^2
	Fe^-	6.9×10^3	FeO^+	3.0×10^3
	FeS^-	6.8×10		
	FeS_2^-	8.8×10^2		
PbS	S^-	7.2×10^6	Pb^+	6.0×10^3
	S_2^-	1.8×10^5	S^+	2.0×10^2
	S_3^-	6.5×10^2		
	Pb^-	7.4×10^2		
	PbS^-	3.5×10^4		
	PbS_2^-	5.5×10^2		
HgS	S^-	5.7×10^5	S^+	5.3×10^3
	Hg^-	B.D.	Hg^+	5.0×10^2
Ti	H^-	7.4×10^4	H^+	3.2×10^2
	Ti^-	1.0×10^4	Ti^+	2.0×10^6
	TiH^-	7.5×10^4	TiO^+	1.2×10^5
			Ti_2^+	1.7×10^3
V	H^-	1.8×10^6	H^+	8.7×10^2
	VH^-	2.3×10^5	VO^+	1.1×10^5
			V_2^+	2.4×10^3
Cr	Cr^-	3.5×10^3	Cr^+	1.7×10^6
			CrO^+	1.7×10^4
Mn	Mn^-	N.D.	Mn^+	6.9×10^5
	MnO^-	2.8×10^3	MnO^+	4.0×10^3
Fe	Fe^-	4.0×10^3	Fe^+	3.8×10^5
	FeO^-	1.3×10^3	FeO^+	1.7×10^3
Co	Co^-	9.2×10^3	Co^+	1.4×10^5
			CoO^+	5.2×10^2
Ni	Ni^-	2.1×10^4	Ni^+	5.8×10^4
			NiO^+	1.7×10^2
Cu	Cu^-	1.5×10^4	Cu^+	6.2×10^4
Zn	Zn^-	B.D.	Zn^+	1.7×10^4

Standard	16.5 keV Cs^+ Secondary ion	cps per 10^{-9} A Cs^+	13.5 keV O^- Secondary ion	cps per 10^{-9} A O^-
GaAs	Ga^-	2.7×10^2	Ga^+	7.6×10^5
	As^-	1.7×10^5	As^+	5.2×10^3
	As^-	1.7×10^5	As^+	5.2×10^3
	$GaAs^-$	2.0×10^4	GaO^+	5.2×10^2
	As^-	4.0×10^3	AsO^+	1.6×10^3
Ge	Ge^-	1.2×10^5	Ge^+	3.2×10^4
	Ge_2^-	3.1×10^4	GeO^+	2.2×10^2
InAs	In^-	82	In^+	9.9×10^5
	As^-	8.3×10^4	As^+	6.5×10^4
	$InAs^-$	5.2×10^3	InO^+	4.3×10^2
	As_2^-	3.6×10^3	AsO^+	1.7×10^2
	$InAs_2^-$	2.9×10^2	$InAs^+$	2.8×10^2
Se	Se^-	3.3×10^5	Se^+	1.4×10^3
	Se_2^-	1.2×10^4	Se^+	1.4×10^3
	Se_3^-	6.8×10^3		
Zr	Zr^-	3.3×10^3	Zr^+	5.1×10^5
	ZrH^-	4.7×10^3	ZrO^+	1.8×10^5
	H^-	6.0×10^4		
Nb	Nb^-	3.1×10^4	Nb^+	4.7×10^5
	NbH^-	1.2×10^5	NbO^+	3.9×10^5
	NbH_2^-	1.8×10^5	NbO_2^+	2.4×10^4
	H^-	7.2×10^5	H^+	5.2×10^2
Mo	Mo^-	1.3×10^3	Mo^+	1.2×10^6
			MoO^+	6.7×10^5
Ru	Ru^-	4.0×10^3	Ru^+	1.6×10^6
			RuO^+	3.2×10^4
Rh	Rh^-	8.3×10^3	Rh^+	7.6×10^5
	Rh^{2-}	1.5×10^3	RhO^+	4.2×10^3
Pd	Pd^-	2.0×10^3	Pd^+	9.4×10^4
	PdH^-	7.0×10^2		
Ag	Ag^-	7.5×10^3	Ag^+	5.9×10^3
			AgO^+	2.4×10^2
Cd	Cd^-	B.D.	Cd^+	1.5×10^3
Sn	Sn^-	9.5×10^3	Sn^+	1.5×10^4
Sb	Sb^-	2.0×10^4	Sb^+	3.2×10^3
			SbO^+	4.7×10^2
Te	Te^-	7.2×10^4	Te^+	1.6×10^3
Hf	Hf^-	B.D.	Hf^+	1.6×10^5
			HfO^+	7.7×10^4
Ta	Ta^-	5.7×10^2	Ta^+	1.1×10^5
	TaH^-	2.4×10^4	TaO^+	2.3×10^5
	H^-	5.2×10^5	TaO_2^+	2.6×10^4
	TaH_2^-	7.0×10^4	H^+	1.0×10^3

Standard	16.5 keV Cs^+ Secondary ion	cps per 10^{-9} A Cs^+	13.5 keV O^- Secondary ion	cps per 10^{-9} A O^-
W	W^-	2.5×10^4	W^+	1.6×10^5
			WO^+	2.2×10^5
Re	Re^-	6.0×10^2	Re^+	2.2×10^5
			ReO^+	1.8×10^4
Os	Os^-	9.7×10^4	Os^+	1.2×10^5
			OsO^+	6.8×10^3
Ir	Ir^-	3.1×10^5	Ir^+	1.1×10^4
Pt	Pt^-	3.0×10^5	Pt^+	6.4×10^2
Au	Au^-	4.4×10^5	Au^+	52
Bi	Bi^-	3.9×10^3	Bi^+	2.6×10^3
			BiO^+	67

B.D.: Barely detected. N.D.: Not detected.

APPENDIX 9

Effects of primary ion beam energy and incidence on sputter yields and secondary ion yields from semiconductor materials

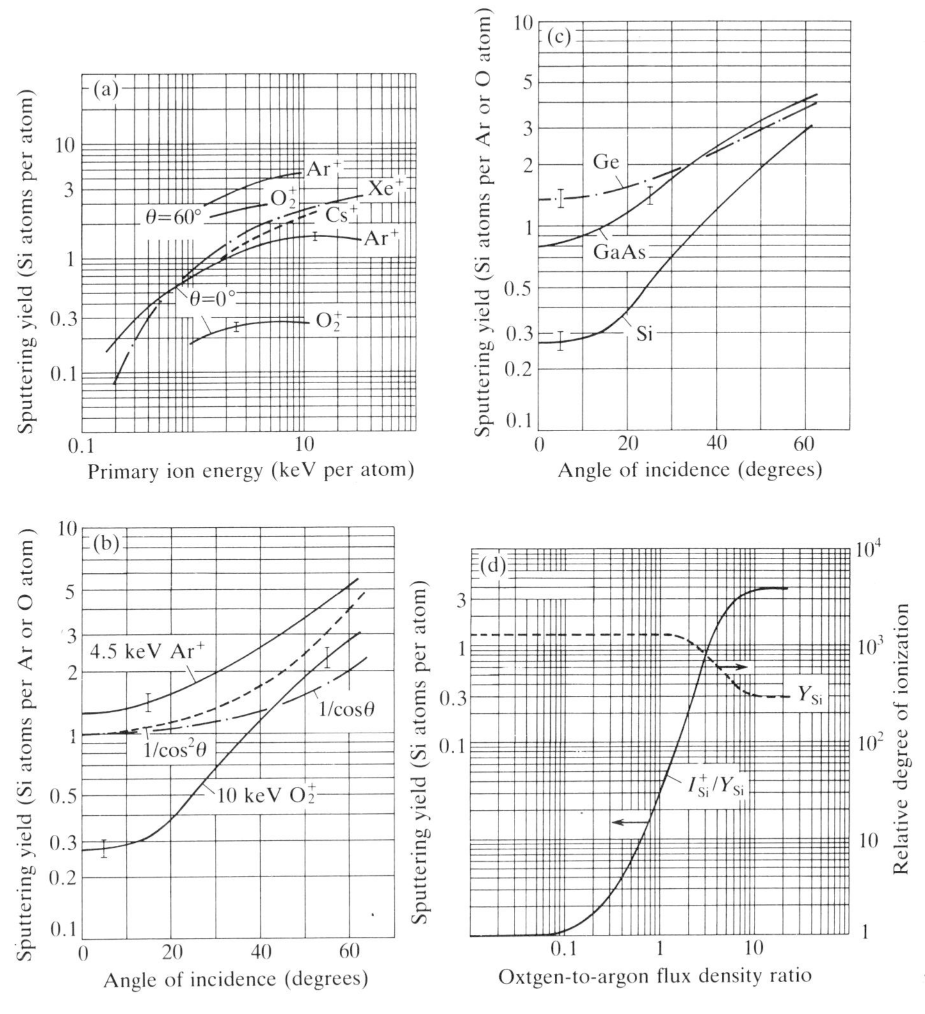

FIG A.9.1 (a) Impact-energy dependence of the sputtering yield of silicon bombarded with various primary ions at normal beam incidence ($\theta = 0°$) or oblique incidence ($\theta = 60°$). (b) Sputtering yield of silicon vs. the angle of incidence of argon or oxygen ions. Angular variations of the form $1/\cos\theta$ and $1/\cos^2\theta$ are shown for comparison. (c) Sputtering yields of 10 keV oxygen-bombarded germanium, gallium arsenide, and silicon vs. the angle of beam incidence. (d) Sputtering yield and fractional ion yield of silicon bombarded with 5 keV argon ions at increasing oxygen partial pressure. Normal beam incidence; j = constant; O_2 bleed-in.

Sputter yields

These data are reproduced by kind permission of K. Wittmaack and Atomica Technische Physik GmbH, Munich, a division of Perkin-Elmer (taken from K. Wittmaack (1982) *Radiats*: *Effects*, **63** , 205; (1983), *Nucl. Inst. Met. Phys. Res.*, **218**, 307; (1984) ibid, **B2**, 674.)

Experimentally, sputter yields vary significantly depending on the primary beam parameters. The following plots, Fig. A9.1 (a)–(d), are a useful guide to the effect of the primary beam parameters on the sputter yields from the semiconductor materials—silicon, germanium and gallium arsenide.

Secondary ion yields

Secondary ion yields are a function of the ionization probability of the element in the particular matrix and the transmission of the instrument. Figure A9.2(a)–(d) shows how the fractional ion yields—the number of secondary ions detected per sputtered atom—vary experimentally as a function of primary beam parameters. The measured values of these yields were obtained using an Atomica Ion Mircoprobe and measurements on other instruments will vary from them depending on their transmission functions; however, the character of the ion yield variation with the primary beam parameters should be qualitatively similar.

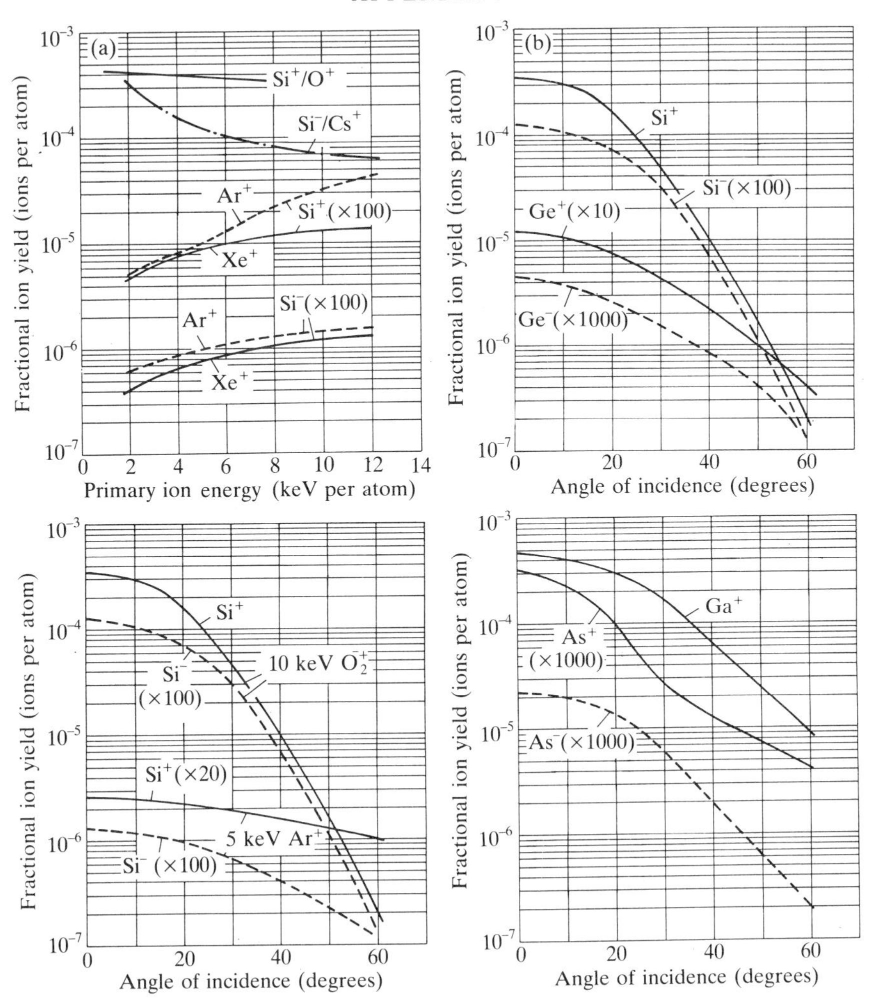

FIG A.9.2 (a) Impact-energy dependence of the fractional ion yield of Si^+ and Si^- ions sputtered from silicon under bombardment with different primary ion species (normal beam incidence). (b) Angular dependence of the fractional ion yield of Si^+ and Si^- ions sputtered from silicon under impact of oxygen or argon ions. (c) Angular dependence of the fractional ion yield of Si^+, Si^+, Ge^+, and Ge^- sputtered from silicon and germanium, respectively. (d) Angular dependence of the fractional ion yield of Ga^+, As^-, and As^- sputtered from gallium arsenide under 10 keV oxygen bombardment.

INDEX

adhesion *see* polymer analysis
adsorption
  adsorbate structure 171–5
  adsorbate–adsorbate interactions 173–5
  amino acids on metal surfaces 163–7
  CO on metal surfaces 162
  coverage, *see* surface coverage measurements
  energetics 169
  ethene 175–7
  NO on metal surfaces 162–3
  on organic surfaces
    PET 182
    phthalocyanines 182
  Rhodamine 6G on PET; *see also* PDMS 293–4
atomic weights 301–3
Auger electron spectroscopy (AES) 1, 149,
  with SSIMS 151, 152, 154, 234

binary collision approximation, *see under* sputtering
biological materials 7

cascade mixing *see under* depth profiling
catalyst analysis
  Cr-silica catalysts 238–9
  Pd catalysts, CO oxidation 239–40, 178–80
  zeolites 236–8
  see also reactions, surface 236–40
cationization 30, 69
^{252}Cf plasma desorption, *see* PDMS
charging of sample 5
  atom bombardment 25, 189
  consequences 191–3
  depth profiling 140–5
  electron neutralization 26, 189, 190–1
  ion bombardment 24–5, 88, 187, 259
  see also electron neutralization
chemical structure 5
cluster ions 5
  adsorbates 28; *see also* adsorption
  formation
    atomic combination model (ACM) 49, 53, 66, 150
    direct emission model (DEM) 49, 66
    inorganic 28
    organic 28–30; *see also* organic materials analysis
    post ionization 287
    recombination of atoms, *see* ACM
    see also laser desorption mass spectrometry; sputtering models, molecular dynamics; unimolecular decomposition
  inorganic
    surface crystallinity 27, 151
    see also oxidation of metals; reactions, surface
  inorganic oxides, dependence on bond dissociation energy 27–8 *see also* valence model
  isotope pattern calculation of 304
  matrix effects 30–1
cluster ions
  dependence, electronic state of solid 7
  see also sputtering, of cluster particles
cluster ion yield
  dependence on primary energy 30
  influence of surface damage *see under* damage
collision cascades
  cross section, electronic stopping 41–2
  cross section, nuclear stopping 37–9
  damage profile 42–3
  penetration profile 41–2
collision induced dissociation 202
composite materials 217–19
  imaging of 268–69
crater base effects, *see* depth profiling, uneven etching
crater edge effects, *see* ion detection

damage
  FAB induced 199–201
  function of magnification 250–1
  function of primary beam parameters 196–9
  imaging 249–52

damage—*contd.*
ion induced, organic cluster ion yield 31, 195–8
monolayer lifetime 31, 249–50
surface crystallinity *see under* Cluster ions - inorganic
see also collision cascades, damage profile; erosion rate; static SIMS conditions
data aquisition 114–15, 246–248
depth profile 4, 10
concentration scale 116–19
SNMS, *see* SNMS
depth profiling
cascade mixing 111
depth calibration 115
depth resolution 109, 130–40
beam induced mixing 137
influence of uneven etching 131–6
experimental conditions 119–23
instrumental drift 122–3
mass interference, *see under* mass interference
multi-layer structures 124–35, 140, 146–7
sputter rate 116
strengths and weaknesses of 106
see also charging of sample
depth resolution, *see* depth profiling
desorption ionization 28–29, 68–9
dimethyl phthalate, *see* spectra
dynamic SIMS
analysis conditions 4, 10, 92
experimental arrangement 110–15

electron bombardment post ionization, *see* SNMS
electron flood gun 89
electron neutralization 6
ToF spectrometers 89
non-conducting samples 88, 107
see also charging of sample
electron spectroscopy 1
electron stimulated ion emission (ESIE) 144, 190–1, 194, 259–60
electronic materials analysis
cleaning procedures for III-V materials 223–9
phthalocyanines 183
solders 229–231
electronic materials, imaging of 264–267
energy dispersive X-ray analysis (EDAX) 244
energy distribution of secondary ions 22–3, 112, 274
charging effects 187–8
energy distribution of sputtered atoms 14–15, 274
erosion rate 305–6
ethyl cellulose, *see* spectra

fast atom bombardment (FAB) 6
beam generation 89–91
imaging *see under* imaging SIMS
induced damage, *see* damage
fingerprint spectra, *see* spectral interpretation
fly's eye, imaging of 268

GaAs, *see* spectra
glow discharge mass spectrometry 280
gold, *see* spectra

high resolution electron energy loss spectroscopy (HREELS) with SSIMS 171–7
hydroxy propyl cellulose, *see* spectra

image contrast
crystallographic 256
isotopic 256
material 255–6
primary beam 258–9
topographic 253–5
voltage 256–8
imaging SIMS 5, 77, 92, 244
aquisition protocol 259–61
fast atom bombardment 90, 269
lateral resolution 267, 251; *see also under* ion beams
three-dimensional 260
ToF 269
ion beams
cold cathode source 83
dynamic SIMS requirements 75–6, 110–12
gas phase electron bombardment source 81–3
generation of 75
lateral resolution 78, 85
influence of aberrations 78, 80–1
liquid metal 4, 78–9, 83–4
rastering 76
SIMS imaging 76, 246
SSIMS requirements 75
surface ionization source 84–5
Wien filters 80
see also erosion rate
ion detection
crater edge effects 76
electronic gating 76, 113–14, 122
optical gating 76, 114
ion formation, *see under* cluster ions; secondary ion formation

ion microprobe 4, 244
ion microscope 4, 77, 246
ion velocities 307–9
ionization probability 17
  dependence on
    ionization potential 17–18
    primary particle 20–2
    sample material 18–20, 107–8
isotope patterns, calculation for cluster ions. 304

laser desorption/ionization mass spectrometry 7, 86
  instrumentation 295–6
  ion formation process 295, 297
lead phthalocyanine, *see* spectra
linear collision cascade theory *see under* sputtering
liquid metal ion sources, *see* ion beams
low energy electron diffraction (LEED) 149, 151

MARLOWE, *see under* sputtering, binary collision approximation
mass interference 127–9
  alleviation by sample bias 129–30
mass spectrometers
  comparison quadrupole and ToF 193–4
  magnetic sector 96–7
  mass resolution, transmission 101, 119
  quadrupole, instrumental parameters for SIMS 310–11
  quadrupole 94–6
  time of flight (ToF) 97–101
molecular dynamics model of sputtering, *see under* sputtering
molecular ion emission, *see* cluster ions, organic 28–30
molybdenum, oxidized, *see* spectra
monolayer lifetime 10, 31, 305
MS-MS 30, 201–2
  instrumentation 92, 202
multi-layer structures, *see* depth profiling
multi-photon non-resonant ionization 286–90; *see also* SNMS
multi-photon resonant ionization 282–6; *see also* SNMS

optical coating, *see* polymer analysis
organic materials analysis 7; *see also* cluster ions, organic formation
oxidation
  of alloys 159–60
  CO, *see* reactions, surface
  of metals 156–9, 181
  imaging 263–4
oxides
  cluster ion generation, *see* cluster ions
  surface analysis 231–6
  surface reactivity 233–4
  surface segregation 234–6

pharmaceutical drug delivery systems, *see* polymer analysis
plasma desorption mass spectrometry (PDMS) 7, 290–5; *see also* adsorption
plasticizers, *see* polymer analysis
polydimenthyl siloxane (PDMSO), *see* spectra
polycarbonate, *see* spectra
polyethylene terephthalate (PET) *see* spectra
polyhydroxy ethyl methacrylate (PHEMA), *see* spectra
polymer analysis
  adhesion 216
  co-polymers 215–16
  drug delivery systems 221–3
  end groups 214
  optical coatings 216
  plasticizers 223–5
  surface treatment 219–20
  see also composite materials; spectra
polymethyl methacrylate (PMMA), *see* spectra
polystyrene, *see* spectra
polytetrafluoro ethylene (PTFE), *see* spectra
post ionization, *see* SNMS 7
potassium bromide, *see* spectra
primary beam effects, *see under* damage; sputtering

quantification
  difficulties 20
  implanted standards 107–8, 116, 123, 125
  of data 123, 146–147; *see also* SNMS
  theoretical model 69–70

reactions, surface 175–81
  decomposition of ethene 177–8
  decomposition of methanol 180–1
  Fischer–Tropsch reaction 180
  hydrogen isotope exchange 180
  oxidation of CO 178–80
  reduction of NO 180
reflection absorption infra-red spectroscopy (RAIRS) with SSIMS 171–3
resistive anode encoder 247
ruthenium (0001), *see* spectra

scanning auger microprobe (SAM) 244
scanning electron microscopy (SEM) 244
scanning SIMS 244; *see also* imaging SIMS
secondary ion emission, mechanism 2
secondary ion formation
comparison, SIMS, PDMS and laser desorption/ionization 28, 292, 295–7
models
bond breaking 63–4
cluster ions
desorption ionization 68–9
lattice fragmentation 66–7
molecular 64–5
nascent ion molecule 64–6
local thermal equilibrium (LTE) theory 69–70
perturbation 62
surface excitation 63
see also ionization probability
secondary ion yield 329–35; *see also* cluster ion yield; sensitivity
sensitivity (XPS) 1
sensitivity
imaging 248–9
instrumental transmission 23–4
secondary ion yield 23–4
useful ion yield 107, 118–20, 124
silicon oxide, *see* spectra
spatial resolution, *see under* ion beams
spectra (SIMS)
diethyl phthalate 225
dimethyl phthalate 224
ethyl cellulose 224
GaAs 276
gold 277
hydroxy propyl cellulose 222
lead phthalocyanine 183
molybdenum - oxidized 313
polydimethyl siloxane 320–21
polyhydroxy ethyl methacrylate 208
polycarbonate 218
polyethylene terephthalate 6, 205
polymethyl methacrylate 32, 208, 316–17
polystyrene 200
polytetrafluoro ethylene 319
potassium bromide 314–15
ruthenium (0001) 153
silicon oxide 266
spectra (SNMS) GaAs 276
gold 277
PMMA 289
spectral interpretation 201–11
commonly observed ions 323–26
inorganic materials 207–13
organic materials 204–7
sputter rate 9–17, 116
sputtered neutral mass spectrometry (SNMS) 7
depth profile 280, 282, 286
electron bombardment post ionization 273–9
electron plasma post ionization 279–82
instrumentation 274–5, 284
laser induced post ionization 282–90
quantification 272–3, 278, 281, 288
relative sensitivity factors 278
single photon ionization 287, 289
technique comparison 291
sputtering
of cluster particles 17
definition 2
electronic 61, 292–3
induced mixing 16, 137
ion explosion 62
models
ACM, DEM *see under* cluster ions, formation
binary collision approximation (BCA) 56–8
cluster formation 61
linear collision cascade theory 36–48; assumptions 44–5; predicted sputter yields 44–7
molecular dynamics 48–56; assumptions 49
of cluster formation,
adsorbates 52–4
clean metals 51–2
single knock-on 36
slow collisional 36
spike regime 36
unimolecular decomposition (RRKM) theory 61
preferential 14–15
yield dependence on angle of incidence 11
dependence on primary energy and mass 10–12, 327–8, 333–5
dependence on sample atomic number 12–14; crystallinity 12; topography 12
static SIMS (SSIMS)
angle resolved 173
analysis conditions 10, 31, 92, 149–50, 186, 195
protocol for optimization 240, 310–12
monolayer lifetime 10, 31
with AES 151, 152, 154, 234
with HREELS 171–7
with RAIRS 171–3
with TD 162, 166–9
with XPS 160, 180, 229
surface chemical structure
metals 150–3
thin metal films 154–6

surface coverage measurements 156, 166–8
surface crystallinity, *see under* cluster ions
surface potential, *see* charging of sample

tandem mass spectrometry, *see* MS-MS
thermal desorption (TD) 162, 166–9
thin metal films, growth of 155–6
three-dimensional images, *see* imaging SIMS

valence model 211–13

Wehner spots 51
Wien filters, *see under* ion beams

X-ray photoelectron spectroscopy (XPS) 1, 149
  with SSIMS 160, 162, 180, 229